HUMAN IMPACTS *on* SEALS, SEA LIONS, *and* SEA OTTERS

HUMAN IMPACTS *on* SEALS, SEA LIONS, *and* SEA OTTERS

Integrating Archaeology and Ecology in the Northeast Pacific

Edited by

Todd J. Braje and Torben C. Rick

UNIVERSITY OF CALIFORNIA PRESS
Berkeley Los Angeles London

University of California Press, one of the most distinguished university presses in the United States, enriches lives around the world by advancing scholarship in the humanities, social sciences, and natural sciences. Its activities are supported by the UC Press Foundation and by philanthropic contributions from individuals and institutions. For more information, visit www.ucpress.edu.

University of California Press
Berkeley and Los Angeles, California

University of California Press, Ltd.
London, England

Library of Congress Cataloging-in-Publication Data

Human impacts on seals, sea lions, and sea otters : integrating archaeology and ecology in the Northeast Pacific / edited by Todd J. Braje and Torben C. Rick.
p. cm.
Includes bibliographical references and index.
ISBN 978-0-520-26726-8 (cloth : alk. paper)
1. Marine mammal remains (Archaeology)—Northwest Coast of North America. 2. Seals (Animals—Effect of human beings on—Northwest Coast of North America—History 3. Sea otter—Effect of human beings on—Northwest Coast of North America—History. 4. Hunting, Prehistoric—Northwest Coast of North America. 5. Paleoecology—Northwest Coast of North America. I. Braje, Todd J., 1976– II. Rick, Torben C.

CC79.5.M35H86 2010
930.1—dc22

2010034645

Cover image: View of St. Paul Island by Louis Choris, from *Voyage Pittoresque Autour du Monde* (1822). Alaska State Library, Louis Choris, Photograph Collection (P139-48).

CONTENTS

CONTRIBUTORS

MATTHEW W. BETTS
Canadian Museum of Civilization
Gatineau, Quebec, Canada
matthew.betts@civilization.ca

TODD J. BRAJE
San Diego State University
San Diego, California
tbraje@mail.sdsu.edu

SUSAN J. CROCKFORD
Pacific Identifications, Inc.
Victoria, BC, Canada
sjcrock@shaw.ca or *scrock@uvic.ca*

BRENDAN J. CULLETON
University of Oregon
Eugene, Oregon
bculleto@uoregon.edu

ROBERT L. DELONG
Alaska Fisheries Service Center
Seattle, Washington
Robert.Delong@noaa.gov

JON M. ERLANDSON
University of Oregon
Eugene, Oregon
jerland@uoregon.edu

S. GAY FREDERICK
Pacific Identifications, Inc.
Victoria, BC, Canada
gay.frederick@viu.ca

DIANE GIFFORD-GONZALEZ
University of California, Santa Cruz
Santa Cruz, California
dianegg@ucsc.edu

WILLIAM R. HILDEBRANDT
Far Western
Davis, California
billh@farwestern.com

ERICA HILL
University of Alaska Southeast
Juneau, Alaska
erica.hill@uas.alaska.edu

TERRY L. JONES
California Polytechnic State University
San Luis Obispo, California
tljones@calpoly.edu

SHAWN LARSON
Seattle Aquarium
Seattle, Washington
shawn.larson@seattle.gov

VERONICA LECH
Idaho State University
Pocatello, Idaho
veronica.lech@gmail.com

ROBERT J. LOSEY
University of Alberta
Edmonton, AB, Canada
robert.losey@ualberta.ca

R. LEE LYMAN
University of Missouri
Columbia, Missouri
lymanr@missouri.edu

HERBERT D.G. MASCHNER
Idaho State Unversity
Pocatello, Idaho
maschner@isu.edu

IAIN MCKECHNIE
University of British Columbia
Vancouver, BC, Canada
ii@interchange.ubc.ca

SARAH MELLINGER
University of California, Santa Barbara
Santa Barbara, California
sarah_mellinger@umail.ucsb.edu

MADONNA L. MOSS
University of Oregon
Eugene, Oregon
mmoss@oregon.edu

JUDITH F. PORCASI
University of California, Los Angeles
Los Angeles, California
porcasi1@verizon.net

TORBEN C. RICK
Smithsonian Institution
Washington, D.C.
rickt@si.edu

ADRIAN R. WHITAKER
Far Western
Davis, California
adie@farwestern.edu

REBECCA J. WIGEN
Pacific Identifications, Inc.
Victoria, BC, Canada
rjwigen@uvic.ca

1

People, Pinnipeds, and Sea Otters of the Northeast Pacific

Torben C. Rick, Todd J. Braje, and Robert L. DeLong

MARINE MAMMALS, SUCH AS polar bears, sea otters, seals, sea lions, and walruses, are an extraordinary group of organisms, many of which maintain a link to both aquatic and terrestrial habitats. Often highly intelligent with sophisticated communication systems, marine mammals are a fundamental component of marine ecosystems around the world (Berta et al. 2005; Riedman 1990). Human interaction with seals, sea otters, and other marine mammals spans millennia and the entire globe (Erlandson 2001; Etnier 2007; Hildebrandt and Jones 1992; Klein and Cruz-Uribe 1996; Lyman 1995; McNiven and Beddingfeld 2008; Monks 2005a; Nagaoka 2002; Stringer et al. 2008). Containing large amounts of meat, oil, ivory, and other important raw material and dietary resources, pinnipeds (seals, sea lions, and walruses) and sea otters have been hunted and scavenged by people in the northeastern Pacific for much of the Holocene or earlier (Braje 2010; Broughton 1994, 1999; Burton et al. 2001, 2002; Colten 2002; Corbett et al. 2008; Crockford and Frederick 2007; Domning et al. 2007; Erlandson al. 2005; Etnier 2002a, 2002b, 2004, 2007; Gifford-Gonzalez et al. 2005; Hildebrandt and Jones 1992, 2002; Jones and Hildebrandt 1995; Lyman 1991, 1995, 2003; Moss et al. 2006; Newsome et al. 2007; Porcasi et al. 2000; Rick et al. 2009). The large terrestrial breeding colonies of some pinnipeds have made them highly susceptible to human hunting in the distant and near past. During the fur and oil trade of the 18th and 19th centuries, for example, human hunting of pinnipeds and sea otters decimated marine mammal populations, causing the extirpation of local populations of sea otters (*Enhydra lutris*), Guadalupe fur seals (*Arctocephalus townsendi*), and northern elephant seals (*Mirounga angustirostris*) and the extinction of the Steller's sea cow (*Hydrodamalis gigas*; Ellis 2003; Ogden 1941; Scammon 1968). Archaeological data demonstrate that human impacts on marine mammals in the more distant past were also substantial and suggest that much remains to be learned about the prehistory of these animals.

With the recognition that human occupation of the world's coastlines has great time depth (Bailey 2004; Erlandson 2001), research on ancient human impacts and influences on the biogeography, behavior, and ecology of pinnipeds, sea otters, and other marine organisms has increased. The shifting baselines phenomenon (Pauly 1995)—the notion that modern and historical ecosystems have been fundamentally altered by humans and that ecological baselines or benchmarks change dramatically over time—has prompted researchers from a variety of disciplines to emphasize the importance of archaeological and historical data in the management and restoration of marine ecosystems around the world (Erlandson and Rick 2010; Jackson et al. 2001; Pinnegar and Engelhard 2008; Rick and Erlandson 2008). Faunal remains found in archaeological sites contain vast amounts of information on the composition, biogeography, and abundance of animals that were present in the past, as well as changing climatic conditions, human subsistence practices, and similarities and differences with the present day (see Lyman 2006; Lyman and Cannon 2004). Although the potential of these data remains underexplored, several recent studies have demonstrated the utility of archaeology for informing contemporary conservation issues, including the historical ecology of seals, sea lions, and sea otters (Lyman 2003; Etnier 2004, 2007; Hildebrandt and Jones 2002; Newsome et al. 2007; Rick et al. 2009).

Using historical ecology as a baseline (Crumley 1994), this volume brings together researchers from a variety of disciplines seeking to integrate archaeology, history, and ecology to understand the ancient and modern interactions between pinnipeds, sea otters, people, and marine ecosystems. Transcending disciplinary boundaries, the studies in this volume use ancient DNA (aDNA), stable isotopes, zooarchaeology, and other analytical techniques to understand the influence of ancient peoples on the biogeography, breeding behavior, and abundance of seals and sea lions over the last several millennia in the northeastern Pacific. While focused on pinnipeds and sea otters, these studies serve as a model for how archaeological data can significantly enhance the contemporary management and conservation of marine and terrestrial ecosystems.

Believing that the past is the key to understanding the present, we also recognize that documenting the interactions between ancient marine mammals and humans requires a firm grasp of the modern ecology of these organisms. Following a brief discussion of archaeology and human environmental impacts, this chapter provides an overview of the ecology of pinnipeds and sea otters in the northeastern Pacific, emphasizing the historical and modern transformations of these various populations (Figure 1.1).

ARCHAEOLOGY, HUMAN ENVIRONMENTAL IMPACTS, AND MARINE MAMMALS

Archaeologists have long investigated the interactions between humans and the environments in which they lived, often seeking to understand the impacts of environmental and climate change on the evolution of human society. More recently, archaeologists and anthropologists have debated the effects of ancient human subsistence, foraging, hunting, and other activities on the ecosystems where people resided (Hames 2007; Kay and Simmons 2002; Kirch 2005; Redman 1999). Assessing the environmental impacts of hunter-gatherers and other small-scale societies has been particularly contentious, with some scholars arguing that these peoples were inherent conservationists and environmental managers and others arguing they lack conservation principles and were generally self-centered and individualistic (see Alvard 1998; Hunn et al. 2003; Hames 2007; Hayashida 2005; Smith and Wishnie 2000). Although hotly contested, recent research suggests that ancient peoples from small-scale hunter-gatherers to stratified agriculturalists had

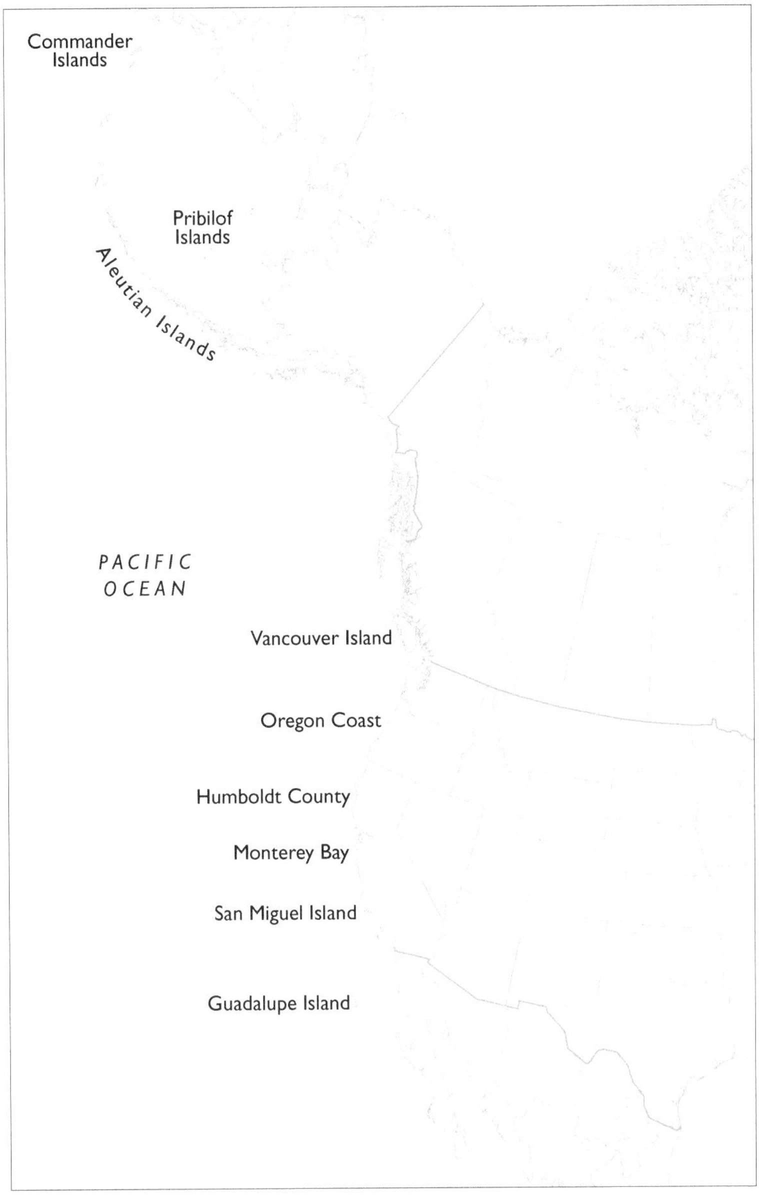

FIGURE 1.1. Western North America and the northeastern Pacific showing the location of many areas discussed in the book, important pinniped breeding areas (Pribilof, San Miguel, and Guadalupe Islands), and the Commander Islands, the last known historical respite for sea cows.

a significant influence and impact on the environment (Kirch 2005; Kay and Simmons 2002; Redman 1999; Rick and Erlandson 2009). Ancient environmental interactions were extremely diverse, ranging from degradation to enhancement, raising significant questions about the structure and function of ancient and modern ecosystems.

Archaeological analyses of pinniped and sea otter remains have played an important role in broader debates over ancient human environmental impacts, providing important lessons for conservation biology (Etnier 2007; Lyman 2006, this volume). During the last two decades, archaeologists have debated the effects of ancient human hunting on northeastern Pacific

Coast seals, sea lions, and sea otters (Etnier 2007; Hildebrandt and Jones 1992; Jones and Hildebrandt 1995; Lyman 1988, 1995, this volume; Moss et al. 2006; Newsome et al. 2007). These and other studies document significant changes in the distribution, abundance, and breeding behavior of pinnipeds and sea otters through time. For example, elephant seals are currently among the most numerous pinnipeds in coastal California, but few archaeological sites contain their remains, suggesting they were rare or absent for much of the last 10,000 years (see Braje et al., this volume; Hildebrandt and Jones 2002; Lyman, this volume; Rick et al. 2010). Similarly, Guadalupe fur seals are currently rare in coastal waters north of Baja, but they are found in archaeological sites as far north as British Columbia and Washington and are among the most common pinnipeds in coastal California archaeological sites (Braje et al., this volume; Etnier 2002b; Lyman, this volume; Moss et al. 2006; Rick et al. 2009). Explored in detail throughout this volume, these and other patterns appear to result from a variety of factors (e.g., climate change and human predation) that vary through time and across space.

With contemporary marine and terrestrial ecosystems around the world in a state of crisis, the studies in this volume use detailed investigations of the archaeology and historical ecology of northeastern Pacific pinnipeds and sea otters to inform contemporary problems. Restoring, managing, and regulating contemporary ecosystems and fisheries is a challenging and often political task that requires carefully balancing the diverse interests of the public, scientists, fishermen, policy makers, and numerous other stakeholders. To help inform these often difficult and contentious issues, the research presented in this volume relies on long-term reconstructions of ancient human interactions with sea otters and pinnipeds to focus on a series of key questions. How did ancient peoples from southern California to Alaska influence and impact ancient pinniped and sea otter abundance, biogeography, and behavior, and what can this tell us about ancient human–environmental interactions more generally? How have humans affected the location and breeding behaviors of pinniped and sea otter populations? What were the effects of ancient climatic changes on pinnipeds, sea otters, and human society, and how do we differentiate between the effects of human impacts and natural climatic changes? What was the role of ancient human technological change, social systems, ideology, and ritual in guiding the types of marine mammals that were taken? How can various analytical techniques, such as aDNA and stable isotope analyses, be integrated with traditional archaeological data to provide a more comprehensive understanding of ancient pinniped and sea otter populations? How can reconstructions of ancient human, pinniped, and sea otter interactions that span millennia inform the contemporary management of these organisms and broader marine ecosystems?

To address these questions, the chapters in this volume present up-to-date archaeological analyses and syntheses of specific regions throughout the northeastern Pacific. Many of these studies note significant differences between ancient and modern pinniped and sea otter abundance, biogeography, and behavior, challenging conventional wisdom about the structure and composition of contemporary pinniped and sea otter populations. In the final chapter, we use the case studies presented in this volume to revisit these questions and explore the challenges presented when archaeological data are incongruent with the modern situation.

ECOLOGY, BEHAVIOR, AND DISTRIBUTION

During the Holocene, eight major species of pinniped, sea otter, and sea cow were exploited by ancient humans throughout much of the northeastern Pacific (Table 1.1). This includes four otariids (eared seals), two phocids (true seals), a sea otter, and an extinct sea cow. The four major otariid species on the Pacific Coast from southern California to the Aleutians include two sea lions (*Eumetopias jubatus* [Steller

TABLE 1.1

Summary of Modern Distribution and Status of Pinnipeds, Sea Otters, and Sea Cows of the Northeastern Pacific

SPECIES/COMMON NAME	RANGE (MODERN/HISTORIC)[a]	COMMENT	STATUS
MUSTELID			
Enhydra lutris (sea otter)	Hokkaido, Japan, around North Pacific Rim to Baja California, Mexico	Primarily aquatic but hauls out on land	Extirpated from much of range; listed (ESA) as threatened in south-western Alaska and California
OTARIID			
Arctocephalus townsendi (Guadalupe fur seal)	Islas San Benitos and Isla de Guadalupe, Baja California, Mexico, to California Channel Islands	Breeds on land in low-density terrestrial colonies	Extirpated from much of its range and population is small but expanding; listed (ESA) as threatened
Callorhinus ursinus (northern fur seal)	Channel Islands, North Pacific Rim to Robben Island in the Sea of Okhotsk	Pelagic and migratory, but breeds in large terrestrial colonies	Population appears to be declining; listed (MMPA) as depleted
Eumetopias jubatus (Steller sea lion)	Sea of Okhotsk, North Pacific Rim to Channel Islands	Breeds in large terrestrial colonies	Population has declined dramatically; listed (ESA) as endangered in western Alaska and threatened from eastern Alaska through California
Zalophus californianus (California sea lion)	Cabo San Lucas, Baja California, Mexico, to the Farrallon Islands	Breeds in large terrestrial colonies	Population is expanding; is near carrying capacity
PHOCID			
Mirounga angustirostris (northern elephant seal)	Cape Arago, Oregon to Cabo San Lazaro, Baja California, Mexico	Large body size, highly migratory, and breeds and molts in terrestrial colonies	Population reduced to near extinction in 19th century but has recovered and is expanding
Phoca vitulina (harbor seal)	Japan, North Pacific Rim to Isla Asuncion, Baja California, Mexico	Mates in water and very wary of humans when on land	Population expanding, but declining in parts of Alaska

[a] Ranges have changed and been altered during historical and modern times. These ranges include historical distributions. All data are based on sources cited in text. Nonbreeding ranges for a few of the species are larger than the breeding ranges provided here.

sea lion] and *Zalophus californianus* [California sea lion]) and two fur seals (*Callorhinus ursinus* [northern fur seal] and *Arctocephalus townsendi* [Guadalupe fur seal]). Two phocids range widely in the Northeastern Pacific: *Mirounga angustirostris* (northern elephant seal) and *Phoca vitulina* (harbor seal). The sea otter, Steller's sea cow (Domning et al. 2007), and walrus (*Odobenus rosmarus*) were the other similar marine mammal species in the northeastern Pacific. A few other seals (e.g., ribbon seals [*Histriphoca fasciata*] and spotted seals [*Phoca largha*]) also venture into the far northern parts of the Pacific. The ecology of some of these seals, as well as walruses, is covered in depth in chapters by Hill (Chapter 3) and Crockford and Frederick (Chapter 4) and is not repeated here.

Pinnipeds and sea otters have long evolutionary histories. The earliest well-documented fossil pinnipeds come from the late Oligocene, roughly 27 to 25 million years ago (MYA), but there is evidence that pinnipeds may be as old as 29 MYA (Berta et al. 2005:27). Morphologic and molecular analyses suggest that pinnipeds are likely monophyletic and most closely aligned with arctoid (raccoons, weasels, bears, etc.) carnivores, especially mustelids (weasels) and ursids (bears; Berta et al. 2005; Jefferson et al. 2008). The earliest phocids may appear as early as 29–23 MYA, with well-dated specimens appearing around 15 MYA (Berta et al. 2005). Fossil otariids currently appear around 11 MYA in the northeastern Pacific. The modern sea otter *(Enhydra)* arose more recently than the pinnipeds around 1–3 MYA in the North Pacific (Berta et al. 2005:104). Human exploitation of these species likely extends back to initial human colonization of North America during the terminal Pleistocene, with definitive archaeological evidence of interaction during the Early to Middle Holocene and of intensive harvest during the Late Holocene (see Hildebrandt and Jones 1992).

The contemporary distribution and breeding behaviors of pinnipeds and sea otters are constantly changing due to a variety of cultural and environmental variables. As the remaining chapters in this volume show, this is something that has occurred in deeper archaeological time as well. The modern and historical behaviors of these animals form an important backdrop for interpreting the archaeological data.

OTARIIDS

There are currently about 16 species of otariids around the world, with four found in the northeast Pacific (Jefferson et al. 2008). The Steller sea lion is the largest of the North Pacific otariids and displays marked sexual size dimorphism, with males up to 325 cm in length and 1120 kg in mass and females up to 291 cm in length and 350 kg in mass (Loughlin et al. 1987). During modern and recent historical times, Steller sea lions ranged from Hokkaido, Japan, around the northern Pacific Rim, and down to the California Channel Islands (Loughlin et al. 1987; Riedman 1990). The population has been declining through portions of its range in recent decades, but Steller sea lions still breed on islands in the Bering Sea, the Aleutians, the Gulf of Alaska, southeast Alaska, British Columbia, Oregon, and in California south to Año Nuevo Island and populations have been increasing from Oregon to southeast Alaska since the 1970s. The species is currently listed by the International Union for the Conservation of Nature (IUCN) as endangered (Jefferson et al. 2008), and it is listed under the U.S. Endangered Species Act (ESA) as endangered in the western portion of the range (Central Gulf of Alaska, westward) and as threatened in the eastern portion of the range (Central Gulf of Alaska to Central California). Steller sea lions feed on fish, cephalopods, and crustaceans, as well as occasionally on harbor seals, northern fur seals, and possibly sea otters (Loughlin et al. 1987).

Generally regarded as nonmigratory, Steller sea lions are known to disperse fairly widely several months after breeding between late May and early July. Consequently, Hildebrandt and Jones (1992) and other archaeologists have referred to Steller sea lions, the other otariids, and northern elephant seals as "migratory breeders" or, more recently, as "terrestrial breeders" because of the large breeding groups they establish on land (Jones et al. 2004) and their seasonal abundance that would make them more susceptible to human predation (see Etnier 2007 for a critique). Steller sea lions are polygynous, mating and pupping on land in territories defended by males. Females give birth to a single pup in May or June and begin a typical otariid pattern, feeding at sea and returning to the rookery to nurse the pup. In the case of Steller sea lions, females leave the rookery in the late afternoon or evening, feed overnight, and return to the rookery the following day to nurse the pup. On many rookeries females leave accompanied by the pup when it is several weeks to a couple months old and move to distant haul-out sites. Females may wean pups when 6 to 11 months of age, but some continue to nurse their young for 2 or more years, forgoing annual reproduction. Alternative terrestrial haul-out sites are used by juvenile animals during the breeding season and by all age and sex groups during the remainder of the year. Many rookery islands are not occupied during winter months. Their large breeding colonies and haul-out sites make them susceptible to human hunters on land. They have been identified in archaeological sites throughout the northeastern Pacific but are generally most abundant in sites in northerly, cooler waters (e.g., Lyman 2003). Archaeological Steller sea lion remains from the Aleutians have also been used to evaluate the role of climate forcing versus human predation for recent and ancient declines of the population, with these studies generally favoring climate forcing (Trites et al. 2007).

California sea lions are a relatively large, sexually size dimorphic otariid, with males averaging 225 cm and 325 kg and females 180 cm and 110 kg (Orr and Helm 1989). Often found on offshore rocks and secluded beaches, they range from the Tres Marianas Islands, Mexico, throughout the Sea of Cortez, and north to the Gulf of Alaska (Jefferson et al. 2008). They breed on islands in the Sea of Cortez, along the Pacific coast of Baja California, northward to the major breeding populations in the California Channel Islands and the Farallon Islands. Like Steller sea lions, they are regarded as nonmigratory, although some age and sex groups of California sea lions display lengthy seasonal movements (Riedman 1990) and feed on a variety of prey, including fishes and squids. The California sea lion may have dropped to as few as 1500 individuals following historic exploitation but have since recovered to about 240,000 animals (Jefferson et al. 2008; Carretta et al. 2007). DeLong and Melin (2002) noted that some 80,000 California sea lions currently breed on California's San Miguel Island (and this population has continued to increase during this decade [DeLong, unpublished data]). The IUCN lists the California sea lion as of least concern, and they are regarded as within the Optimum Sustainable Population Level under the U.S. Marine Mammal Protection Act (Carretta et al. 2007).

California sea lions are polygynous and breed and pup on land in territorial breeding colonies that are highly susceptible to human predation and other activities. Female sea lions give birth to pups in May and June and alternate terrestrial nursing and multiday marine foraging trips for 6 to 11 months, when the pup is weaned. Thus females and pups are found on the rookery island all year. Some juvenile animals remain on haul-out sites on rookery islands all through the year, but some portion of the juvenile population migrate northward to central and northern California where they alternate marine foraging and terrestrial hauling activity on offshore rocks and islands; in recent times they have hauled out in urban areas such as Monterey Bay and San Francisco Bay. Adult males visit the rookery islands during the pupping and breeding season from May through July. Many adult males establish and maintain reproductive territories on rookeries, but a near-equal number of adult males appear to spend the breeding season hauled out on island beaches that are not rookeries. Following the breeding season, all adult males migrate northward to haul-out sites all along the coast from central California through British Columbia, Canada. California sea lions are fairly common in archaeological sites in California, especially in Late Holocene sites on the Channel Islands and southern California coast (Braje 2010; Hildebrandt and Jones 1992; Porcasi et al. 2000; Rick 2007; Walker et al. 2002).

Northern fur seals are also sexually size dimorphic, with males up to 210 cm in length and from 180 to 270 kg and females up to 150 cm and from 43 to 50 kg (Jefferson et al. 2008; Orr and Helm 1989). Fur seals are highly migratory, ranging from southern California around the northern Pacific Rim to Honshu, Japan. Adults breed on terrestrial rookeries in reproductive territories. Following a 4-to-5-month breeding season, they go to sea, with the majority of adults and most juveniles remaining at sea and feeding pelagically across the North Pacific until the following breeding season, when they again come ashore. Most juvenile animals from 4 months through 3 years of age remain at sea year round. In the fourth year of life, many of the animals come ashore for the first time since birth, females to breed and males to haul out, and begin developing the terrestrial skills needed to obtain and maintain reproductive territories. Most males are not successful in obtaining a territory until 8 to 10 years of age. Northern fur seals prey on a variety of fish and squid found on the continental shelf and over deep oceanic waters. Northern fur seals were heavily affected by historical human exploitation, with the population currently thought to be around 1,000,000 and declining. They are listed by the IUCN as vulnerable and as a depleted species under the Marine Mammal Protection Act (Jefferson et al. 2008; Carretta et al 2007).

A highly polygynous species, northern fur seals breed from mid-June to August. Northern fur seals breed today on isolated islands that are generally distant from humans, bears, and other predators (e.g., Commander, Pribilof, and Bogoslof Islands and San Miguel Island; see DeLong and Melin 2002). Females give birth to a single pup, are bred 6–7 days later, and begin to alternate multiday marine feeding trips and terrestrial nursing cycles, which continue until the pup is weaned at 4 to 5 months of age. Following weaning, females and pups go to sea, and females do not return until the next breeding season. Archaeological data from California, Washington, and elsewhere are interpreted by some to indicate that northern fur seals may have been more common prehistorically in mainland areas and may have even bred on some portions of the California and Washington mainland (see Etnier 2002a; Gifford-Gonzalez et al. 2005; Newsome et al. 2007). Alternatively, fur seals may have been harvested at sea in coastal waters by early humans using marine technology developed in parts of the eastern Pacific coast during the Holocene, and/or as sick or injured animals that washed ashore (see Gifford-Gonzales et al. 2005 for a discussion and Moss and Erlandson 1995 for discussion of technologies). The presence of northern fur seal bones, including pups in archaeological contexts, has prompted considerable debate among a variety of researchers as to whether these represent sick or injured animals that washed ashore, were animals taken at sea, or if fur seal breeding and migration patterns were dramatically different during the Late Holocene than the modern and historical era (see Gifford-Gonzalez et al. 2005). These debates are revisited in this volume.

Like other otariids, Guadalupe fur seals are sexually dimorphic, with females averaging 120 cm in length and 40 to 50 kg and males 180 cm in length and 160 to 170 kg (Belcher and Lee 2002). Guadalupe fur seals are also nonmigratory, although they are known to disperse to the Channel Islands and other areas (Belcher and Lee 2002; DeLong and Melin 2002). Prior to the 19th century, Guadalupe fur seals ranged from San Miguel Island southward through Isla de Guadalupe, Islas San Benitos, and Isla Cedros, Baja California, Mexico (Rice 1998; Belcher and Lee 2002; Repenning et al, 1971). They currently breed only on Isla de Guadalupe and Islas San Benitos (Carretta et al. 2007), although a single pup was born on San Miguel Island in 1997 (Melin and DeLong 1999). Recent archaeological data from the Ozette site in northwestern Washington yielded the remains of several Guadalupe fur seals dated to the last 500 years, suggesting they once ranged considerably further to the north (Etnier 2002b). They generally feed on squid, bony fish, and crustaceans. Over 12,000 Guadalupe fur seals were thought to exist in 2003 (Gallo-Reynoso et al. 2005), and the IUCN currently lists the population as vulnerable and it is listed as threatened under the U.S. Endangered Species Act.

Guadalupe fur seals are polygynous and give birth to pups and breed on rocky beaches and in caves (Belcher and Lee 2002). Females give birth to a single pup in June and July and are bred several weeks later. They then alternate nursing the pup for 1 to 2 days and feeding at sea away from the rookery for several days until pups are weaned at 8 to 11 months of age. Guadalupe fur seal females are also very common in archaeological sites from the Late Holocene California Channel Islands (Braje 2010; Porcasi et al. 2000; Rick 2007; Rick et al. 2009; Walker et al. 2002), suggesting they were much more abundant prior to historical depletion.

PHOCIDS

Some 19 species of phocid are found around the world, with only two species ranging widely in the northeastern Pacific (Jefferson et al. 2008). The northern elephant seal is sexually dimorphic and the largest of all the North Pacific pinnipeds, with males up to 410 cm and 2300 kg and females 280 to 300 cm and 600 to 800 kg (Stewart and Huber 1993). Pushed to the brink of extinction in the 1800s, a few animals were identified on Isla de Guadalupe off Baja California, Mexico, in the late 1800s. The population has since recovered dramatically, with some 50,000 animals breeding on San Miguel Island alone (DeLong and Melin 2002) and in excess of 120,000 in U.S. waters (Carretta et al. 2009). They have low genetic diversity, probably resulting from the historical population collapse (Stewart and Huber 1993). Currently, they range widely in the North Pacific, foraging from the Gulf of Alaska and the Aleutians southward to the Oregon and California coast and the Channel Islands and Baja California, where they pup and breed. The IUCN lists the northern elephant seal as least concern (Jackson et al. 2008); they are not threatened or endangered under the ESA, and they might be approaching the Optimum Sustainable Population (OSP) (Carretta et al. 2007). They are a highly migratory species, preying on squid and fish. They are known to dive to 1000 meters or more and make two yearly migrations totaling at least 21,000 km (DeLong and Stewart 1991; Stewart and DeLong 1995).

Elephant seals breed on land in large reproductive territories. Pups are born in January and February, are nursed for about 30 days, and are weaned by the departure of the female following breeding. Pups then remain on the island for up to 3 months, gradually developing swimming skills, and finally go to sea not to return until they haul out to molt 1 year later. As is the case in all phocids, elephant seals spend the majority of the molt period, which lasts about 4 to 6 weeks, on land (or ice in the case of some of the ice seals). Although they are large and conspicuous marine mammals that breed and haul out on land, they are fairly rare in archaeological assemblages. They have been found in a limited number of sites in the northeastern Pacific primarily dated to the last 1500 years but usually occur in very low frequencies (Rick et al. 2010).

The harbor seal is a small phocid. Although males are generally larger than females, they are not sexually dimorphic and are about 150 to 190 cm and 60 to 150 kg (Jefferson et al. 2008; Orr and Helm 1989). Their geographic range includes islands and the coast in the North Pacific from Japan to central Baja California, and they are also present in the southern Bering Sea. Harbor seals rarely venture far from the water, and when on land they frighten easily when humans are near. They forage on fish, cephalopods, and crustaceans. Estimating harbor seal populations is difficult, but Jefferson et al. (2008) suggest that there are likely some 300,000 to 500,000 animals, with Carretta et al. (2007) estimating the California population at least 31,600 animals, the Oregon and Washington stock at around 22,380, with at least 12,844 in Washington's inland waters. The Alaska population is estimated at 180,000 animals (Angliss and Allen 2009). The IUCN categorizes the harbor seal as a species of least concern (Jefferson et al. 2008), and they are not listed as endangered or threatened under the ESA (Carretta et al. 2007).

Like all the Pacific coastal pinnipeds, harbor seal females give birth to a single pup on land. They then nurse the pup for about 4 to 5 weeks before weaning. Harbor seals mate in the water from February to October and are promiscuous to weakly polygamous (Jefferson et al. 2008). Harbor seals do not migrate and may not move away from the rookery and haul-out areas, spending some portion of all weeks (in some cases days) of the year on land. These and other factors led Hildebrandt and Jones (1992) to call harbor seals and sea otters "residential" and later "aquatic breeders" (Jones et al. 2004), arguing that they would generally be much more difficult to acquire through hunting than otariids. Harbor seals are fairly common in northeastern Pacific archaeological sites but are rarely a dominant species (Lyman 2003; Porcasi et al. 2000; Walker et al. 2002). Harbor seals remain a valued resource to Alaska Natives and in excess of 2000 are harvested there each year (Angliss and Allen 2009).

SEA OTTER

The smallest of the North Pacific marine mammals, sea otters are members of the family Mustellidae, which includes some 67 species, only two of which are marine (Jefferson et al. 2008). There are two subspecies of sea otters that occur in the North Pacific archeological record, *E.l. kenyoni* is distributed from the western Aleutian Islands across the North Pacific and south currently to Washington. The second, *E.l. nereis,* is currently found only in central California but was prehistorically distributed from central Baja California north probably through at least Oregon (Rice 1998; Valentine et al. 2008; Wilson et al. 1991). Like the harbor seal, sea otters have limited sexual size dimorphism, with males averaging 126 to145 cm in length and 21 to 45 kg and females 107 to 140 cm and 14 to 33 kg (Estes 1980). Sea otters spend most or all of their lives in the water. They are often solitary but can rest in groups and will only venture on land in very protected areas or when sea otter abundance is high (Estes 1980). They are generally nonmigratory, but seasonal movements of long distances have been documented. Sea otters were pushed to the brink of extinction during the fur trade and were extirpated in numerous areas. Prior to the fur trade, they appear to have ranged from Baja California, around the North Pacific Rim, and into Hokkaido, Japan (Estes 1980). According to the IUCN, the sea otter is currently endangered throughout its range. The California sea otter and the Southwest Alaska otter populations are listed as threatened under the ESA. The remaining two Alaska stocks (Southeast and South Central) and the Washington stock are not listed.

They forage on invertebrates, especially sea urchins, mussels, crab, and abalones, as well as on fish in some areas (Estes 1980). They are a fundamental component of kelp forest ecosystems and help regulate sea urchin and other grazing invertebrate populations. During the fur trade and in modern times, the decimation of sea otters allowed sea urchin populations to skyrocket, resulting in kelp barrens in numerous areas (Simenstad et al. 1978; Steneck et al. 2002). Identified in San Miguel Island archaeological sites as old as 9000 years (Erlandson et al. 2005), sea otters are often abundant in Aleutian, Californian, and other archaeological sites (see Corbett et al. 2008). Sea otters are polygynous and mate and breed in the water. Although sea otters commonly haul out on beaches in Alaska, it is generally considered by archeologists that sea otters would have likely been captured in water and, compared to many pinnipeds and sea cows, would have been difficult prey (Hildebrandt and Jones 1992). Simenstad et al. (1978), Corbett et al. (2008), and Erlandson et al. (2005) have suggested that prehistoric reductions in sea otters from ancient human hunting, albeit on different scales than the fur and oil trade, may have resulted in significant changes in ancient nearshore kelp forest ecosystems (see also Jones et al. this volume).

HISTORICAL TRANSFORMATIONS

All the northeastern Pacific marine mammal species described in this chapter underwent dramatic declines during the 18th and 19th centuries (Ellis 2003). As human hunters moved to new areas of the globe to acquire these animals for the commercial fur trade, populations were dramatically reduced, pushed to the brink of extinction or to local extirpation, or went completely extinct in the case of Steller's sea cows. Some of this depletion was extremely rapid, with the sea cow apparently extinct just a few decades after its discovery by Europeans.

Bodkin (2000) estimates that between 500,000 and 900,000 sea otters were killed from 1742, when commercial harvest began, until 1910, when a final, unsuccessful hunt was staged. Historical commercial hunting of seals and cetaceans began even earlier than that of sea otters. Intensive harvest of whales began at least in the 16th century and continued well into the 20th century. The oil of many whales and various dolphins has been a prized commodity for a number of household and industrial purposes, including the production of lamp oil, soap, leather dressing, steelmaking, and textile sizing (Ellis 2003:237). The historic decimation of cetacean populations is staggering, with thousands killed each year for at least four centuries and commercial harvest of many whale and dolphin species continuing today.

Fur seals and sea lions have endured a similar history. Throughout North Pacific waters, Guadalupe fur seals, northern fur seals, California sea lions, and Steller sea lions were all slaughtered by the thousands for their fur, blubber, and "trimmings" beginning in the early to mid-1700s (Ellis 2003:161–178). The trimmings trade was especially wasteful, targeting only the genitals of bull seals for medicinal and aphrodisiacal purposes, the gall bladder for a type of medicine, and the whiskers for opium pipe cleaners or toothpicks (Busch 1985:201–202). Northern elephant seals, lacking the luxurious pelts of the fur seal, were targeted for their thick blubber, which was boiled down into oil.

This historic blitzkrieg dramatically altered marine mammal populations in the northeastern Pacific and beyond. Nearly every species that survived suffered significant historic population size reductions and, in some cases, decreases in genetic variability (genetic bottlenecks). Mitochondrial DNA (mtDNA) studies, for example, have demonstrated that all extant populations of sea otters have suffered at least one population bottleneck (Riedman and Estes 1990) and show low levels of genetic variability (Bodkin et al. 1999; Lidicker and McCollum 1997). A similar pattern of low genetic diversity exists among northern elephant seals and Guadalupe fur seals (Bernardi et al. 1998; Bonnell and Selander 1974), especially when compared to DNA of prehistoric specimens (Hoelzel et al. 2002; Weber et al. 2004).

In 1972, the Marine Mammal Protection Act was passed, providing protection to all marine mammals in U.S. waters regardless of their status. The result has been a recovery of many species to levels that may be comparable to those prior to intensive historical predation (Jefferson et al. 2008), but the dearth of historical records prior to the mid 19th century makes archaeology one of the few sources of data for reconstructing prehistoric pinniped biogeography. Northern elephant seals and California sea lions have fared particularly well, with Guadalupe fur seals also increasing in population. In some parts of their range, Steller sea lions are endangered and have undergone declines throughout the southern part of their range in recent decades and have stopped breeding on San Miguel Island completely (DeLong and Melin 2002; Trites and Donnelly 2003), but they have been increasing from Oregon to southeast Alaska, where they are listed as threatened, and in Canada they are listed as a species of special concern. The Northern fur seal population recovered to nearly 2 million animals after reaching a low point in 1910. But since 1970 the population has declined by half due to unknown causes. Finally, sea otters are absent from much of their original range, have undergone population declines in southwestern Alaska, and translocations have been met with varying degrees of success (see Hatfield 2005).

ARCHAEOLOGICAL IMPLICATIONS AND VOLUME OVERVIEW

These historical transformations, including dramatic population declines, recoveries, and rebounds, range expansions and contractions, and other shifts provide an important backdrop for interpreting the archaeological record of pinnipeds, sea otters, and other marine mammals in the northeastern Pacific. Archaeology is poised to provide crucial information on long-term ecology, biogeography, and human interactions with pinnipeds and sea otters. As Monks (2005b) noted, however, compared to other organisms (i.e., terrestrial mammals) marine mammals have received relatively limited attention from zooarchaeologists. In the case of pinnipeds and sea otters, this appears to be changing, especially in the northeastern Pacific, as several studies since the 1990s have provided archaeological information on pinnipeds and otters (e.g., Crockford and Frederick 2007; Etnier 2002a, 2007; Gifford-Gonzalez et al. 2005; Hildebrandt and Jones 1992; Jones and Hildebrandt 1995; Lyman 1995, 2003; Moss et al. 2006; Newsome et al. 2007; Rick et al. 2009, 2010).

With each new study it is clear that much remains to be learned about pinnipeds and sea otters in the more distant past, especially as some archaeological data appear to be at odds with modern ecological data or at least raise questions about modern pinniped distributions and behavior. Data on northern fur seals from the northeastern Pacific, for example, have been used to support the hypothesis that breeding colonies of northern fur seals may have existed on the mainland of central California, a pattern that, if correct, is dramatically different from the present or historical situation (Gifford-Gonzalez et al. 2005; Newsome et al. 2007). Work at the Ozette site in Washington, moreover, documented Guadalupe fur seals in deposits dated to the last 500 years, showing a much larger range than is known from the modern era (Etnier 2002b). Hunting of sea otters in the Aleutians and California Channel Islands by ancient Native peoples also suggests that human hunters may have caused localized impacts on kelp forests in the distant past (Corbett et al. 2008; Erlandson et al. 2005; Simenstad et al. 1978).

The chapters in this volume deepen our knowledge by providing information on ancient pinnipeds, sea otters, and sea cows throughout the northeastern Pacific. In total, the volume comprises the most comprehensive, interdisciplinary study on the archaeology of pinnipeds and sea otters from this region and serves as a model for other historical ecological studies elsewhere in the world.

Thirteen chapters from the Aleutian Islands to the southern California Channel Islands cover a variety of different species and draw on diverse methodological perspectives to yield empirical and theoretical insights. In Chapter 2, Lyman provides a comprehensive analysis of the history of paleoecological research on northeastern Pacific pinnipeds and sea otters. After this chapter, ten regional case studies from the northeastern Pacific are presented. In Chapter 3, Hill explores human exploitation of walrus in the far North Pacific along the Bering Strait. In Chapter 4, Crockford and Frederick present their research on fur seal breeding patterns and changing sea ice and other environmental variables in Alaska. This is followed by Betts, Maschner, and Lech's results (Chapter 5 and Chapter 6) from their research in the Aleutians (Betts et al. and Lech et al., respectively). In Chapter 7, McKechnie and Wigen summarize the history of pinniped exploitation on the coast of southern British Columbia. We then move to the Oregon and Washington coasts, where Moss and Losey (Chapter 8) analyze the regional archaeological record of pinnipeds and sea otters. Whitaker and Hildebrandt (Chapter 9) examine aspects of the northern California record of pinniped hunting and human prestige economies. In Chapter 10, Gifford-Gonzalez provides detailed information on her long-term research on fur seals in central California. Jones et al. (Chapter 11) then provide insight into the long-term dynamics of human hunting of sea otters in central California, including ancient DNA and stable isotope data. This is followed by Braje et al.'s (Chapter 12) overview of pinniped and sea otter hunting on the California Channel Islands. Finally, Braje and Rick provide an analysis, critique, and suggestions for further research in Chapter 13.

ACKNOWLEDGMENTS

Discussions with Jon Erlandson, Mike Etnier, Mike Glassow, Bill Hildebrandt, Terry Jones, Doug Kennett, Sharon Melin, Madonna Moss, Seth Newsome, Tony Orr, Tom Wake, Phil Walker, and others have greatly increased our understanding of the archaeology and biology of pinnipeds and sea otters. We thank all the contributors for their efforts in meeting our deadlines, writing insightful chapters, and helping bring this volume to fruition. Mark Raab and Andrew Trites provided important comments that greatly improved this chapter and the rest of the volume.

REFERENCES CITED

Alvard, M. A.
1998 Evolutionary Ecology and Resource Conservation. *Evolutionary Anthropology* 7:62–74.
Angliss, R. P., and B. M. Allen
2009 Alaska Marine Mammal Stock Assessments, 2008. NOAA Technical Memorandum NMFS-AFSC-193.
Bailey, G.
2004 World Prehistory from the Margins: The Role of Coastlines in Human Evolution. *Journal of Interdisciplinary Studies in History and Archaeology* 1:39–50.
Belcher, R. L., and T. E. Lee, Jr.
2002 *Arctocephalus townsendii. Mammalian Species* 700: 1–5.
Bernardi, G., S. R. Fain, J. P. Gallo-Reynoso, A. L. Figeuroa-Carranza, B. J. Le Boeuf
1998 Genetic variability in Guadalupe fur seals. *Journal of Heredity* 89:301–305.
Berta, A., J. L. Sumich, and K. M. Kovacs
2005 *Marine Mammals: Evolutionary Biology*. Second Edition. Academic Press, San Diego.
Bodkin, J. L.
2000 Sea Otters. *Alaska Geographic* 27(2):74–92.
Bodkin, J. L., B. E. Ballachey, M. A. Cronin, K. T. Scribner
1999 Population Demographics and Genetic Diversity in Remnant and Translocated Populations of Sea Otters. *Conservation Biology* 13:1378–1385.
Bonnell, M. L., and R. K. Selander
1974 Elephant Seals: Genetic Variation and Near Extinction. *Science* 184:908–909.
Braje, T. J.
2010 *Modern Oceans, Ancient Sites: Archaeology and Marine Conservation on San Miguel Island, California*. University of Utah Press, Salt Lake City.
Broughton, J. M.
1994 Declines in Mammalian Foraging Efficiency during the Late Holocene San Francisco Bay, California. *Journal of Anthropological Archaeology* 13:371–401.
1999 *Resource Depression and Intensification during the Late Holocene, San Francisco Bay: Evidence from the Emeryville Shellmound Vertebrate Fauna*. University of California Anthropological Records 32.
Burton, R. K., D. Gifford-Gonzalez, J. J. Snodgrass, and P. L. Koch
2002 Isotopic Tracking of Prehistoric Pinniped Foraging and Distribution along the Central California Coast: Preliminary Results. *International Journal of Osteoarchaeology* 12:4–11.
Burton, R. K., J. J. Snodgrass, D. Gifford-Gonzalez, T. Guilderson, T. Brown, and P. L..Koch
2001 Holocene Changes in the Ecology of Northern Fur Seals: Insights from Stable Isotopes and Archaeofauna. *Oecologia* 128:107–115.
Busch, B. C.
1985 *The War against the Seals: A History of the North American Seal Fishery*. Kingston, Ontario: McGill-Queen's University Press.
Carretta, J. V., K. A. Forney, M. M. Muto, J. Barlow, J. Baker, B. Hanson, and M. S. Lowry
2007 *U.S. Pacific Marine Mammal Stock Assessments: 2006*. U.S. Dep. Commer. NOAA Tech. Memo. NOAA-TM-NMFS-SWFSC-398.
Carretta, J. V., K. A. Forney, M. S. Lowry, J. Barlow, J. Baker, D. Johnston, B. Hanson, R. L. Brownell, Jr., J. Robins, D. K. Mattila, K. Ralls, M. M. Muto, D. Lynch, and L. Carswell
2009 *U.S. Pacific Marine Mammal Stock Assessments: 2009*. U.S. Dep. Commer. NOAA Tech. Memo. NOAA-TM-NMFS-SWFSC -453.
Colton, R. H
2002 Prehistoric Marine Mammal Hunting in Context: Two Western North American Examples. *International Journal of Osteoarchaeology* 12:12–22.
Corbett, D. G., D. Causey, M. Clementz, P. L. Koch, A. Doroff, C. Lefebre, and D. West
2008 Aleut Hunters, Sea Otters, and Sea Cows: 3000 Years of Interactions in the Western Aleutian Islands, Alaska. In *Human Impacts on Ancient Marine Ecosystems: A Global Perspective*, edited by T. C. Rick and J. M. Erlandson, pp. 43–75. University of California Press, Berkeley.
Crockford, S. J., and S. G. Frederick
2007 Sea Ice Expansion in the Bering Sea during the Neoglacial: Evidence from Archaeozoology. *The Holocene* 17:699–706.
Crumley C. L. (editor)
1994 *Historical Ecology: Cultural Knowledge and Changing Landscapes*. School of American Research Press, Santa Fe, New Mexico.

DeLong, R. L., and S. R. Melin
2002 Thirty Years of Pinniped Research at San Miguel Island. In *The Fifth California Islands Symposium*, edited by D. Brown, K. Mitchell, and H. Chaney, pp. 401–406. Santa Barbara Museum of Natural History, Santa Barbara.
DeLong, R. L., and B. Stewart
1991 Diving Patterns of Northern Elephant Seal Bulls. *Marine Mammal Science* 7:369–384.
Domning, D. P., J. Thomason, and D. G. Corbett
2007 Steller's Sea Cow in the Aleutian Islands. *Marine Mammal Science* 23:976–983.
Ellis, R.
2003 *The Empty Ocean*. Island Press/Shearwater Books, Washington D.C.
Erlandson, J. M.
2001 The Archaeology of Aquatic Adaptations: Paradigms for a New Millennium. *Journal of Archaeological Research* 9:287–350.
Erlandson, J. M., and T. C. Rick
2010 Archaeology Meets Marine Ecology: The Antiquity of Maritime Adaptations and Human Impacts on Marine Fisheries and Ecosystems. *Annual Review of Marine Science* 2, 231–251.
Erlandson, J. M., T. C. Rick, J. A. Estes, M. H. Graham, T. J. Braje, and R. L. Vellanoweth
2005 Sea Otters, Shellfish, and Humans: 10,000 Years of Ecological Interaction on San Miguel Island, California. In *Proceedings of the Sixth California Islands Symposium*, edited by D. Garcelon and C. Schwemm, pp. 9–21. National Park Service Technical Publication CHIS-05-01, Institute for Wildlife Studies, Arcata, California.
Estes, J. A.
1980 *Enhydra lutris. Mammalian Species* 133:1–8.
Etnier, M. A.
2002a *The Effects of Human Hunting on Northern Fur Seal* (Callorhinus ursinus) *Migration and Breeding Distributions in the Late Holocene*. PhD dissertation, University of Washington, Seattle.
2002b Occurrences of Guadalupe Fur Seals (*Arctocephalus townsendi*) on the Washington Coast over the Past 500 Years. *Marine Mammal Science* 18:551–557.
2004 The Potential of Zooarchaeological Data to Guide Pinniped Management Decisions in the Northeastern Pacific. In *Zooarchaeology and Conservation Biology*, edited by R. L. Lyman and K. Cannon, pp. 88–102. University of Utah Press, Salt Lake City.
2007 Defining and Identifying Sustainable Harvests of Resources: Archaeological Examples of Pinniped Harvests in the Northeastern Pacific. *Journal for Nature Conservation* 15:196–207.
Gallo-Reynoso, J. P., B. J. Le Boeuf, A. L. Figueroa-Carranza, and M. O. Maravilla-Chávez
2005 Los Pinnípedos de Isla Guadalupe. In *Isla Guadalupe, Restauración y Conservación*, edited by K. Santos del Prado and E. Peters, pp. 170–201. Instituto Nacional de Ecología, Mexico.
Gifford-Gonzalez, D., S. D. Newsome, P. L. Koch, T. P. Guilderson, J. J. Snodgrass, and R. K. Burton
2005 Archaeofaunal Insights on Pinniped–Human Interactions in the Northeastern Pacific. In *The Exploitation and Cultural Importance of Sea Mammals*,edited by G. Monks, pp. 19–38. Oxbow Books, Oxford.
Hames, R.
2007 The Ecologically Noble Savage Debate. *Annual Review of Anthropology* 36:177–190.
Hatfield, B. B.
2005 The Translocation of Sea Otters to San Nicolas Island: An Update. In *Proceedings of the Fifth California Islands Symposium*, edited by D. Garcelon and C. Schwemm, pp. 473–475. National Park Service Technical Publication CHIS-05-01, Institute for Wildlife Studies, Arcata, California.
Hayashida, F. M.
2005 Archaeology, Ecological History, and Conservation. *Annual Review of Anthropology* 34:43–65.
Hildebrandt, W. R., and T. L. Jones
1992 Evolution of Marine Mammal Hunting: A View from the California and Oregon Coasts. *Journal of Anthropological Archaeology* 11:360–401.
2002 Depletion of Prehistoric Pinniped Populations along the California and Oregon Coasts: Were Humans the Cause? In *Wilderness and Political Ecology: Aboriginal Influences on the Original State of Nature*, edited by C. E. Kay and R. T. Simmons, pp. 72–110. University of Utah Press, Salt Lake City.
Hoelzel, A. R., R. C. Fleischer, C. Campagna, B. J. Le Bouef, and G. Alvord
2002 Impact of a Population Bottleneck on Symmetry and Genetic Diversity in the Northern Elephant Seal. *Journal of Evolutionary Biology*. 15:567–575.
Hunn, E. S., D. R. Johnson, Priscilla N. Russell, and T. F. Thornton
2003 Huna Tlingit Traditional Environmental Knowledge, Conservation, and the Management of a "Wilderness" Park. *Current Anthropology* 44 (Supplement): S79–S103.
Jackson, J. B. C., M. Kirby, W. Berger, K. Bjorndal, L. Botsford, B. Bourque, R. Bradbury, R. Cooke, J. Erlandson, A. James, J. Estes, T. Hughes, S. Kidwell, C. Lange, H. Lenihan, J. Pandolfi, C. Peterson, R. Steneck, M. Tegner, and R. Warner

2001 Historical Overfishing and the Recent Collapse of Coastal Ecosystems. *Science* 293:629–637.

Jefferson, T. A., M. A. Weber, and R. Pitman
2008 *Marine Mammals of the World: A Comprehensive Guide to Their Identification.* Academic Press, San Diego.

Jones, T. L., and W. R. Hildebrandt
1995 Reasserting a Prehistoric Tragedy of the Commons: Reply to Lyman. *Journal of Anthropological Archaeology* 14:78–98.

Jones, T. L., W. R. Hildebrandt, D. J. Kennett, and J. F. Porcasi
2004 Prehistoric Marine Mammal Overkill. *Journal of California and Great Basin Anthropology* 24: 69–80.

Kay, C. E., and R. F. Simmons, editors
2002 *Wilderness and Political Ecology: Aboriginal Influences and the Original State of Nature.* University of Utah Press, Salt Lake City.

Kirch, P. V.
2005 Archaeology and Global Change: The Holocene Record. *Annual Review of Environment and Resources* 30:409–440.

Klein, R. G., and K. Cruz-Uribe
1996 Exploitation of Large Bovids and Seals at Middle and Later Stone Age Sites in South Africa. *Journal of Human Evolution* 31:315–334.

Lidicker, W. Z., and F. C. McCollum
1997 Genetic Variation in California Sea Otters. *Journal of Mammalogy* 78:417–425.

Loughlin, T. R., M. A. Perez, and R. L. Merrick
1987 *Eumetopias jubatus. Mammalian Species* 283:1–7.

Lyman, R. L.
1988 Zoogeography of Oregon Coast Marine Mammals: The Last 3000 Years. *Marine Mammal Science* 4:247–264.
1991 *Prehistory of the Oregon Coast: The Effects of Excavation Strategies and Assemblage Size on Archaeological Inquiry.* Academic Press, San Diego.
1995 On the Evolution of Marine Mammal Hunting on the West Coast of North America. *Journal of Anthropological Archaeology* 14:45–77.
2003 Pinniped Behavior, Foraging Theory, and the Depression of Metapopulations and Nondepression of a Local Population on the Southern Northwest Coast of North America. *Journal of Anthropological Archaeology* 22:376–388.
2006 Paleozoology in the Service of Conservation Biology. *Evolutionary Anthropology* 15:11–19.

Lyman, R. L., and K. Cannon (editors)
2004 *Zooarchaeology and Conservation Biology.* University of Utah Press, Salt Lake City.

McNiven, I. J., and A. C. Beddingfield
2008 Past and Present Marine Mammal Hunting Rates and Abundances: Dugong (*Dugong dugon*) evidence from Dabangai Bone Mound, Torres Strait. *Journal of Archaeological Science* 35:505–515.

Melin, S. R., and R. L. DeLong
1999 Observations of a Guadalupe fur seal (*Arctocephalus townsendi*) female and pup at San Miguel Island, California. *Marine Mammal Science* 15:885–888.

Monks, G. G. (editor)
2005a *The Exploitation and Cultural Importance of Sea Mammals.* Oxbow Books, Oxford.

Monks, G. G.
2005b Discussion: Sea Mammals in Zooarchaeology AD 2002. In *The Exploitation and Cultural Importance of Sea Mammals,* edited by G. Monks, pp. 167–173. Oxbow Books, Oxford.

Moss, M. L., and J. M. Erlandson
1995 Reflections on North American Pacific Coast Prehistory. *Journal of World Prehistory* 9:1–45.

Moss, M. L., D. Y. Yang, S. D. Newsome, C. F. Speller, I. McKechnie, A. D. McMillan, R. J. Losey, and P. L. Koch
2006 Historical Ecology and Biogeography of North Pacific Pinnipeds: Isotopes and Ancient DNA from Three Archaeological Assemblages. *Journal of Island and Coastal Archaeology* 1: 165–190.

Nagaoka, L.
2002 The Effects of Resource Depression on Foraging Efficiency, Diet Breadth, and Patch Use in Southern New Zealand. *Journal of Anthropological Archaeology* 21:419–442.

Newsome, S. D., M. A. Etnier, D. Gifford-Gonzalez, D. L. Phillips, M. van Tuinen, E. A. Hadly, D. P. Costa, D. J. Kennett, T. P. Guilderson, and P. L. Koch
2007 The Shifting Baseline of Northern Fur Seal Ecology in the Northeast Pacific Ocean. *Proceedings of the National Academy of Sciences* 104: 9709–9714.

Ogden, A.
1941 *The California Sea Otter Trade 1784–1848.* University of California Publications in History 26. University of California Press, Berkeley.

Orr, R. T., and R. C. Helm
1989 *Marine Mammals of California.* University of California Press, Berkeley.

Pauly, D.
1995 Anecdotes and the Shifting Baselines Syndrome of Fisheries. *Trends in Ecology and Evolution* 10(10):430.

Pinnegar, J. K., and G. H. Engelhard
2008 The "Shifting Baseline" Phenomenon: A Global Perspective. *Review of Fish Biology and Fisheries* 18:1–16.
Porcasi, J. F., T. L. Jones, and L. Mark Raab
2000 Trans-Holocene Marine Mammal Exploitation on San Clemente Island: A Tragedy of the Commons Revisited. *Journal of Anthropological Archaeology* 19:200–220.
Redman, C.
1999 *Human Impact on Ancient Environments.* University of Arizona Press, Tucson.
Repenning, C. A., R. S. Peterson, and C. L. Hubbs
1971 Contributions of the Systematics of the Southern Fur Seals, with Particular Reference to the Juan Fernandez and Guadalupe Species. In Antarctic Pinnipedia, edited by W. H. Burt, pp. 1–34. *Antarctic Research Series* 18:1–226.
Rice, D. W.
1998 *Marine Mammals of the World: Systematics and Distribution.* Special Publication No. 4. The Society for Marine Mammalogy.
Rick, T. C.
2007 *The Archaeology and Historical Ecology of Late Holocene San Miguel Island.* Cotsen Institute of Archaeology, University of California, Los Angeles.
Rick, T. C., and J. M. Erlandson (editors)
2008 *Human Impacts on Ancient Marine Ecosystems: A Global Perspective.* University of California Press, Berkeley.
Rick, T. C., and J. M. Erlandson
2009 Coastal Exploitation. *Science* 325:952–953.
Rick, T. C., R. L. DeLong, J. M. Erlandson, T. J. Braje, D. J. Kennett. T. L. Jones, T. A. Wake, and P. L. Walker
2009 A Trans-Holocene Archaeological Record of Guadalupe Fur Seals on the California Coast. *Marine Mammal Science* 25:487–502.
Rick, T. C., R. L. DeLong, J. M. Erlandson, T. L. Jones, T. J. Braje, J. E. Arnold, M. R. Des Lauriers, W. R. Hildebrandt, D. J. Kennett, and T. A. Wake
2010 Where Were the Northern Elephants Seals? Archaeology and Holocene Biogeography of *Mirounga angustirostris. The Holocene,* in review.
Riedman, M.
1990 *The Pinnipeds: Seals, Sea Lions, and Walruses.* University of California Press, Berkeley.
Riedman, M. L., and J. A. Estes
1990 The Sea Otter (*Enhydra lutris*): Behavior, Ecology, and Natural History. *Biological Report* 90(14). Washington D. C.: U.S. Department of the Interior, Fish and Wildlife Service.
Scammon, C. M.
1968 *The Marine Mammals of the Northwestern Coast of North America.* Originally published in 1874. Dover, New York.
Simenstad, C.A., J. A. Estes, and K. W. Kenyon
1978 Aleuts, Sea Otters, and Alternate Stable-State Communities. *Science* 200:403–411.
Smith, E. A., and M. Wishnie
2000 Conservation and Subsistence in Small-Scale Societies. *Annual Review of Anthropology* 29:493–524.
Steneck, R., M. Graham, B. Bourque, D. Corbett, J. M. Erlandson, and J. Estes
2002 Kelp Forest Ecosystems: Biodiversity, Stability, Resilience, and Their Future. *Environmental Conservation* 29:436–459.
Stewart, B. S., and R. L. DeLong
1995 Double Migrations of the Northern Elephant Seal, *Mirounga angustirostris. Journal of Mammalogy* 76:196–205.
Stewart, B. S., and H. R. Huber
1993 *Mirounga angustirostris. Mammalian Species* 449:1–10.
Stringer, C. B., J. C. Finlayson, R. N. E. Barton, Y. Fernandez-Jalvo, I. Cacares, R. C. Sabin, E. J. Rhodes, A. P. Currant, J. Rodriguez-Vidal, F. Giles-Pacheco, and J. A. Riquelme-Cantal
2008 Neanderthal Exploitation of Marine Mammals in Gibraltar. *Proceedings of the National Academy of Sciences* 104: 14319–14324.
Trites, A. W., and C. P. Donnelly
2003 The Decline of Steller Sea Lions *Eumetopias jubatus* in Alaska: A Review of the Nutritional Stress Hypothesis. *Mammal Review* 33: 3–28.
Trites, A. W. A. J. Miller, H. D. G. Maschner, M. A. Alexander, S. J. Bograd, A. Capotondi, K. O. Coyle, E. Di Lorenzo, T. C. Royer, E. J. Gregr, C. E. Grosch, B. P. Finney, L. Fritz, G. L. Hunt, J. Jahncke, N. B. Kachel, H. Kim, C. Ladd, N. J. Mantua, C. Marzban, W. Maslowski, D. J. Neilson, J. E. Overland, S. R. Okkonen, K. L. Reedy-Maschner, J. X. L. Wang, and A. J. Winship
2007 Bottom-Up Forcing and the Decline of Steller Sea Lions in Alaska: Assessing the Ocean Climate Hypothesis. *Fisheries Oceanography* 16:46–67
Valentine, K., D. A. Duffield, L. E. Patrick, D. R. Hatch, V. L. Butler, R. L. Hall, and N. Lehman
2008 Ancient DNA Reveals Genotypic Relationships among Oregon Populations of the Sea Otter (*Enhydra lutris*). *Conservation Genetics* 9:933–938.
Walker, P. L., D. J. Kennett, T. Jones, and R. L. Delong

2002 Archaeological Investigations of the Point Bennett Pinniped Rookery on San Miguel Island. In *Proceedings of the Fifth California Islands Symposium*, edited by D. Brown, K. Mitchell, and H. Chaney, pp. 628–632. Santa Barbara Museum of Natural History, Santa Barbara.
Weber, D. S., B. S. Stewart, and N. Lehman
2004 Genetic Consequences of a Severe Population Bottleneck in the Guadalupe Fur Seal (*Arctocephalus townsendi*). *Journal of Heredity* 95(2):144–153.
Wilson, D. E., M. A. Bogan, R. L. Brownell, Jr., A. M. Burdin, and M. K. Maminov
1991 Geographic Variation in the Sea Otters, *Enhydra lutris*. *Journal of Mammalogy* 72:22–36.

2

A History of Paleoecological Research on Sea Otters and Pinnipeds of the Eastern Pacific Rim

R. Lee Lyman

TODAY IT IS NOT UNUSUAL to pick up the latest issue of an archaeology journal such as *American Antiquity*, or *Journal of Archaeological Science*, or the like, and to find in the table of contents an article on a topic in zooarchaeology. It is more difficult to find a piece on the zooarchaeology of marine mammals. It is equally difficult to find an article on some prehistoric aspect of marine mammals based on zooarchaeological remains in a natural history journal such as *Marine Mammal Science* or *Oecologia*. These informal observations prompt some musings. Has the history of zooarchaeological research on North Pacific pinnipeds and sea otters (*Enhydra lutris*) been unique, or has it been but a portion of the much larger history of zooarchaeology in general? Have trends in analysis of pinniped and sea otter remains tracked analytical trends in zooarchaeology in general? In this chapter, I provide some initial answers to these questions. In particular, I report and comment on the history of paleoecological research on pinniped and sea otter remains recovered from archaeological sites along the northeastern Pacific coast. I conclude with some observations on potentially significant research topics that have as yet been little explored.

METHODS, MATERIALS, AND CAVEATS

To write the history that follows, I read the published literature only and did not examine unpublished archival records nor did I delve into the extensive unpublished (and thus basically inaccessible) grey literature resulting from CRM projects. Exceptions include several unpublished doctoral dissertations, copies of which I happen to have. Taxonomic abundance data are presented as the number of identified specimens (NISP; Grayson 1984; Lyman 2008). Discussion is limited to the pinnipeds and sea otters; whales and other cetaceans are not considered.

For purposes of this chapter, the northeastern Pacific Rim includes, from south to north,

Human Impacts on Seals, Sea Lions, and Sea Otters: Integrating Archaeology and Ecology in the Northeast Pacific, edited by Todd J. Braje and Torben C. Rick.

TABLE 2.1

Pinniped and Sea Otter Nomenclature for the Northeastern Pacific Ocean

TAXON	COMMON NAME
Enhydra lutris	Sea otter
Phoca vitulina	Harbor seal
Mirounga angustirostris	Northern elephant seal
Callorhinus ursinus	Northern fur seal
Arctocephalus townsendi	Guadalupe fur seal
Eumetopias jubatus	Steller (northern) sea lion
Zalophus californianus	California sea lion

NOTE: Taxonomy after King (1983).

the California, Oregon, Washington, British Columbia, and southeastern Alaskan coasts, both those coasts along the open ocean and those facing the inside passage. The groups of pinnipeds and sea otters discussed comprise seven species (Table 2.1). I mention the paleontological record where pertinent, and focus on the archaeological record of these creatures because that is where the bulk of the pertinent data originates. I mostly restrict discussion to aspects of paleoecology—biogeography, rookery locations, and the like—indicated by the remains of marine mammals. There have been many discussions of the human economic implications of the marine mammal remains recovered from archaeological contexts along the eastern Pacific Rim (e.g., Calvert 1980; Cannon 1991; Colton 2002; Hildebrandt 1981, 1984b; Huelsbeck 1983; Stewart and Stewart 1996; Wigen and Stucki 1988). I do not review these discussions here, as many of the other chapters in this volume cover this topic in detail.

Given the focus of this chapter, it is appropriate to note that natural historian W. J. Wintemberg (1919:63) pointed out nine decades ago that zooarchaeological remains would "give valuable aid to zoology." In particular, Wintemberg noted that zooarchaeological remains would reveal much about prehistoric ranges of taxa, long-extinct and thus zoologically unknown species might be revealed by such remains, skeletal pathologies of taxa might be better documented with zooarchaeological materials, and temporal changes in the size of animals might be revealed by prehistoric faunal remains. Some but not all of these revelations have occurred via study of marine mammal remains recovered from archaeological sites along the northeastern Pacific Rim.

BEGINNINGS

William Healy Dall (1877) excavated several sites in the Aleutian Islands during the late 19th century. He listed the sea mammals represented by faunal remains he found: northern fur seal (*Callorhinus ursinus*), Steller sea lion (*Eumetopias juabutus*), *Phoca* (two species), and walrus (*Odobenus rosmarus*) (Dall 1877:74). Dall was a naturalist of the first rank (Merriam 1927); he did not need to indicate how he identified the remains nor did he list the frequencies of the remains he found. This was typical of the time (Reitz 1993; Robison 1987)—what are pejoratively referred to as "species lists" were the typical result of zooarchaeological work in the late 19th and early 20th centuries, if any such work was done at all. Dall did not mention any paleoecological implications of the remains he reported. This is not surprising for the simple reason that naturalists were still learning about the modern zoological world in the late 19th century and thus did not know what might be unusual in the prehistoric record.

The report on northeastern Pacific Rim marine mammals that most people know about and that can be taken as the seminal modern zooarchaeological study was undertaken by Gretchen M. Lyon in the 1930s. Lyon (1935, 1937) described remains from the Point Mugu shell mound in Ventura County, southern California. She was a zoologist in the natural history sense. She studied living amphibians, mammals, and birds; she illustrated the work of other zoologists; and she did some paleontology (published

TABLE 2.2

NISP and MNI Data for Pinnipeds and Sea Otters Reported by Lyon (1937)

SPECIES	NISP	MNI	PERCENT OF MNI
Enhydra lutris	304	31	13.7
Zalophus californianus	145	19	8.3
Eumetopias jubatus	12	4	1.7
Callorhinus ursinus	57	12	5.2
Arctocephalus townsendi	1557	152	66.9
Phoca vitulina	18	4	1.7
Mirounga angustirostris	21	5	2.2

NOTE: Taxonomy is updated.

under her married name; Burleson 1941, 1948). In her report on the Point Mugu materials Lyon not only provided quantitative data in the form of NISP and MNI (Minimum Number of Individuals) per species (Table 2.2), but she described taxonomically diagnostic morphometric features of some of the bones she identified. Lyon's use of NISP and MNI followed work by paleontologists who were her contemporaries (e.g., Stock 1929). Lyon's description of taxonomically diagnostic anatomical features mirrored some of the seminal efforts of individual researchers to present what can loosely be termed skeletal key-like information that could be used by other archaeologists (e.g., Brainerd 1939). At the time, such information was new and worthy of publication. Lyon's work preceded by more than a decade the more widely known (among archaeologists) efforts of Theodore White (e.g., 1952, 1953a, 1953b), a paleontologist who studied zooarchaeological remains recovered from sites on the Plains and who is typically credited with introducing the concept of MNI to archaeologists. Credit likely should be given to White because he published in *American Antiquity* and other archaeological venues whereas Lyon published her work in zoological journals that were seldom read by archaeologists until about 1980.

Lyon (1937:163) believed that the remains she described reflected "a changing picture of marine life," and thus, as a biologist, she set a precedent that would not soon be mimicked by zooarchaeologists. Lyon discussed the "past and present status of the species" represented by the faunal remains from Point Mugu. She noted, for example, that though the sea otter was rare in the site area today and that it had obviously been hunted by prehistoric people, the abundance of its remains indicated that it had been "formerly abundant" in the area and that it was "likely the white man" who had decimated the population in the late 18th century and throughout the 19th century (Lyon 1937:163, 164). Lyon noted regarding the northern fur seal that no remains of males had been found; only remains of females were in the collection, and this matched expectations based on the modern migratory habits of this species. The large number of Guadalupe fur seal (*Arctocephalus townsendi*) remains (Table 2.2) was surprising to Lyon (1937:164), who noted that this species was thought to be extirpated "north of the Mexican line." Significantly, Lyon (1937:165) reported that she had examined a small sample of bones collected from a shell midden located at Yachats in Lincoln County, Oregon, and those remains included several specimens of female Guadalupe fur seal. Lyon (1937:165) took those specimens as evidence that this species had once been found that far north, adding that samples from other sites were necessary "to establish with certainty the northern limit of the range of this species." This research avenue would not be exploited until more then 60 years later. Finally, Lyon found that the age-sex demography of the sample suggested that rookeries had been exploited. In offering her biogeographic and biological inferences, Lyon was holding to the natural history tradition in which she was trained.

BIOGEOGRAPHY, DEMOGRAPHY, AND ROOKERIES

Lyon's observations regarding differences between modern and prehistoric distributions of taxa set a standard that would not be diverged from for nearly 65 years. Thirty years after Lyon's report was published, Carl E. Gustafson (1968a, 1968b) noted that his work with the Ozette site zooarchaeological materials from the northwestern tip of Washington State's Olympic Peninsula revealed abundant remains of northern fur seal. Gustafson was a zoologist by training, though throughout his career he was Washington State University's zooarchaeologist in the Department of Anthropology. Gustafson perceived no difference in the relative abundance of fur seal remains from precontact to postcontact time among the remains from Ozette, and this suggested to him that some 2000 years of human predation had not had an impact on the local population. The abundance of male northern fur seals relative to the abundance of females was about 1:1 throughout the stratigraphic sequence, but males were absent from the historic record. Gustafson (1968b:51) attributed this demographic shift to "a change in the migratory pattern of male fur seals." Gustafson's work is noteworthy because it represents an early zooarchaeological study in which an estimate of the ontogenetic age of individual organisms based on tooth development was used. Gustafson (1968b:50) submitted teeth to the "Bureau of Commercial Fisheries, Division of Marine Mammals, for exact age determinations" based on "annual growth rings in the canines (Scheffer 1950)," but in the absence of those data he "constructed relative age categories based on the size of the root canal, which becomes smaller as dentine is deposited with increasing age." Gustafson's research was published in *Science,* yet it seems to have had no more (or less) impact on paleozoology than did Lyon's less widely circulated report. The article was rarely cited by paleozoologists over the next three decades, and many of the insights Gustafson provided were neither replicated nor evaluated in light of other data until early in the 21st century.

A decade after Gustafson's work, Phillip Walker and Steven Craig (1979:50), two archaeologists, reiterated Wintemberg's (1919) six-decades-old statement: zooarchaeological remains can, they said, "provide information concerning the biology of animal species." They described remains from a site on San Miguel Island (part of California's Channel Islands), noting that remains of Guadalupe fur seal were much more abundant relative to the remains of other pinnipeds than would be expected given modern abundances of the taxa in the area. They attributed this difference in observed versus expected abundances to historic "commercial sealing activities" having decimated the metapopulation, and noted that their data matched those described by Gretchen Lyon (Walker and Craig 1979:53). Walker and Craig published their research results in *California Fish and Game,* a regional journal unlikely to have been read by many archaeologists. The interests of archaeologists at the time largely concerned those of artifact-centric culture history or culture process-centric processual archaeology; the paleoecological implications of faunal remains were of little interest to the majority of archaeologists because prior to about 1970 they were asking questions that did not require paleoecological data (Lyman 2007a, 2007b; Lyman et al. 1997). This would change as the population of archaeologists grew and archaeologists diversified and specialized their interests (Reitz 1993; Trigger 2006).

By the end of the 1970s, zooarchaeology had become an important and potentially autonomous research endeavor (Reitz 1993; Reitz and Wing 1999). The likely catalysts for this development were equal parts of processual archaeology's quest for materialist and functionalist explanations including economic variables (see for example Graham's [1979] and Lundelius's [1974] synopses of late Pleistocene paleomammalogy in volumes devoted to Paleo-Indian archaeology), the federal government's mandates for protection of archaeological resources via recovery and analysis of all cultural materials, and a growing population of archaeologists such that intradisciplinary specialization was not only possible but predictable (O'Brien et al. 2005; Reitz 1993). Zooarchaeology along the northeastern Pacific Rim tracked these developments, including an increased rate of publication.

William Hildebrandt (1981, 1984b), an archaeologist, identified and tallied the remains of sea otters and four taxa of pinnipeds recovered from six sites on the northern California coast (ΣNISP = 746) for his doctoral research. Hildebrandt (1984a) observed that the zooarchaeological remains he described "may indicate a former deviation from [the historically documented] pattern" of northern fur seals not frequenting the coast of northern California. Because northern fur seal remains comprised two-thirds of the marine mammal remains from the site of Stone Lagoon, Hildebrandt (1984a:29) hypothesized that prehistoric hunters had obtained these animals from the nearby offshore Redding Rock and that this location "may have been used heavily by the northern fur seal as a hauling ground." Hildebrandt cited Lyon's (1937) work, but not Gustafson's (1968a, 1968b). Importantly, Hildebrandt published his hypothesis in a natural history journal rather than an archaeological journal. It was, however, a regional journal with limited circulation, and thus it is likely that few marine mammalogists were aware of Hildebrandt's significant observations.

Archaeologist Donald Clark (1986:39) effectively reiterated Wintemberg's (1919) notation when he remarked that a "highly effective mode of long-term biological sampling is the analysis of kitchen middens or refuse deposits left at ancient habitation sites." Clark interpreted changing abundances of remains of northern fur seal in archaeological sites on Kodiak Island, Alaska, as indicating that this species had been abundant near the southeastern side of the island during the late prehistoric/earliest historic period. Both prior to and subsequent to a high abundance of northern fur seal remains relative to other pinnipeds, northern fur seal remains were not very abundant. Clark (1986:42) was unsure of the cause of the "blip" in the abundance of fur seal remains, but safely concluded that it could not be presumed that prehistoric maritime ecosystems were stable.

In the late 1980s, zooarchaeologist Lee Lyman (1988, 1989, 1991) reported on remains of sea otters and pinnipeds from three sites on the coast of Oregon (ΣNISP = 3235). He used the sex and age demography indicated by the pinniped remains to determine whether prehistoric hunters were exploiting haul-outs or rookeries. Lyman (1988) inferred the local presence of rookeries for northern fur seal and Steller sea lion. Rookeries for northern fur seal were historically unknown along the Oregon coast, and at least one and perhaps two Steller sea lion rookeries Lyman inferred were historically not documented. Lyman (1988) attributed differences between the prehistoric and historic records to 19th-century commercial exploitation of pinnipeds and the decimation of local populations. His results were published in an international marine mammalogy journal with the explicit purposes of illustrating the precise nature of commercial exploitation on marine mammal populations and contributing pertinent data to biological conservation and management decisions. Such applied zooarchaeological research is increasing with respect to both terrestrial mammals (references in Lyman 2006) and marine mammals (Murray 2008).

In the 1990s, the long-awaited reports on the late-prehistoric archaeological site at Ozette, Washington, were published (Huelsbeck 1994). Only a sample of the mammal remains was described, but that sample was an order of magnitude larger than any other sample from the eastern Pacific (ΣNISP = >48,000). Unfortunately, remains of the two species of sea lions—*Eumetopias jubatus* and *Zalophus californianus*—were not distinguished. The only biological observation offered was that the abundances of the taxa were "very similar to those observed by Gustafson" in a much smaller sample (Huelsbeck 1994:27). A few years later, zooarchaeologist Mike Etnier (2002a, 2002b) studied a sample of the pinniped mandibles from Ozette and found 34 specimens of Guadalupe fur seal mixed with 1374 specimens of northern fur seal. Etnier (2002a:555) drew three conclusions: (1) because all of the Guadalupe fur seal remains seemed to originate from individuals that were a year or less in age, this species likely had no rookeries in the area; (2) despite the historic range of the species being limited to southern California and Baja California, it had until late prehistoric times been found as far north as northern Washington; and (3) this species may have ranged northward outside its modern range coincident with El Niño events. Etnier (2002b) used demographic data to conclude that previously unidentified northern fur seal rookeries had existed prehistorically on or near the Washington coast. Finally, Etnier (2002b) found no evidence that prehistoric human predation had influenced the local population of northern fur seals. He concluded that historic commercial exploitation had altered the biogeography of the species (see also Etnier 2007).

Zooarchaeologist Diane Gifford-Gonzalez and colleagues (Gifford-Gonzalez et al. 2005) summarized the zooarchaeological record for pinnipeds along the northeastern Pacific coast. They noted that remains of northern fur seals were much more abundant prehistorically throughout the area than would be expected given the modern abundances and migratory behaviors of the species. They also pointed out that Steller sea lion rookeries as well as northern fur seal rookeries existed prehistorically in places where there was no historical record of rookeries for these species, but cautioned that additional research was necessary to confirm the occurrence of uniquely prehistoric rookeries. Most recently, archaeologist Torben Rick and colleagues (Rick et al. 2009) compiled and summarized all zooarchaeological data for Guadalupe fur seals on the California coast. The early-19th-century distribution of this species was poorly known given it had nearly been exterminated by the late 19th century. Its zooarchaeological remains indicate it was relatively abundant in southern California, particularly south of 35°N latitude, during the late Holocene (last 3500 years). In producing syntheses, Gifford et al. (2005) and Rick et al. (2009) mimic a trend in zooarchaeology in general. By the end of the 20th century, paleozoological (zooarchaeological and paleontological) data were sufficiently abundant that syntheses of information for numerous taxa in many areas could be written (e.g., Grayson 2005; Lyman 2004).

Pinnipeds are categorized as marine mammals, yet individuals of various taxa are known to occasionally ascend rivers, likely in pursuit of prey such as anadromous fish. This means that remains of marine taxa will potentially be recovered from riverine sites in freshwater settings. So far as I know, only Lyman and colleagues (Lyman et al. 2002) have examined this phenomenon with respect to eastern North Pacific pinnipeds. They found that harbor seals occurred in the lower reach of the Columbia River prehistorically, virtually since the beginning of the Holocene 10,000 years ago. This could be a critical bit of knowledge with respect to damming rivers and industrial alteration of estuaries. Thus far, wildlife managers have focused only on the impacts of such activities to anadromous fish. It may become necessary to monitor as well the impacts of anthropogenic activities on pinnipeds that ascend rivers. Such monitoring may, however, be a long time coming because California sea lions have been observed for the past decade or so preying upon salmon and steelhead—two economically valued fish—below Bonneville Dam on the Columbia River. In light of these observations, state fish and wildlife agencies in Washington, Oregon, and Idaho requested and received federal authorization in early 2008 to remove these pinnipeds by capture and relocation if possible, but by lethal means if necessary (Washington Department of Fish and Wildlife, 2008). The zooarchaeological record suggests pinnipeds have ascended the Columbia River for millennia, so one must wonder if lethal removal is ecologically wise and really the best option, especially in light of the dim future that has been forecast for marine mammals the world over (e.g., Anderson 2001).

CHEMISTRY, GENETICS, AND EXTIRPATION

By the end of the 20th century, zooarchaeology had entered what can be informally labeled the archaeometry stage. Animal bones were not just identified and their macroscopic features recorded; sometimes the chemistry, and sometimes the genetics of particular animal bones, were recorded (Reitz and Wing 1999). Study of eastern Pacific pinnipeds and sea otters followed suit.

ISOTOPES

One of the first studies of the chemistry of archaeological pinniped remains focused on the northern fur seal. Mammalogist Robert Burton, zooarchaeologist Diane Gifford-Gonzalez, and their colleagues (Burton et al. 2001, 2002) studied the stable isotopes of fur seal remains from Monterey Bay, California. They found that the carbon and nitrogen isotopes of the bones indicated that, unlike in the 20th century, the northern fur seal foraged off the coast of central and northern California in the past. The ontogenetic age of some individuals was estimated based on the size of the dentary and indicated that fur seals considerably younger than 3 months of age were represented. Isotopic analysis indicated these individuals had not yet been weaned when they died. Together, these data confirmed earlier suggestions that northern fur seals had reproduced in lower latitudes prior to the 18th century than they did today.

A second study of bone chemistry quickly followed. Etnier (2004a:99) noted that the stable isotopes of the northern fur seal bones from the Ozette site fell midway between those of modern Alaska and the archaeological specimens from California reported by Burton et al. (2001, 2002). In light of demographic data suggesting a local rookery had existed, Etnier (2004a:99) concluded that the isotope data indicated that the Ozette northern fur seals "maintained a foraging pattern distinctly different than those of the California and Alaska populations."

Archaeologist Madonna Moss and colleagues (Moss et al. 2006) examined the stable carbon and nitrogen isotopes evident in northern fur seal specimens from a site in southeastern Alaska and a site on the Oregon coast. They found that although specimens from both sites indicated the fur seals "were feeding offshore," they could not distinguish specimens at either site from modern northern fur seals from the Pribilof Islands of southwestern Alaska (Moss et al. 2006:179). In 2007, the collaborative efforts of several biologists and archaeologists provided resolution to Moss et al.'s (2006) conundrum. Newsome et al. (2007) were able to sort several collections of fur seal remains, both modern and prehistoric, into three geographic groups. Female northern fur seals from central and southern California have the highest isotope values; individuals from northern Oregon, Washington, British Columbia, southeastern Alaska, and the eastern Aleutians have intermediate values; and individuals from the western Aleutian Islands have the lowest isotope values. Newsome et al. (2007:9710) conclude that these distinctions "confirm that prehistoric northern fur seal from California were not immigrants from northern waters but instead were year-round residents." Isotope values also indicate that the northern fur seals in the geographically intermediate group "weaned at a much older age than their modern Bering Sea counterparts." Newsome et al.'s (2007:9711) suggestion that older weaning age is likely a result of less selective pressure from long-duration severe winter weather in southern latitudes relative to northern latitudes has recently received some very suggestive but not quite conclusive confirmation that has implications for conservation biology (Lea et al. 2009). Study of isotopes to detect migration patterns of terrestrial mammals have also been recently undertaken (e.g., Hughes 2004), and though not based on stable isotope analyses, other evidence of prehistoric migration of ungulates has been noted by conservation biologists (Berger et al. 2006). Thus, study of zooarchaeological remains of marine mammals is tracking zooarchaeological research in general.

GENES

Historical records indicate that some populations of eastern Pacific pinnipeds were decimated and others were exterminated by historic commercial exploitation. This well-documented fact provides geneticists with a testable hypothesis—the degree of genetic diversity in descendants of decimated populations should be less than the genetic diversity of descendant populations that did not experience a bottleneck. Similarly, prehistoric, prebottleneck populations should display relatively greater genetic diversity than modern, postbottleneck populations. Biologist Shawn Larson and colleagues (Larson et al. 2002a, 2002b) found exactly these test implications among eastern Pacific sea otters. Both extant and prehistoric sea otter populations displayed low variability in mtDNA, but sea otter remains from the Ozette site indicate that the population prior to the fur trade had more variation than all tested extant populations.

When Lyman (1988) reported on the phenotypic variation of Oregon's prehistoric sea otters, he implied that some of that variation was the result of genetic variation. The editors of *Marine Mammal Science* requested that discussion of the possible genetic implications of the phenotypic variation of prehistoric Oregon sea otters and of the likely results of transplanting the wrong phenotype (and by implication the wrong genotype) of sea otter to the Oregon coast both be omitted from the manuscript that eventually became Lyman (1988). In the original, unpublished manuscript, Lyman suggested that the Alaskan sea otters transplanted to the Oregon coast in the 1970s may have been doomed from the start because of their phenotypic (and implied genetic) adaptation to a high-latitude habitat and attendant dietary differences from sea otters in more southern latitudes.

Subsequent to Lyman's (1988) research, a collaborative project was undertaken between Oregon biologists and zooarchaeologists who studied ancient DNA extracted from archaeological specimens recovered from Oregon sites. These researchers (Valentine et al. 2008) found that the genetic composition of Oregon's prehistoric sea otters best matches the genetic composition of modern Californian sea otters rather than the Alaskan sea otter population that was exploited for individuals that were moved to other areas along the northeastern Pacific coast.

Moss et al. (2006) examined ancient DNA in a specific attempt to distinguish populations of fur seals among remains recovered from a site in southeastern Alaska (Cape Addington), a site on the west coast of Vancouver Island (Ts'ishaa), and a site on the Oregon coast. Although they found that prehistoric northern fur seals tended to have "a much higher genetic variability" than modern populations, the limited modern data prompted them to conclude that they could not determine "the extent of recent genetic bottlenecks" (Moss et al. 2006:181). Similarly, Moss et al. (2006) could not determine if their specimens represented distinct, prehistorically unique local breeding populations. This sort of research mimics that of others studying paleozoological remains of terrestrial taxa (e.g., Pusch et al. 2003 and references therein).

EXTIRPATION—TIMING AND CAUSE

Historical documents suggest that populations of many northeastern Pacific Rim pinniped and sea otter taxa were nearly extirpated by commercial and bounty exploitation during the 19th and early 20th centuries. Some archaeologists argue that prehistoric hunters drove many populations toward extinction (Hildebrandt and Jones 1992, 2002; Jones and Hildebrandt 1995; Porcasi et al. 2000). Not surprisingly, what we have found is that extirpation is historically contingent; it is population and location specific (Etnier 2002b; Gifford et al. 2005; Lyman 2003; Newsome et al. 2007). Available data suggest that central and southern California populations of fur seals and sea lions were extirpated 800 or more years ago whereas more northern populations of these taxa were extirpated only in the last 200 or so years (Etnier 2002b; Gifford et al. 2005; Lyman 2003).

It has been suggested that terrestrial climatic change in southern California prompted intensified exploitation and eventual decimation of pinniped populations (Colten and Arnold 1998). Detailed paleoclimatic records (e.g., Arnold and Tissot 1993; Jones and Kennett 1999) are required, as are tight chronological controls of zooarchaeological data, in order to establish that climatic variables had causative roles in the decimation of marine mammal populations (e.g., Trites et al. 2007). Coincidentally, it must be demonstrable that humans did not play a significant role in depressing populations of prey animals. Study of the timing of the extirpation of local populations of sea mammals will continue, and debate over the cause—was it natural or was it anthropogenic?—will also continue, just as it is has for the terminal Pleistocene extinctions of North American mammals (e.g., Fiedel and Haynes 2004; Grayson and Meltzer 2003). Hopefully, identifying a human cause of extinctions will not become evidence used for political purposes as it has for the overkill hypothesis regarding Pleistocene extinctions (Grayson and Meltzer 2004).

A good way to sum up the discussion thus far concerns an ontological point basic to modern ecology. When studying the history of multiple taxa, no matter how related they might be in an ecological or phylogenetic sense, each taxon's history will be more or less independent of every other taxon's history. This ontology, known as the "individualistic hypothesis," grew from botanist Henry A. Gleason's (1926) observations (Nicolson 1990). It's general acceptance among ecologists grew in part from increasing amounts of paleontological data that contradicted notions of static interspecies associations and interlinked ecologies and histories thereof (e.g., Hewitt 2000; King and Graham 1981). Hanson and Kusmer (2001) exemplify this taxon-specific approach, examining all faunal collections in the Strait of Georgia region to determine if sea otters were historically absent from the area as a result of historic overhunting or if some environmental factor precluded use of the area by this marine mustelid. They found no evidence of the former, and postulated that the sea otter's ability to detect and avoid paralytic shellfish toxins caused them to avoid the area. What distinguishes Hanson and Kusmer's analysis is their choice of a very narrow research question concerning a single taxon in a geographically limited area, and their intensive analysis of all available data that bears on their research question. This is not to denigrate or discourage multitaxon analyses, but rather to emphasize that different taxa have different physiologies, ecologies, and the like, making the individualistic species model the most viable initially. Further, my comments are not meant to deny that the presence or the absence of particular marine mammal taxa could have significant cascading ecological effects on littoral biotic communities; interestingly, one of the seminal studies to demonstrate this built its case on zooarchaeological data (Simenstad et al. 1978). Large zooarchaeological samples are necessary for studies of prehistoric species interactions, and those samples must span sufficient temporal durations and occur in sequent, sufficiently short-duration assemblages that the signal of the ecological cascade effects is not muted. Sadly, few samples with all of these characteristics presently exist.

MORPHOMETRICS

The potential implications of zooarchaeological data for wildlife management were particularly evident in Lyman's (1988) examination of phenotypic differences between prehistoric and modern sea otter populations. Lyman (1988) noted that Oregon's prehistoric sea otters displayed some characteristics of 20th-century Alaskan sea otters and other characteristics of historic California sea otters. These characteristics ranged over simple qualitative traits such as the angle of the ascending ramus of the mandible, to quantitative features such as size and shape of teeth. In light of a recent failure of efforts to re-establish an Oregon sea otter population with transplanted Alaskan sea otters, Lyman suggested that further study of prehistoric sea otter remains might prove informative to wildlife managers and conservation biologists. The morphometry of prehistoric sea otters has not been pursued and remains a wide-open research avenue. This is not unusual; bone size tends to be minimally exploited in zooarchaeology generally, except for the use of allometric relationships between bone size and body size for purposes of estimating biomass (e.g., Reitz et al. 1987).

Mike Etnier (2002b, 2004b) examined the size of male Alaskan northern fur seals in the 20th century and found that the fur seals collected between 1911 and 1920 were larger than individuals collected from 1940 through 1953. He attributed this shift in size to coincident changes in the density of the fur seal population. When the population was more dense, intraspecific competition was greater and fur seals tended to grow slower and be smaller as adults; when the population was less dense, intraspecific competition was less and fur seals grew more rapidly and were larger as adults. Etnier (2004b:1624) suggested that "long-term data on relative population levels [might be provided by] paleontological or archaeological samples." In his unpublished dissertation, Etnier (2002b) reported that the Ozette northern fur seals were smaller than the Alaskan individuals throughout the archaeological sequence. The archaeological specimens were smaller either because the prehistoric Washington population was denser than the 20th-century Alaskan population, or because the two populations were not only distinct in terms of size but latitudinally distinct as well (Etnier 2002b:227). This issue has not yet been resolved. The correlation of population density and individual body size is a phenomenon that is beginning to be regularly used in the zooarchaeology of terrestrial mammals (e.g., Wolverton 2008).

Etnier (2002b, 2004b; Newsome et al. 2007) also used bone size to develop growth curves that allow determination of the ontogenetic age of northern fur seals. Such determinations are more exact than earlier ones based on much smaller samples of known-age individuals for other pinniped species (Lyman 1991). Assessment of ontogenetic age is critical to evaluation of whether or not rookeries may have been near archaeological sites that produce remains of immature pinnipeds (Lyman 1988). Using bone or tooth size (e.g., crown height) to estimate ontogenetic age is not unusual in zooarchaeology in general (e.g., Klein et al. 1981; Munson 1984).

A unique use of morphometric data is found in Crockford et al.'s (2002) suggestion that a previously unknown and now-extinct species of fur seal occupied the Barkley Sound–Cape Flattery area of southwestern Vancouver Island and the northwestern Olympic Peninsula of Washington state. Ethnohistoric data they consulted indicate a form of fur seal that was not only unique in terms of its reproductive behaviors but also its pelage. These data in combination with slight differences in the morphology of archaeological remains of newborn fur seals relative to modern comparative specimens suggested to Crockford et al. (2002) that a unique, now-extinct species of fur seal occurred around the western end of the Strait of Juan de Fuca until recent historic times. Crockford et al. wisely note the small size of their sample and suggest testing the validity of the proposed unique taxon with ancient DNA. Crockford et al.'s use of morphometric data to identify previously unknown extinct taxa is part of a long history of such in zooarchaeology and paleontology (Mead et al. 2000 and references therein).

The paucity of morphometric studies on northeast Pacific sea otters and pinnipeds mirrors the paucity of such studies in North American zooarchaeology in general. A difference exists, however, in the fact that most morphometric research away from the Pacific Rim has involved converting bone size to biomass (e.g., Emerson 1978; Purdue 1987; Reitz et al. 1987). No such algorithms have yet been developed for pinnipeds along the northeastern Pacific Rim.

DISCUSSION

Moss et al. (2006) point out that they misidentified several specimens of pinniped and detected the mistakes only when ancient DNA revealed the errors. They attribute the misidentifications to the "fragmentary and juvenile [character of the archaeological] skeletal remains" and the lack of "comparative specimens [of known taxonomy] representing the full range of morphological pinniped species" (Moss et al. 2006:179). This episode in the history of zooarchaeological study of northeastern Pacific Rim marine mammals tracks similar episodes with terrestrial fauna and marine fishes and prompts the same conclusion—extensive comparative collections of known taxonomy are mandatory to accurate identifications of zooarchaeological remains (e.g., Driver 1992; Gobalet 2001; Lyman 2005).

Two variables exacerbate the identification problem along the northeastern Pacific Rim. First, with respect to pinnipeds, relative to the terrestrial interior, few natural history museums are located in coastal settings where skeletons of sea mammals are likely to be housed. The land-locked comparative collection I visit most often—the University of Kansas Museum of Natural History, where the world famous mammalogist E. Raymond Hall (1981) once worked—has thousands of mammal skeletons but only a couple marine mammal skeletons. The second variable that influences our ability to correctly identify remains of pinnipeds is that marine mammals are, by and large, now under strict international protection, so building comparative skeletal collections presents unique logistical problems above and beyond the more typical one of traveling to a comparative collection of sufficient size to adequately facilitate identification.

Correct taxonomic identification is critical to any zooarchaeological endeavor. Consider, for example, the walrus, a pinniped taxon that has not often been associated with the Northwest Coast culture area. More than 15 years ago, Harington and Beard (1992) reported the recovery of a walrus skeleton from Vancouver Island. A radiocarbon date on bone collagen from the skeleton assayed at greater than 40,000 BP (see also Harington 1984:516). Four paleontologists recently reported geological trace fossils of walrus in Willapa Bay, southwestern Washington (Gingras et al. 2007). These trace fossils may be more than 190,000 years old (Gingras et al. 2007). Nevertheless, the recovery of walrus remains from Vancouver Island, southwestern Washington, and even from San Francisco Bay (Harington 1984) indicates that paleozoologists should keep their eyes and their minds open when identifying pinniped remains from the Northwest Coast. Do not allow modern biogeography to bias your taxonomic-search grid (Driver 1992). Even if walrus were extirpated on the Northwest Coast prior to the arrival of humans, it is particularly important to not completely exclude them from consideration given that prehistoric peoples utilized fossil bone (e.g., Nelson et al. 1986).

Part of the solution to difficulties with taxonomic identification resides in adequate reporting (Butler and Lyman 1996), in particular, describing the morphometric criteria used to distinguish taxa, sexes, and ontogenetic cohorts (Lyman 2005). Such reporting would allow zooarchaeologists to evaluate identifications made by others; this is what paleontologists do, and for good reason. But with respect to reporting the morphometric criteria used to make taxonomic identifications, zooarchaeological research on northeastern Pacific pinnipeds is no different than that anywhere else in the world; basically, very few people report the criteria they use to make identifications. Editorial concerns with per-page publication costs may be a limiting factor, but if so, then we simply haven't done our job in terms of convincing editors of the necessity of describing the criteria we used to identify a particular taxon. In fact, this lacuna in our reporting is mirrored in archaeology generally. Can anyone tell me where the definitive morphometric characteristics of a Clovis projectile point are published? To be sure, many points given the name "Clovis" have been described, but the necessary and sufficient attributes a specimen must possess in order to be given the name "Clovis" are not generally agreed upon nor are they well known (e.g., Howard 1990).

Accurate and well reported taxonomic identification is critical to many kinds of

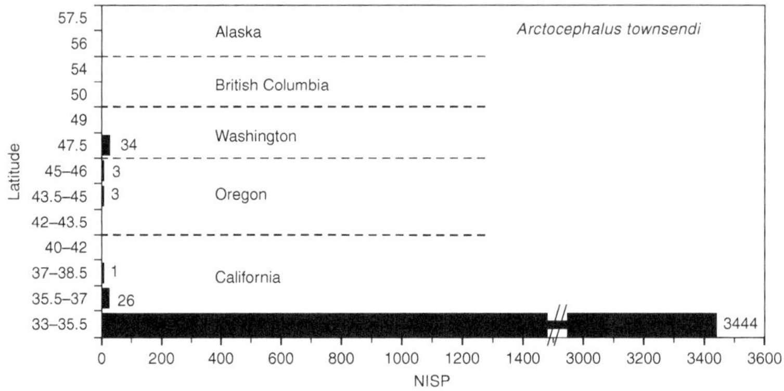

FIGURE 2.1. Frequency (NISP) of Guadalupe fur seal (*Arctocephalus townsendi*) remains relative to latitude. The absence of remains from 40 to 43.5 degrees is likely the result of failure to identify Guadalupe fur seal remains and to distinguish them from northern fur seal (*Callorhinus ursinus*) remains. Data from Table 2.3.

paleobiological analyses. Consider the likely artifact of incorrect identification of fur seal as *Callorhinus ursinus* rather than *Arctocephalus townsendi* apparent in Figure 2.1. One would expect that the biogeographic border for a taxon would not be abrupt in terms of relative abundance of that taxon but rather geographically gradual (Brown and Lomolino 1998). Data on all sea otter and pinniped remains identified between Kodiak Island and the California-Mexico border were compiled to generate Figure 2.1. The Guadalupe fur seal's frequency distribution should gradually taper off north of the southern California area given the occasional presence of vagrant individuals (e.g., Gaston 1996). Instead, there are no remains of this species explicitly reported for the area between southern California and central Oregon; fur seal remains are reported as just that—"fur seal"—without species designation (e.g., Hildebrandt and Jones 1992). The only reason Guadalupe fur seals are found in central and northern Oregon is because Gretchen Lyon (1937) reported them in the former area based on skeletal morphology and Moss et al. (2006) identified them in the latter area based on ancient DNA. I predict that if extant collections from northern California and southern Oregon are reexamined with a critical eye, remains of Guadalupe fur seal will be detected.

The data on which Figure 2.1 is based are given in Table 2.3. While compiling those data, it was found that many such data were not reported in a manner useful to this sort of biogeographic analysis. Rather than reporting pinniped remains as to genus or species represented, several authors simply reported "phocids," or "fur seals," or "sea lions." Each of these folk taxa is polytypic—each comprises more than one genus or species. Such reporting may suffice for assessing prehistoric human subsistence, but it simply won't allow detailed paleoecological and paleobiological research. Taxonomically ambiguous reporting is, sadly, a characteristic of zooarchaeological research in non-coastal contexts as well. For better or worse, we are keeping pace with the discipline at large.

Zooarchaeological research on eastern North Pacific sea otters and pinnipeds focusing on biological issues has centered on the northern fur seal, likely because remains of this species have been so often out of place biogeographically relative to modern times—something recognized initially by Gretchen Lyon 70 years ago and still the center of attention today. Lyon also noted the unexpected abundance of Guadalupe fur seals in southern California, something only recently pursued with intensity (Rick et al. 2009). Ignoring the likely bias

TABLE 2.3

Summary of Taxonomic Abundance Data (NISP) by Latitude.
List Reads from North to South.

AREA (LATITUDE)	*E. l.*	*P. v.*	*M. a.*	*C. u.*	*A. t.*	*E. j.*	*Z. c.*	DATA SOURCE
Kodiak Isl. (57.5)	8	664	0	354	0	19	0	Clark (1986)
Angoon (57.5)	476	166	0	0	0	9	0	Moss (1989)
Cape Addington (56)	n.d.	97	0	20	0	52	0	Moss et al. (2006)
Prince Rupert (54)	1915	564	0	44	0	59	0	Stewart and Stewart (1996)
Queen Charlottes (53)	719	662	0	104	0	108	0	Orchard and Clark (2005)
Hesquiat Harbor (50)	158	124	1	329	0	41	31	Calvert (1980)
Ts'ishaa (49)	n.d.	43	1	250	0	19	1	Moss et al. (2006)
N.W. Washington (47.5)	45	57	7	1923	0	42	0	Friedman (1976)
Ozette (47.5)	501	377	2	47,296	34	10	1	Huelsbeck (1994), except *A. t., E. j., Z. c.* data from Etnier (2002b)
N. Oregon (45–46)	724	317	0	186	3	584	29	Lyman (1995), Colten (2002), Moss et al. (2006), and Minor et al. (2008), except *A. t.* data from Moss et al. (2006), and Moss et al. (2006) do not report data for *E. l.*
Central Oregon (43.5–45)	222	264	0	135	3	1047	42	Lyman (1995), except *A. t.* data from Lyon (1937)
S. Oregon (42–43.5)	329	1332	3	73	0	146	22	Lyman (1995)
N.N. Calif. (40–42)	483	116	0	259	0	975	250	Hildebrandt and Jones (1992) and Whitaker (2008)
S.N. Calif. (38.5–40)	185	155	1	93	0	311	126	Hildebrandt and Jones (1992), Wake and Simmons (2000), and Whitaker (2008)
N. Cent. Calif. (37–38.5)	5057	549	0	312	1	296	304	Simmons (1992) and Broughton (1999), except *A. t.* from Rick et al. (2009)
S. Cent. Calif. (35.5–37)	744	103	0	292	26	9	26	Hildebrandt and Jones (1992), except *A. t.* from Rick et al. (2009)
S. Calif. (33–35.5)	1609	275	51	128	3444	22	364	Lyon (1937), Walker and Craig (1979), Colten and Arnold (1998), Porcasi et al. (2000), Walker et al. (2002), and Jones et al. (2008), except *A. t.* from Rick et al. (2009)

NOTE: Latitude is approximate. Data not reported, n.d.

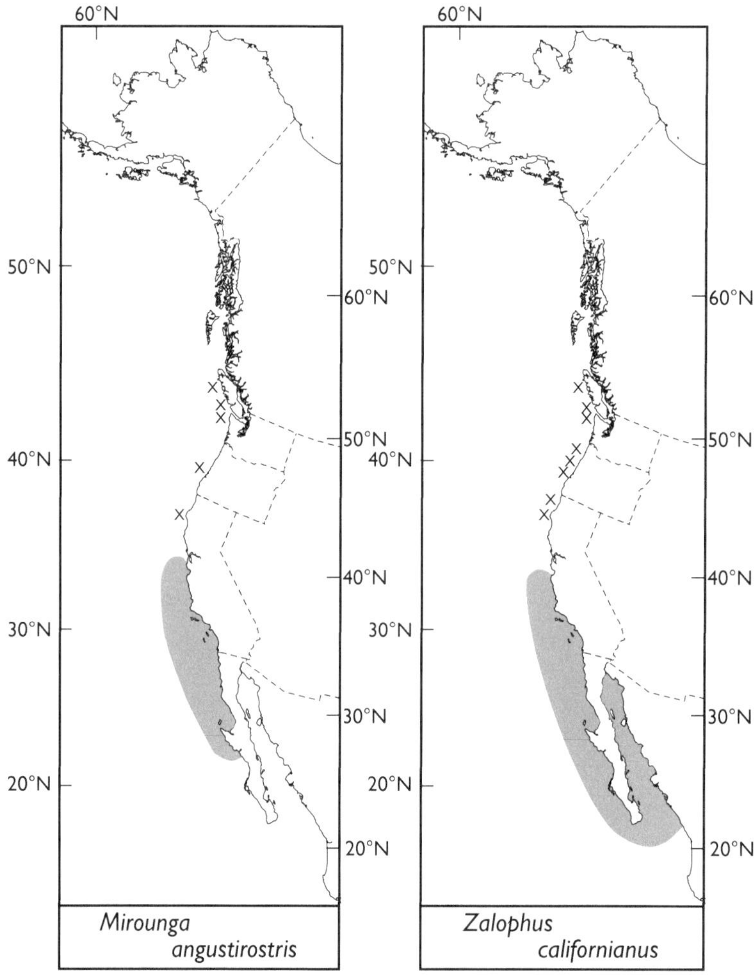

FIGURE 2.2. Modern breeding ranges of northern elephant seal (*Mirounga angustirostris*) and of California sea lion (*Zalophus californianus*) (shaded), and locations of archaeological remains of each (x) outside of the breeding range. Based on data in Table 2.3.

with respect to taxonomic identification of Guadalupe fur seal remains, the biogeographic implications of Figure 2.1 are that during the late Holocene (last 3500 years), this species, like today, seldom ventured north of the latitude of the Channel Islands (33–34°N). Although a historic, commercial-exploitation–related population bottleneck for this species is likely, the prehistoric record suggests relative stasis in its distribution and, perhaps, migratory habits. And this is not the only taxon whose prehistoric remains suggest stasis.

By stasis I mean relative to, particularly, northern fur seals. Despite the commercial exploitation of all pinniped taxa, there are few biogeographic differences between what we know of the 20th century and what the prehistoric record implies for some taxa. This is particularly evident with northern elephant seals (*Mirounga angustirostris*) and California sea lions (*Zalophus californianus*). Today both species breed in latitudes south of the vicinity of San Francisco Bay at about 37°N (King 1983:23, 125), but both are occasionally observed in waters to the north of that bay (Figure 2.2; references in Lyman 1988). Reports of northern elephant seals establishing a more northern breeding and pupping location or rookery

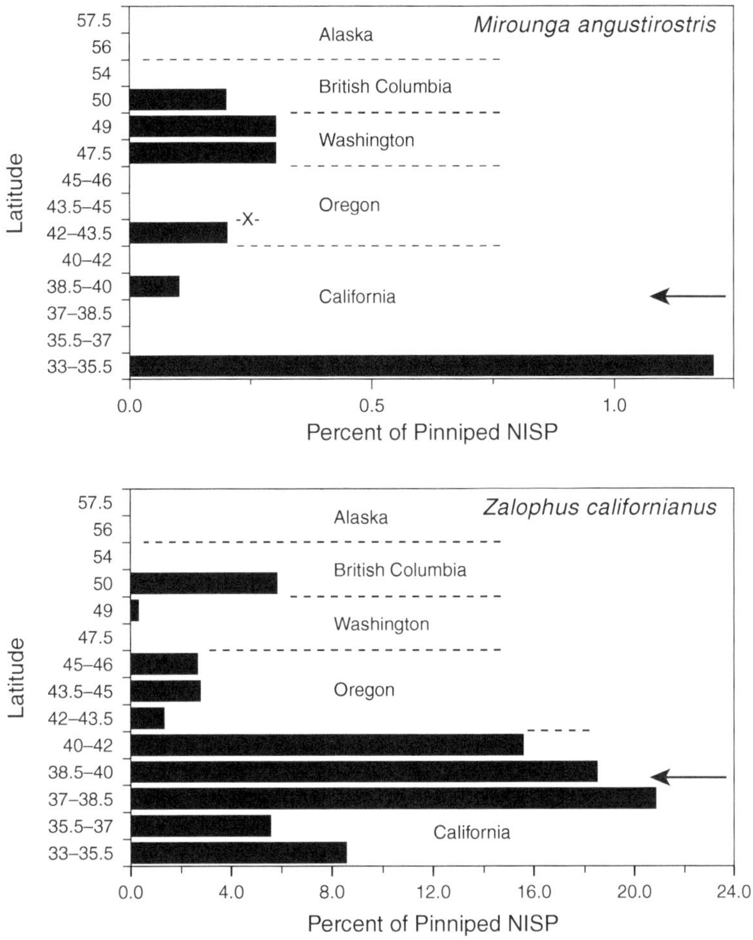

FIGURE 2.3. Relative frequency (percent of all pinniped remains) of northern elephant seal (*Mirounga angustirostris*) and of California sea lion (*Zalophus californianus*) remains relative to latitude. Arrows denote the northern limit of the modern breeding range (see Figure 2.2); "x" in the *Mirounga angustirostris* graph denotes approximate latitude of a newly established rookery. Data from Table 2.3.

(Hodder et al. 1998) at approximate latitude 43.2°N may represent recolonization of previously abandoned rookeries or invasions. This species today migrates north to south-central Alaska (Stewart 1997; see also Crocker et al. 2006). Only study of the paleozoological record will clarify which possibility applies in the case of northern elephant seals (Campbell 1987) and in the case of California sea lions (Bigg 1987).

Zooarchaeological remains of northern elephant seals suggest they occasionally could be found north of 37°N prehistorically, but were only abundant south of that latitude just as they are today, perhaps because that is where they were accessible to humans. Remains of both taxa are abundant in sites within the modern geographic breeding range relative to remains of all other pinniped taxa, but they are relatively rare in sites in more northern latitudes (Figure 2.3). Why these taxa may not have been noticeably affected by historic commercial exploitation whereas sea otters and northern fur seals were markedly influenced—the latter two were extirpated from large expanses of their prehistoric ranges—remains an open question. Part of the answer may reside in the fact that only sea otters and northern fur seals were exploited for their furs; the other taxa were exploited for

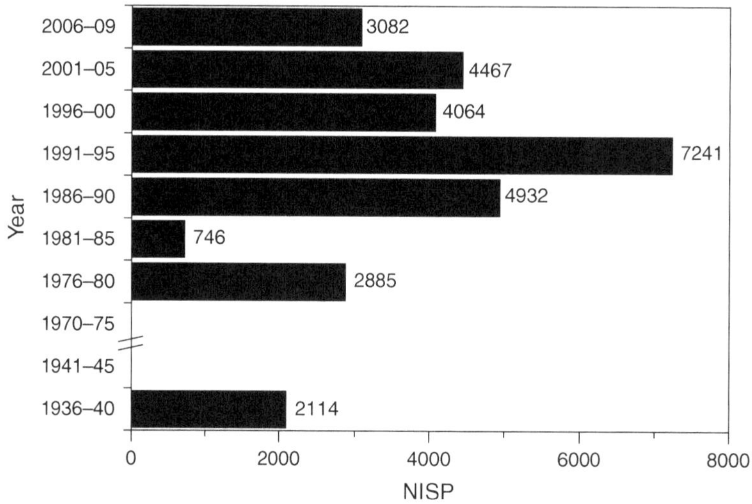

FIGURE 2.4. NISP of sea otter and pinniped remains reported over the last 70 years. The trend to increase over time is real, but the trend to decrease after 1995 may be the result of unpublished data that has not been widely circulated.

other reasons. Whatever the case, this difference between taxa with respect to the degree to which their populations were affected highlights the earlier point regarding the Gleasonian model of taxonomically individualistic histories.

CONCLUSION

By and large, zooarchaeological research on eastern North Pacific sea otters and pinnipeds seems to be closely tracking analytical trends in more land-locked loci. By way of conclusion, there is another arena where study of marine mammal remains is tracking study of terrestrial mammal remains. Analysis of zooarchaeological remains from the North Pacific coastal zone began slowly, with an early, relatively large sample. It was some years before the next sample was studied, but once that second sample was described, sample sizes per 5-year bin have, in general, increased (Figure 2.4). The apparent decrease in NISP over the past decade or so is likely an artifact of the fact that many data, particularly those in the CRM-generated grey literature, have not yet entered the published record (a prime example is Rick et al. 2009). This, too, tracks the character of how the general archaeological record is known in the literature (e.g., Lyman 1997).

There are few documented paleontological remains of any of the taxa listed in Table 2.1 (see Harington et al. 2004 and Ray 2008 for notable exceptions), so I have focused here on what zooarchaeological remains can tell us about the paleoecology of marine mammals. The problems of studying the marine mammal paleorecord of the northeastern Pacific Rim are not insurmountable. The questions driving research on those mammals are interesting and significant. Let us hope that the next 70 years will be as equally exciting as the first 70 years.

ACKNOWLEDGMENTS

I thank Torben Rick and Todd Braje for inviting me to participate, for comments on an early draft, and for time to update the discussion.

REFERENCES CITED

Anderson, P. K.
2001 Marine Mammals in the Next One Hundred Years: Twilight for a Pleistocene Megafauna? *Journal of Mammalogy* 82:623–629.
Arnold, J. E., and B. N. Tissot
1993 Measurement of Significant Marine Paleotemperature Variation Using Black Abalone Shells from Prehistoric Middens. *Quaternary Research* 39:390–394.

Berger, J., S. L. Cain, and K. M. Berger
2006 Connecting the Dots: An Invariant Migration Corridor Links the Holocene to the Present. *Biology Letters* 2:528–531.
Bigg, M. A.
1987 Status of the California Sea Lion, *Zalophus californianus,* in Canada. *Canadian Field Naturalist* 102:307–314.
Brainerd, G. W.
1939 An Illustrated Field Key for the Identification of Mammal Bones. *Ohio State Archaeological and Historical Quarterly* 48:324–328.
Broughton, J. M.
1999 *Resource Depression and Intensification During the Late Holocene, San Francisco Bay.* Anthropological Records 32. University of California, Berkeley.
Brown, J. H., and M. V. Lomolino
1998 *Biogeography,* 2nd edition. Sinauer, Sunderland, Mass.
Burleson, G. M. Lyon
1941 A Miocene Sea Lion from Lomita, California. *University of California Publications in Zoology* 47(2):23–41.
1948 A Pliocene Pinniped from the San Diego Formation of Southern California. *University of California Publications in Zoology* 47(10):247–253.
Burton, R. K., D. Gifford-Gonzalez, J. J. Snodgrass, and P. L. Koch
2002 Isotopic Tracking of Prehistoric Pinniped Foraging and Distribution along the Central California Coast: Preliminary Results. *International Journal of Osteoarchaeology* 12:4–11.
Burton, R. K., J. J. Snodgrass, D. Gifford-Gonzalez, T. Guilderson, T. Brown, and P. L. Koch
2001 Holocene Changes in the Ecology of Northern Fur Seals: Insights from Stable Isotopes and Archaeofauna. *Oecologia* 128:107–115.
Butler, V. L., and R. L. Lyman
1996 Taxonomic Identifications and Faunal Summaries: What Should We Be Including in Our Faunal Reports? *SAA Bulletin* 14(1):22.
Calvert, S. G.
1980 *A Cultural Analysis of Faunal Remains from Three Archaeological Sites in Hesquiat Harbor, B.C.* Doctoral dissertation, University of British Columbia, Vancouver.
Campbell, R. R.
1987 Status of the Northern Elephant Seal, *Mirounga angustirostris,* in Canada. *Canadian Field Naturalist* 101:266–270.
Cannon, A.
1991 *The Economic Prehistory of Namu.* Simon Fraser University, Department of Archaeology, Publication No. 19. Burnaby, British Columbia.
Clark, D. W.
1986 Archaeological and Historical Evidence for an 18th-Century "Blip" in the Distribution of Northern Fur Seal at Kodiak Island, Alaska. *Arctic* 39:39–42.
Colton, R. H.
2002 Prehistoric Marine Mammal Hunting in Context: Two Western North American Examples. *International Journal of Osteoarchaeology* 12:12–22.
Colten, R. H., and J. E. Arnold
1998 Prehistoric Marine Mammal Hunting on California's Northern Channel Islands. *American Antiquity* 63:679–701.
Crocker, D. E., D. P. Costa, B. J. Le Boeuf, P. M. Webb, and D. S.Houser
2006 Impact of El Niño on the Foraging Behavior of Female Northern Elephant Seals. *Marine Ecology Progress Series* 309:1–10.
Crockford, S. J., S. G. Frederick, and R. J. Wigen
2002 The Cape Flattery Fur Seal: An Extinct Species of *Callorhinus* in the Eastern North Pacific? *Canadian Journal of Archaeology* 26:152–174.
Dall, W. H.
1877 On the Succession of Shell-Heaps of the Aleutian Islands. *Contributions to North American Ethnology* 1:41–91.
Driver, J. C.
1992 Identification, Classification and Zooarchaeology. *Circaea* 9(1):35–47.
Emerson, T. E.
1978 A New Method for Calculating the Live Weight of the Northern White-Tailed Deer from Osteoarchaeological Material. *Mid-Continental Journal of Archaeology* 3:35–44.
Etnier, M.
2002a Occurrences of Guadalupe Fur Seals (*Arctocephalus townsendi*) on the Washington Coast over the Past 500 Years. *Marine Mammal Science* 18:551–557.
2002b *The Effects of Human Hunting on Northern Fur Seal* (Callorhinus ursinus) *Migration and Breeding Distributions in the Late Holocene.* Doctoral dissertation, University of Washington, Seattle.
2004a The Potential of Zooarchaeological Data to Guide Pinniped Management Decision in the Eastern North Pacific. In *Zooarchaeology and Conservation Biology,* edited by R. L. Lyman and K. P. Cannon, pp. 88–102. University of Utah Press, Salt Lake City.
2004b Reevaluating Evidence of Density-Dependent Growth in Northern Fur Seals (*Callorhinus ursinus*) Based on Measurements of Archived Skeletal Specimens. *Canadian*

Journal of Fisheries and Aquatic Sciences 61: 1616–1626.
2007 Defining and Identifying Sustainable Harvests of Resources: Archaeological Examples of Pinniped Harvests in the Eastern North Pacific. *Journal for Nature Conservation* 15:196–207.
Fiedel, S., and G. Haynes
2004 A Premature Burial: Comments on Grayson and Meltzer's "Requiem for Overkill." *Journal of Archaeological Science* 31:121–131.
Friedman, E.
1976 *An Archaeological Survey of Makah Territory: A Study in Resource Utilization.* Doctoral dissertation, Washington State University, Pullman.
Gaston, K. J.
1996 Species Richness: Measure and Measurement. In *Biodiversity: A Biology of Numbers and Difference,* edited by K. J. Gaston, pp. 77–113. Blackwell Science, Oxford.
Gifford-Gonzalez, D., S. D. Newsome, P. L. Koch, T. P. Guilderson, J. J. Snodgrass, and R. K. Burton
2005 Archaeofaunal Insights on Pinniped–Human Interactions in the Northeastern Pacific. In *The Exploitation and Cultural Importance of Sea Mammals,* edited by G. Monks, pp. 19–38. Oxbow Books, Oxford.
Gingras, M. K., I. A. Armitage, S. G. Pemberton, and H. E. Clifton
2007 Pleistocene Walrus Herds in the Olympic Peninsula Area: Trace-Fossil Evidence of Predation by Hydraulic Jetting. *Palaios* 22:539–545.
Gleason, H. A.
1926 The Individualistic Concept of the Plant Association. *Bulletin of the Torrey Botanical Club* 53:7–26.
Gobalet, K. W.
2001 A Critique of Faunal Analysis: Inconsistency among Experts in Blind Tests. *Journal of Archaeological Science* 28:377–386.
Graham, R. W.
1979 Paleoclimates and Late Pleistocene Faunal Provinces in North America. In *Pre-Llano Cultures of the Americas: Paradoxes and Possibilities,* edited by R. L. Humphrey and D. Stanford, pp. 49–69. Anthropological Society of Washington, Washington, D.C.
Grayson, D. K.
1984 *Quantitative Zooarchaeology.* Academic Press, Orlando.
2005 A Brief History of Great Basin Pikas. *Journal of Biogeography* 32:2103–2111.
Grayson, D. K., and D. J. Meltzer
2003 A Requiem for North American Overkill. *Journal of Archaeological Science* 30:585–593.
2004 North American Overkill Continued? *Journal of Archaeological Science* 31:133–136.
Gustafson, C. E.
1968a Historic Change in Migratory Fur Seal Populations: Result of Pelagic Sealing? *Journal of Mammalogy* 49:842.
1968b Prehistoric Use of Fur Seals: Evidence from the Olympic Coast of Washington. *Science* 161: 49–51.
Hall, E. R.
1981 *The Mammals of North America,* 2nd edition. Wiley, New York.
Hanson, D. K., and K. D. Kusmer
2001 Sea Otter (*Enhydra lutris*) Scarcity in the Strait of Georgia, British Columbia. In *People and Wildlife in Northern North America: Essays in Honor of R. Dale Guthrie,* edited by S. C. Gerlach and M. S. Murray, pp. 58–66. British Archaeological Reports International Series 944. Oxford.
Harington, C. R.
1984 Quaternary Marine and Land Mammals and Their Paleoenvironmental Implications—Some Examples from Northern North America. In *Contributions in Quaternary Vertebrate Paleontology: A Volume in Memorial to John E. Guilday,* edited by H. H. Genoways and M. R. Dawson, pp. 511–525. Carnegie Museum of Natural History Special Publication No. 8. Pittsburgh.
Harington, C. R. and G. Beard
1992 The Qualicum Walrus: A Late Pleistocene Walrus (*Odobenus rosmarus*) Skeleton from Vancouver Island, British Columbia, Canada. *Annales Zoologici Fennici* 28:311–319.
Harington, C. R., R. L. M. Ross, R. W. Mathewes, K. M. Stewart, and O. Beattie
2004 A Late Pleistocene Steller Sea Lion (*Eumetopias jubatus*) from Courtenay, British Columbia: Its Death, Associated Biota, and Paleoenvironment. *Canadian Journal of Earth Sciences* 41: 1285–1297.
Hewitt, G.
2000 The Genetic Legacy of the Quaternary Ice Ages. *Nature* 405:907–913.
Hildebrandt, W. R.
1981 *Native Hunting Adaptations on the North Coast of California.* Doctoral dissertation, University of California, Davis.
1984a Archaeological Presence of the Northern Fur Seal (*Callorhinus ursinus*) Along the Coast of Northern California. *The Murrelet* 65: 28–29.
1984b Late Period Hunting Adaptations on the North Coast of California. *Journal of California and Great Basin Anthropology* 6:189–206.

Hildebrandt, W. R., and T. L. Jones
1992 Evolution of Marine Mammal Hunting: A View from the California and Oregon Coasts. *Journal of Anthropological Archaeology* 11: 360–401.
2002 Depletion of Prehistoric Pinniped Populations along the California and Oregon Coasts: Were Humans the Cause? In *Wilderness and Political Ecology: Aboriginal Influences on the Original State of Nature*, edited by C. E. Kay and R. T. Simmons, pp. 72–110. University of Utah Press, Salt Lake City.
Hodder, J., R. Brown, and C. Cziesla
1998 The Northern Elephant Seal in Oregon: A Pupping Range Extension and Onshore Occurrence. *Marine Mammal Science* 14:873–881.
Howard, C.
1990 The Clovis Point: Characteristics and Type Description. *Plains Anthropologist* 35:255–265.
Huelsbeck, D. R.
1983 *Mammals and Fish in the Subsistence Economy of Ozette.* Doctoral dissertation, Washington State University, Pullman.
1994 Mammals and Fish in the Subsistence Economy of Ozette. In *Ozette Archaeological Project Research Reports Volume II: Fauna*, edited by S. R. Samuels, pp. 17–91. Washington State University Department of Anthropology Reports of Investigations No. 66. Pullman.
Hughes, S. S.
2004 Postcontact Changes in the Behavior and Distribution of Rocky Mountain Bighorn Sheep (*Ovis canadensis*) in Northwestern Wyoming. In *Zooarchaeology and Conservation Biology*, edited by R. L. Lyman and K. P. Cannon, pp. 116–135. University of Utah Press, Salt Lake City.
Jones, T. L., and W. R. Hildebrandt
1995 Reasserting a Prehistoric Tragedy of the Commons: Reply to Lyman. *Journal of Anthropological Archaeology* 14:78–98.
Jones, T. L., and D. J. Kennett
1999 Late Holocene Sea Temperatures along the Central California Coast. *Quaternary Research* 51:74–82.
Jones, T. L., J. F. Porcasi, J. Gaeta, and B. F. Codding
2008 The Diablo Canyon Fauna: A Coarse-Grained Record of Trans-Holocene Foraging from the Central California Mainland Coast. *American Antiquity* 73:289–316.
King, F. B., and R. W. Graham
1981 Effects of Ecological and Paleoecological Patterns on Subsistence and Paleoenvironmental Reconstructions. *American Antiquity* 46:128–142.
King, Judith E.
1983 *Seals of the World.* Cornell University Press, Ithaca, NY.
Klein, R. G., C. Wolf, L. G. Freeman, and K. Allwarden
1981 The Use of Dental Crown Heights for Constructing Age Profiles of Red Deer and Similar Species in Archaeological Samples. *Journal of Archaeological Science* 8:1–31.
Larson, S., R. Jameson, J. L. Bodkin, M. Staedler, and P. Bentzen
2002a Microsatellite DNA and mtDNA Variation within and among Remnant and Translocated Sea Otter, *Enhydra lutris*, Populations. *Journal of Mammalogy* 83:893–906.
Larson, S., R. Jameson, M. Etnier, M. Fleming, and P. Bentzen
2002b Loss of Genetic Diversity in Sea Otters (*Enhydra lutris*) Associated with the Fur Trade of the 18th and 19th Centuries. *Molecular Ecology* 11:1899–1903.
Lea, M.-A., D. Johnson, R. Ream, J. Sterling, S. Melin, and T. Gelatt
2009 Extreme Weather Events Influence Dispersal of Naïve Northern Fur Seals. *Biology Letters* 5:252–257.
Lundelius, E. L., Jr.
1974 The Last Fifteen Thousand Years of Faunal Change in North America. In *History and Prehistory of the Lubbock Lake Site*, edited by C. C. Black, pp. 141–160. The Museum Journal 15.
Lyman, R. L.
1988 Zoogeography of Oregon Coast Marine Mammals: The Last 3000 Years. *Marine Mammal Science* 4:247–264.
1989 Seal and Sea Lion Hunting: A Zooarchaeological Study from the Southern Northwest Coast of North America. *Journal of Anthropological Archaeology* 8:68–99.
1991 *Prehistory of the Oregon Coast: The Effects of Excavation Strategies and Assemblage Size on Archaeological Inquiry.* Academic Press, San Diego.
1995 On the Evolution of Marine Mammal Hunting on the West Coast of North America. *Journal of Anthropological Archaeology* 14:45–77.
1997 Impediments to Archaeology: Publishing and the (Growing) Translucency of Archaeological Research. *Northwest Anthropological Research Notes* 31:5–22.
2003 Pinniped Behavior, Foraging Theory, and the Depression of Metapopulations and Nondepression of a Local Population on the Southern Northwest Coast of North America. *Journal of Anthropological Archaeology* 22:376–388.

2004 Late-Quaternary Diminution and Abundance of Prehistoric Bison (*Bison* sp.) in Eastern Washington State, U.S.A. *Quaternary Research* 62:76–85.
2005 Zooarchaeology. In *Handbook of Archaeological Methods*, edited by H. D. G. Maschner and C. Chippendale, pp. 835–870. AltaMira Press, Lanham, Maryland.
2006 Paleozoology in the Service of Conservation Biology. *Evolutionary Anthropology* 15: 11–19.
2007a Archaeology's Quest for a Seat at the High Table of Anthropology. *Journal of Anthropological Archaeology* 26:133–149.
2007b What is the "Process" in Cultural Process and in Processual Archaeology? *Anthropological Theory* 7:217–250.
2008 *Quantitative Paleozoology*. Cambridge University Press, Cambridge.

Lyman, R. L., J. L. Harpole, C. Darwent, and R. Church
2002 Prehistoric Occurrence of Pinnipeds in the Lower Columbia River. *Northwestern Naturalist* 83:1–6.

Lyman, R. L., M. J. O'Brien, and R. C. Dunnell
1997 *The Rise and Fall of Culture History*. Plenum, New York.

Lyon, G. M.
1935 Some Marine Mammals from a Southern California Shellmound. *Journal of Mammalogy* 16:151–152.
1937 Pinnipeds and a Sea Otter from the Point Mugu Shell Mound of California. *University of California at Los Angeles Publications in Biological Sciences* 1:133–168.

Mead, J. I., A. E. Spiess, and K. D. Sobolik
2000 Skeleton of Extinct North American Sea Mink (*Mustela macrodon*). *Quaternary Research* 53:247–262.

Merriam, C. H.
1927 William Healey Dall. *Science* 65:345–347.

Minor, R., R. L. Greenspan, and D. C. Barner
2008 Chinookan Resource Exploitation in the Columbia River Estuary: The View from Indian Point. In *Dunes, Headlands, Estuaries, and Rivers: Current Archaeological Research on the Oregon Coast*, edited by G. L. Tasa and B. L. O'Neill, pp. 37–55. Association of Oregon Archaeologists Occasional Papers No. 8. University of Oregon, Eugene.

Moss, M. L.
1989 *Archaeology and Cultural Ecology of the Prehistoric Angoon Tlingit*. Doctoral dissertation, University of California, Santa Barbara.

Moss, M. L., D. Y. Yang, S. D. Newsome, C. F. Speller, I. McKechnie, A. D. McMillan, R. J. Losey, and P. L. Koch
2006 Historical Ecology and Biogeography of North Pacific Pinnipeds: Isotopes and Ancient DNA from Three Archaeological Assemblages. *Journal of Island and Coastal Archaeology* 1: 165–190.

Munson, P. J.
1984 Teeth of Juvenile Woodchucks as Seasonal Indicators on Archaeological Sites. *Journal of Archaeological Science* 11:395–403.

Murray, M. S.
2008 Zooarchaeology and Arctic Marine Mammal Biogeography, Conservation, and Management. *Ecological Applications* 18:S41–S55.

Nelson, D. E., R. E. Morlan, J. S. Vogel, J. R. Southon, and C. R. Harington
1986 New Dates on Northern Yukon Artifacts: Holocene Not Upper Pleistocene. *Science* 232:749–751.

Newsome, S. D., M. A. Etnier, D. Gifford-Gonzalez, D. L. Phillips, M. van Tuinen, E. A. Hadly, D. P. Costa, D. J. Kennett, T. P. Guilderson, and P. L. Koch
2007 The Shifting Baseline of Northern Fur Seal Ecology in the Northeast Pacific Ocean. *Proceedings of the National Academy of Sciences* 104:9709–9714.

Nicolson, M.
1990 Henry Allan Gleason and the Individualistic Hypothesis: The Structure of a Botanist's Career. *Botanical Review* 56:91–161.

O'Brien, M. J., R. L. Lyman, and M. B. Schiffer
2005 *Archaeology as a Process: Processualism and Its Progeny*. University of Utah Press, Salt Lake City.

Orchard, T. J., and T. Clark
2005 Multidimensional Scaling of Northwest Coast Faunal Assemblages: A Case Study from Southern Haida Gwaii, British Columbia. *Canadian Journal of Archaeology* 29:88–112.

Porcasi, J. F., T. L. Jones, and M. L. Raab
2000 Trans-Holocene Marine Mammal Exploitation on San Clemente Island, California: A Tragedy of the Commons Revisited. *Journal of Anthropological Archaeology* 19:200–220.

Purdue, J. R.
1987 Estimation of Body Weight of White-Tailed Deer (*Odocoileus virginianus*) from Bone Size. *Journal of Ethnobiology* 7:1–12.

Pusch, C. M., M. Broghammer, and N. Blin
2003 Molecular Phylogenetics Employing Modern and Ancient DNA. *Journal of Applied Genetics* 44:269–290.

Ray, C. E.
2008 The Fossil Record of *Eumetopias*. *Marine Mammal Science* 24:251–252.

Reitz, E. J.
1993 Zooarchaeology. In *The Development of Southeastern Archaeology*, edited by J. K. Johnson, pp. 108–131. University of Alabama Press, Tuscaloosa.

Reitz, E. J., I. R. Quitmyer, H. S. Hale, S. J. Scudder, and E. S. Wing
1987 Application of Allometry to Zooarchaeology. *American Antiquity* 52:304–317.

Reitz, E. J., and E. S. Wing
1999 *Zooarchaeology*. Cambridge University Press, Cambridge.

Rick, T. C., R. L. DeLong, J. M. Erlandson, T. J. Braje, T. L. Jones, D. J. Kennett, T. A. Wake, and P. L. Walker
2009 A Trans-Holocene Archaeological Record of Guadalupe Fur Seals (*Arctocephalus townsendi*) on the California Coast. *Marine Mammal Science* 25:487–502.

Robison, N. D.
1987 History of Zooarchaeology. In *The Zooarchaeology of Eastern North America*, 2nd edition, edited by A. E. Bogan and N. D. Robison, pp. 1–26. Tennessee Anthropological Association Miscellaneous Paper No. 12.

Scheffer, V. B.
1950 Growth Layers on the Teeth of Pinnipeedia as an Indication of Age. *Science* 112:309–311.

Simenstad, C. A., J. A. Estes, and K. W. Kenyon
1978 Aleuts, Sea Otters, and Alternate Stable-State Communities. *Science* 200:403–411.

Simons, D. D.
1992 Prehistoric Mammal Exploitation in the San Francisco Bay Area. In *Essays on the Prehistory of Maritime California*, edited by T. L. Jones, pp. 73–103. Center for Archaeological Research at Davis Publication No. 10. University of California, Davis.

Stewart, B. S.
1997 Ontogeny of Differential Migration and Sexual Segregation in Northern Elephant Seals. *Journal of Mammalogy* 78:1101–1116.

Stewart, F. L., and K. M. Stewart
1996 The Boardwalk and Grassy Bay Sites: Patterns of Seasonality and Subsistence on the Northern Northwest Coast, B.C. *Canadian Journal of Archaeology* 20:39–60.

Stock, C.
1929 A Census of the Pleistocene Mammals of Rancho La Brea, Based on the Collections of the Los Angeles Museum. *Journal of Mammalogy* 10:281–289.

Trigger, B. G.
2006 *A History of Archaeological Thought*, 2nd edition. Cambridge University Press, Cambridge.

Trites, A. W., A. J. Miller, H. D. G. Haschner, M. A. Alexander, S. J. Bograd, et al.
2007 Bottom-Up Forcing and the Decline of Steller Sea Lions (*Eumetopias jubatus*) in Alaska: Assessing the Ocean Climate Hypothesis. *Fisheries Oceanography* 16:46–67.

Valentine, K., D. A. Duffield, L. E. Patrick, D. R. Hatch, V. L. Butler, R. L. Hall, and N. Lehman
2008 Ancient DNA Reveals Genotypic Relationships among Oregon Populations of the Sea Otter (*Enhydra lutris*). *Conservation Genetics* 9: 933–938.

Wake, T. A., and D. D. Simons
2000 Trans-Holocene Subsistence Strategies and Topographic Change on the Northern California Coast: The Fauna from Duncan's Point Cave. *Journal of California and Great Basin Anthropology* 22:295–320.

Walker, P. L., and S. Craig
1979 Archaeological Evidence Concerning the Prehistoric Occurrence of Sea Mammals at Point Bennett, San Miguel Island. *California Fish and Game* 65:50–54.

Walker, P. L., D. J. Kennett, T. Jones, and R. L. DeLong
2002 Archaeological Investigations of the Point Bennett Pinniped Rookery on San Miguel Island. In *Proceedings of the Fifth California Islands Symposium*, edited by D. Browne, K. Mitchell, and H. Chaney, pp. 628–632. Santa Barbara Museum of Natural History, Santa Barbara, CA.

Washington Department of Fish and Wildlife
2008 Columbia River Sea Lion Management: Federal Authorization for States to Remove Sea Lions. http.wdfw.wa.gov/conservation/sealions/links.html (accessed Oct. 20, 2008 and Oct. 28, 2010).

Whitaker, A. R.
2008 *The Role of Human Predation in the Structuring of Prehistoric Prey Populations in Northwestern California*. Doctoral dissertation, University of California, Davis.

White, T. E.
1952 Observations on the Butchering Technique of Some Aboriginal Peoples: I. *American Antiquity* 17:337–338.
1953a A Method of Calculating the Dietary Percentage of Various Food Animals Utilized by Aboriginal Peoples. *American Antiquity* 19:396–398.
1953b Studying Osteological Material. *Plains Archaeological Conference News Letter* 6:58–66.

Wigen, R. J., and B. R. Stucki
1988 Taphonomy and Stratigraphy in the Interpretation of Economic Patterns at Hoko River Rockshelter. In *Prehistoric Economies of the Pacific Northwest Coast*, edited by B. L. Isaac, pp. 87–146. Research in Economic Anthropology, Supplement 3. JAI Press, Greenwich, Conn.

Wintemberg, W. J.
1919 Archaeology as an Aid to Zoology. *The Canadian Field-Naturalist* 33:63–72.

Wolverton, S.
2008 Harvest Pressure and Environmental Carrying Capacity: An Ordinal-Scale Model of Effects on Ungulate Prey. *American Antiquity* 73:179–199.

3

The Historical Ecology of Walrus Exploitation in the North Pacific

Erica Hill

THE PACIFIC WALRUS (*Odobenus rosmarus divergens*) is a highly salient species of sea mammal and was a fundamental resource for prehistoric maritime peoples in the North Pacific. For at least 2000 years, inhabitants of the coasts of St. Lawrence Island, Alaska, and the Chukchi Peninsula of Chukotka have relied upon walrus as either a primary or secondary resource. At some highly debated point in time, Eskimo on both sides of the Bering Strait adopted whaling, and in most areas the use of walrus as a resource declined (for discussion of possible dates, see Mason and Gerlach 1995). Modern inhabitants of these regions continue their whaling traditions, focusing on gray and bowhead whales (both baleen whales), as well as the smaller belugas, or "white whales." Also living along the coast of the Chukchi Peninsula at contact were the so-called maritime Chukchi (also called "coastal" or "settled" Chukchi). Though unrelated linguistically to the Eskimo, the Chukchi pursued many of the same subsistence practices as their Eskimo neighbors, including walrusing and whaling. When they arrived on the coast is unknown; however, the Chukchi presence was recorded by mid-18th-century Russian observers (Krupnik 1993a).

On the eastern side of the Bering Strait, Alaska Eskimos included caribou (*Rangifer tarandus*) and ringed (*Phoca hispida*) and bearded (*Erignathus barbatus*) seals in their subsistence economies (Giddings 1964; Larsen and Rainey 1948; Stanford 1976), with walrus and whales most important at sites near sea mammal migration routes. At specific sites along the Alaska coast, such as the western tip of the Seward Peninsula and Point Hope, farther north, Eskimo appear to have followed the same general subsistence trajectory as the coastal populations of Chukotka; that is, a long history of seal and walrus exploitation followed by a shift to whaling. The composition of the zooarchaeological assemblages at these sites is very similar to patterns seen across the strait in Chukotka,

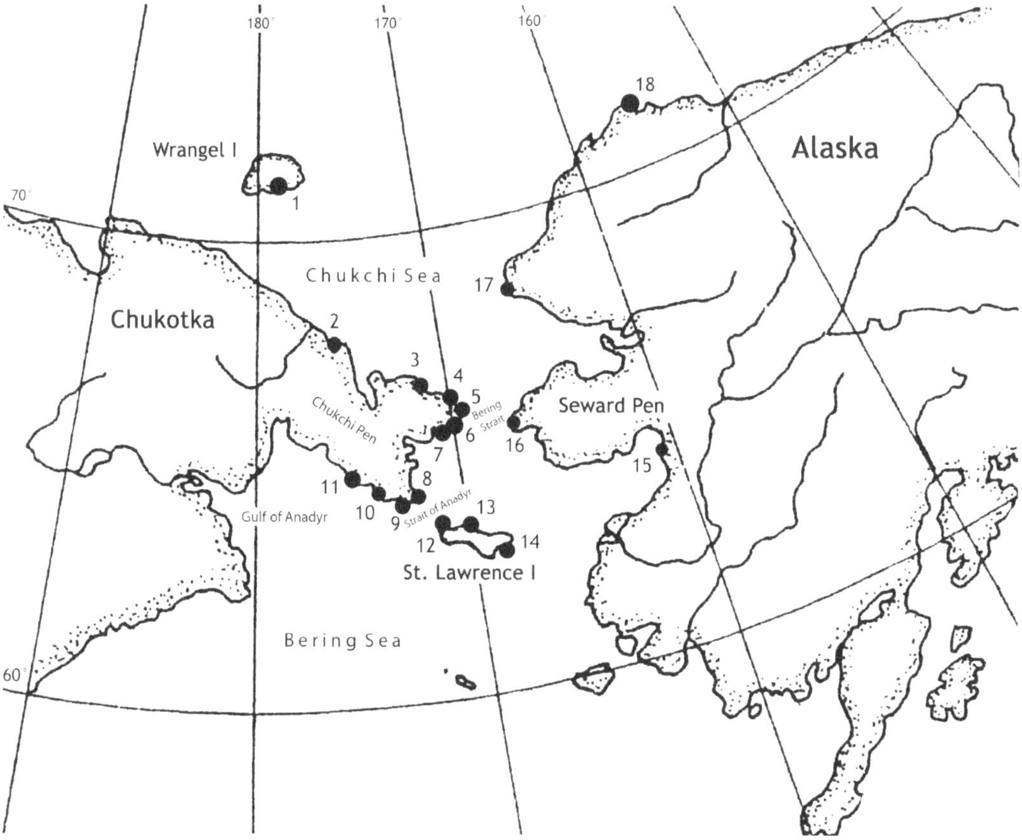

FIGURE 3.1. Key archaeological sites containing walrus located along the coasts of the Bering and Chukchi seas. Compare the site density along the coasts of Chukotka and St. Lawrence Island with the Alaska coast.

WRANGEL ISLAND: 1 = Chertov Ovrag (Devil's Gorge).

CHUKOTKA: 2 = Cape Vankarem; 3 = Seshan; 4 = Inchouan; 5 = Uelen; 6 = Ekven; 7 = Yandygay; 8 = Cape Chaplino; 9 = Kivak (Kiwak); 10 = Sireniki (Sirhenik); 11 = Unanan (Nunligran).

ST. LAWRENCE ISLAND: 12 = Gambell site complex (including Ievoghiyoq, Hillside, Miyowagh, and Seklowaghyaget); 13 = Kukulik; 14 = Punuk Islands.

ALASKA MAINLAND: 15 = Iyatayet and Nukleet; 16 = Kurigitavik Mound (Wales); 17 = Ipiutak (Pt. Hope); 18 = Utqiagvik (Barrow).

but the distribution of sites is significantly different (Figure 3.1) (see Harritt 1995 for a brief overview of prehistoric subsistence in northwest Alaska).

In this chapter, I examine the historical ecology of the Pacific walrus considering the association between human settlements and walrus migration routes along the North Pacific coast on both sides of the Bering Strait. I also discuss how humans adapted their hunting strategies to deal with the habitat preferences of female and juvenile walruses. These observations have major implications for our understanding of how cooperative hunting developed along the coasts of the Bering and Chukchi seas. Hunting by large crews aboard *umiat* (large, open skin boats) is a strategy that has long been associated—exclusively—with whaling. But emerging archaeological evidence has the potential to demonstrate that North Pacific Eskimos around the Bering Strait lived and hunted in large, cooperative kin groups *before* whaling emerged as the preferred subsistence strategy.

By the early first millennium AD, walrus had become one component of a diet that included seabirds, waterfowl, small (phocid) seal, bearded seal, and caribou/reindeer (*Rangifer tarandus*). The last item is highly variable in its dietary role, as *R. tarandus* exploitation is dependent upon access to inland herds. Prior to whaling, most coastal sites are dominated by seal, with walrus forming a significant secondary component, except on St. Lawrence Island, where walrus was the primary food and raw material source. Phocid seals, especially *Phoca hispida*, the ringed seal, are among the smallest pinnipeds exploited by North Pacific coastal foragers. Ringed seals, which in Alaska average 50 kilos, can be taken and butchered expediently by one or two men. Large, cooperative groups are unnecessary, as stealth, patience, and expert knowledge of sea ice are the most important factors in hunting success. In contrast, baleen whales are taken only by hunters on the sea in *umiat* working in large cooperative groups. Abundant ethnohistoric and ethnographic documentation indicates that whaling presents significant challenges to hunters and requires the participation of as many as eight or ten men (e.g., Bockstoce 1976; Ellanna 1988; Sheehan 1985, 1995).

Walrus hunting exists somewhere between these two forms of sea mammal exploitation. As I demonstrate below, the behavior of walrus and the number of men needed to coordinate their harvest required cooperation beyond the household level. Hunters continued to take walrus for hundreds of years, despite the labor and energy costs, because walrus met a number of raw material needs in addition to their contribution to subsistence. In this chapter, I argue that walrusing required a level of organizational complexity that was firmly established among coastal groups before whaling emerged as a preferred subsistence strategy. The ability to harvest large numbers of walrus at key points along the Alaska coast, and at numerous sites across the Bering Strait on the Chukchi Peninsula, positioned Eskimo to switch to whaling once a combination of social and ecological factors made it a less costly and more efficient strategy.

Because whaling has been a major focus of research in both the western and eastern Arctic, the importance of walrus to prehistoric diet, material culture, and social organization has been largely underestimated, with a few notable exceptions (e.g., Rainey 1941). I will therefore discuss sites on both sides of the Bering Strait that contain walrus remains in significant numbers. Sites yielding only walrus artifacts, such as bolas made of walrus teeth, are largely excluded as these items likely represent trade goods. Coastal settlements are consistently located at capes and headlands along the spring migration routes of Pacific walrus. Most of these sites later became centers of bowhead and gray whale exploitation (e.g., Barrow, Point Hope, Wales). As I discuss below, walrus migrations today are associated with the route of a nutrient-rich water mass that presently passes through the Strait of Anadyr, along the coast of Chukotka, and through the Bering Strait. The evidence from prehistoric settlement patterns and associated walrus remains indicates that such a water mass has existed for centuries and that the biomass it carried in the past facilitated walrus migration, just as it does today.

HISTORICAL ECOLOGY AND THE NORTH PACIFIC

The most critical issue in the arctic and subarctic regions of the North Pacific in the 21st century is global climate change and its effects on weather patterns, sea levels, sea ice extent and distribution, coastal erosion, traditional subsistence practices, and the survival of both marine and terrestrial species. As the chapters in this volume demonstrate, human impact on coastal and marine ecosystems stretches back thousands of years. The framework of historical ecology, with its emphasis on active human engagement with the natural world, allows us to study ecosystems of the North Pacific as products of millennia of intensive human use and alteration of coastal and marine landscapes and their associated flora and fauna.

Our understanding of Amazonian ecosystems, for example, has been transformed as archaeologists, biologists, and ecologists recognize the long-term effects of shifting cultivation, anthropogenic fire, and soil enrichment (Heckenberger et al. 2003, 2007; McEwan et al. 2001).

Pacific island ecosystems have also been the subject of research, with archaeological work demonstrating the extraordinary effects the human presence has had on island ecology, landscape, and species diversity (Kirch and Hunt 1997; Steadman 1995). Along the California Coast, zooarchaeological evidence suggests that humans, in concert with paleoenvironmental factors, played a significant role in altering the abundance and distribution of shellfish (Braje et al. 2007) and marine mammal (Glassow 2005) populations. Changes in terrestrial and marine resource availability, in turn, have been shown to affect human settlement and social organization, patterns of resource exploitation, and incidence of violence, especially when exacerbated by rapid population growth (Kennett and Kennett 2000; Walker 1989).

The historical ecology of sea mammals shifts the focus from terrestrial landscapes to those of the coast and to the "waterscapes" of sea ice and the open ocean. In the North Pacific, the waterscape is defined by ice distribution, direction and composition of currents and water masses, concentrations of nutrients, phytoplankton, algae, and sea grasses, and the depth and configuration of the shelves and basins of the sea floor. The Bering and Chukchi seas of the North Pacific differ from many other marine ecosystems in the major role played by sea ice and in the unique conditions created by the properties and actions of three major water masses. These two variables structured human adaptations to the coasts of the Bering and Chukchi seas. Site density and location, diet, and even human social organization are linked to sea ice and to the water masses that make the Bering Sea ecosystem one of the most productive in the world.

Sea ice structures and facilitates human, sea mammal, and seabird exploitation of the entire arctic ecosystem. Far from being a homogenous mass of frozen water, arctic sea ice varies in thickness, extent, snow cover, salinity, and age. It is part of a dynamic system in which Eskimo, and later Chukchi, hunters procured sea mammals. Constant monitoring of ice conditions was, and still is, essential to the survival of subsistence hunters. Cracks, wind direction, lead formation, and floe drift and rotation are all factors critical to the efficient exploitation of sea mammals hauled out on the ice or appearing at breathing holes. Inupiat of northwest Alaska use an extensive focal vocabulary to identify young versus old ice, thick, slush, and rotten ice, gray versus black ice, as well as a number of ice behaviors, such as piling, rafting, and cracking (Nelson 1969).

While sea ice facilitates exploitation of mammalian resources at the highest trophic levels, the flow, salinity, and nutrient composition of the Anadyr, Bering Shelf, and Alaskan coastal water masses affect Bering Sea productivity beginning at the lowest levels (Grebmeier et al. 2006; Springer and McRoy 1993; Springer et al. 1996). Each transports and circulates the nutrients (nitrates, phosphorus, silicon) needed by primary producers, such as phytoplanktons, algae, and sea grasses. These producers are consumed by invertebrates, including bivalves, which in turn are preyed upon by walruses in what is a relatively short food chain. In a very real sense, then, human use of walrus is dependent upon the nutrients transported by water masses. Seabirds such as auklets, fulmars, and phalaropes are similarly dependent upon nutrient availability, and the distribution of some species has been linked to the Anadyr water mass (Elphick and Hunt 1993).

This mass is associated with high phytoplankton biomass in the Gulf of Anadyr and flows northward through the Strait of Anadyr, along the coast of the Chukchi Peninsula, and through the Bering Strait. Both straits, and much of the Bering Sea, are situated on the broad continental shelf, where water depth averages less than 100 meters. Walrus forage for benthic invertebrates in these relatively shallow waters, avoiding the deeper basins. The co-occurrence of the nutrient-rich Anadyr water mass and the availability of walrus prey at appropriate depths means that walrus are able to migrate north in the spring along a highly productive route through the Bering Sea. This route takes them close to headlands and through the Bering Strait, which acts as a bottleneck. Human hunters on both sides of the strait moved to sites along the coast where walrus and seabirds were concentrated, consistently settling on the capes and headlands that allowed them to take their prey most efficiently. In other words, ancient human settlements occur where Anadyr water flows nearest the coast today. Predictably, where the Bering Strait is narrowest, human occupation has been intensive. Both Wales, on the tip of the Seward Peninsula, and the sites surrounding Cape Dezhnev, across the strait in Chukotka, have relatively long occupations with multiple settlements and abundant evidence for exploitation of walrus (Dumond 2000; Giddings 1967:189; Harritt 1995, 2004; Mason 1998).

Similarly, St. Lawrence Island was most intensively occupied at the northwestern end, where the Strait of Anadyr forces the Anadyr water mass close to the coast. Across the strait at Kivak, near Cape Nizmenny, and at a site on Cape Chukchi, there is also evidence of walrus use, in addition to the remains of phocid and bearded seals (Rudenko [1947] 1961:63–66). Once through the Strait of Anadyr, portions of the Anadyr mass deviate to the east and run along the northern coast of St. Lawrence Island, where the accompanying walrus herds supported the prehistoric inhabitants in the past and continue to support modern inhabitants today.

SETTLEMENT PATTERNS AND WALRUS MIGRATION

Eskimo on both sides of the Bering Strait have consistently settled on capes and headlands that provide access to marine resources and views of large ocean expanses since at least Okvik/Old Bering Sea times (Ackerman 1988; Gusev et al. 1999; see Table 3.1). Below, I review coastal sites with evidence for walrus exploitation. As Figure 3.1 shows, the coast of the Chukchi Peninsula is much more intensively settled than the coast of Alaska (Gusev et al. 1999) and reflects the proximity of the Anadyr water mass and the seabirds and sea mammals it supports. High concentrations of walrus remains occur, for example, at the site complex at Gambell (Sivuqaq), which is located in the northwestern corner of St. Lawrence Island. Walrus pass the site through the Strait of Anadyr every year during their spring migration. Multicomponent Gambell, with its Old Bering Sea artifactual evidence, has been intensively occupied since at least the middle of the first millennium (see Gerlach and Mason 1992 for a discussion of dates from St. Lawrence Island) and appears to have been dependent upon walrus for the entire time. Other St. Lawrence sites with large walrus assemblages include Kukulik, located on the north coast of the island on a cape near the town of Savoonga (Collins 1937; Geist and Rainey 1936), and at Okvik, on the Punuk Islands, where the excavator, Froelich Rainey, concluded that people depended almost entirely upon walrus for both subsistence and material culture. He noted, too, that the name "Okvik" was suggested to him by a St. Lawrence Islander, as it meant "place where many walrus haul out" (Rainey 1941:467, 543).

On the Alaska coast, walrus have been recovered at Nukleet on Cape Denbigh (Giddings 1964:37); in smaller numbers at Cape Nome, on the south side of Seward Peninsula (Bockstoce 1979:59); and at Wales, the westernmost point of the Seward Peninsula, where Harritt has uncovered significant evidence of walrus exploitation, including walrus skull features that bear striking similarities to those identified in Chukotka (Dumond 2000; Giddings 1964:189; Harritt 2004; see also Dumond 2000 on Collins' work at Wales). North of the Seward Peninsula, in Kotzebue Sound, Darwent (2006) found evidence for sealing and caribou hunting—not walrusing—at Cape Krusenstern, which at first appears anomalous

TABLE 3.1

Cultural Chronology of North Pacific Coastal Sites

Key Sites for Each Cultural Period Discussed in the Text Are Given in Italics.

CHUKOTKA	ST. LAWRENCE ISLAND	NORTHWEST ALASKA
Okvik/Old Bering Sea (AD 1?–1000) *Cape Dezhnev, Cape Vankarem, Seshan, Uelen*	Okvik/Old Bering Sea (550 BC–AD 1000) *Gambell complex*	Ipiutak (AD 450–750) *Point Hope type site, Cape Krusenstern*
Birnirk (AD 650?–900?) *Uelen, Ekven, Nesh'kan*	Punuk/Birnirk (AD 650–1200) *Ievoghiyoq, Kukulik*	Birnirk (AD 650–1100) *Cape Krusenstern, Kurigitavik (Wales), Utqiagvik (Barrow), Walakpa*
		Punuk (AD 650–1100)
Punuk (AD 800?–1200?)		Thule (AD 1000–contact) *Iyatayet (?), Nukleet (?), Utqiagvik (Barrow)*

NOTE: Data derived from Gerlach and Mason (1992) and Mason (1998). This chronology is speculative, based on dates derived from a number of materials using different methods and correction factors. See Gerlach and Mason (1992) for a detailed discussion of chronology and calibration in the North Pacific region.

given the position of the site on a cape extending into the sound. Biological data indicates, however, that walrus presently take a direct route north from Wales to Point Hope. According to Fay's (1982:7–26) mapping of walrus sightings in the North Pacific between 1930 and 1979, not a single live walrus was sighted in Kotzebue Sound during that 50-year period. The lack of walrus at Krusenstern suggests that walrus followed the same or similar migration route today that they followed in the past. While walrus may occasionally move east along the northern edge of Seward Peninsula, both the archaeological and biological data indicate that this area is not, and has not been in the past, desirable habitat. It has been consistently bypassed by walrus, leading prehistoric Krusenstern inhabitants to rely upon other resources.

Farther north, at the site of Ipiutak on Point Hope, Larsen and Rainey (1948:68) estimated that 23% of bones from 14 houses represented walrus, second in number only to seal. Also recovered were tools of walrus tusk, such as mattocks and shovel blades made of walrus scapulae. A harpoon socket made from a walrus baculum was recovered from the later site of Tigara, located to the southwest along the spit (Larsen and Rainey 1948:87, 176)

At Point Barrow, on the coast of the North Slope of Alaska, walrus remains have also been identified, but not in significant numbers. The Utqiagvik excavations, carried out from 1981 to 1983 prior to new construction, produced few remains, likely because the site was occupied during the late prehistoric and early historic periods, when whaling had become the subsistence focus (Hall and Fullerton 1990). Stanford (1976) recovered walrus at the site of Walakpa, southwest of Barrow, but seal and caribou bone far outnumbered walrus. Walrus teeth, however, were abundant in the form of bird bolas (Stanford 1976); these small, portable, and very durable items appear to have been traded to people at nonwalrusing sites with considerable frequency on both sides of the Bering Strait. Numerous bolas were also recovered by Ford (1959:143), whose excavations at Nunagiak, another site at

FIGURE 3.2. Illustration of walrus hunting on St. Lawrence Island by Henry Elliott (1881:96). Elliott made additional observations on use, abundance, and distribution of walrus on his voyage to the "Seal Islands," i.e., the Pribilof Islands of St. Paul and St. George.

Point Barrow, yielded a shovel made from a walrus scapula. Unfortunately, Ford does not discuss the fauna he excavated at Nunagiak, so it is not possible to determine whether his finds of bolas and a scapula shovel represent goods acquired through trade or whether they were the products of a local walrus hunting tradition.

In contrast to Wales and Point Hope, much of the coast of the Seward Peninsula and Kotzebue and Norton sounds are largely devoid of walrus remains. Certainly part of the zooarchaeological walrus distribution pattern is due to preservation, which is notoriously poor at Cape Nome and at sites on Cape Denbigh, with the possible exceptions of Nukleet and Iyatayet (Giddings 1964:93, 96, 113). Because migration patterns rarely took walrus along the coast of Alaska for any distance, the dearth of walrus and the intensive use of phocid seals at sites such as Cape Krusenstern is likely an anthropogenic pattern, rather than a product of taphonomy.

With the exception of the Alaska sites discussed above—Wales, Point Hope, a handful of additional Alaskan sites, and possibly Sledge Island (Bockstoce 1979:13), off the southern coast of Seward Peninsula—walrus are more numerous at sites along the coast of Chukotka. This pattern is matched by recent observations that walrus are more common in the western Bering Sea and around St. Lawrence Island than they are on the eastern, Alaska side. The relative impoverishment of the eastern coastal waters that we see today (Springer and McRoy 1993; Springer et al. 1996), therefore, appears to reflect a historical pattern. Early naturalists such as Henry Elliott (Figure 3.2) recorded immense numbers of walrus near St. Lawrence Island (Elliott 1881). An engraving from James Cook's 1778 voyage through the Bering Strait and into the Chukchi Sea (Figure 3.3) was accompanied by the observation that "[sea horses] lie upon the ice in herds of many hundreds, huddling like swine, one over the other; and they roar very loud . . . [and give] us notice . . . of the ice" (Cook 1784:42).

On the western side of the Bering Strait, the northernmost Russian site to yield walrus remains thus far is Chertov Ovrag on Wrangel

FIGURE 3.3. Engraving based on observations in the Bering Strait made during the third voyage of James Cook, 1778 (Cook 1784:facing page 40).

Island (Gerasimov et al. 2006:203). The densest concentrations of sites yielding walrus remains, however, are located along the northern and eastern coasts of mainland Chukotka. These sites, like those across the strait in Alaska, rim the Bering Sea shelf, much of which comprised ancient Beringia. The shallow shelf, which averages less than 100 m in depth, is attractive to walrus in terms of both sea floor depth and prey availability. Ringed seal appears to have been heavily exploited at many of these sites (Savinetsky et al. 2004), with walrus representing a smaller proportion of the fauna. It is possible, given the abundance of walrus bone and tusk artifacts relative to those of seal, which occur rarely, that the raw materials provided by walrus in the form of blubber, hide, teeth, and tusk were more important than the contribution walrus made to the diet. In other words, the intensive use of all parts of the walrus produced a pattern at archaeological sites that underrepresents their importance in Eskimo economies. The focus on proportions of taxa in dietary terms alone may, therefore, be somewhat misleading, especially considering that bone artifacts are often analyzed separately from fauna interpreted as the residue of food processing and consumption. Orekhov ([1987] 1999:186), for example, estimated that at certain points in Chukotkan prehistory, 90 percent of all bone artifacts were made of walrus tusk or sea mammal bone. If these materials are excluded from consideration in subsistence reconstructions, our understanding of the actual contributions of different taxa to the economy will be skewed, overrepresenting species whose remains were not considered suitable for artifact manufacture.

Although some excavations have occurred along the coast of the Chukchi Peninsula, the bulk of Russian archaeological work has taken the form of either reconnaissance or excavations at major cemetery sites, such as Chini, Uelen, and Ekven (Mason 1998). Absolute dates are unavailable for most sites identified on surveys, especially sites located south of Cape Dezhnev (Gerlach and Mason 1992; Mason 1998), as the primary form of data analysis has been qualitative and descriptive, rather than quantitative and analytical. Small sites have been, until recently, assigned to cultural periods on the basis of harpoon head morphology and described in the broadest terms, often accompanied by assertions regarding site function, subsistence focus, and ritual behavior. Excavators such as Dikov ([1974] 2002, [1977] 2003), Orekhov ([1987] 1999), and Rudenko ([1947] 1961) generally noted the presence and concentration of mammalian faunal remains, including walrus, but rarely provided information on the proportions of seal to walrus, or caribou to sea mammal.

Relying exclusively upon these early reports would give the impression that the contribution of birds to the diet and material culture of coastal Eskimo was minimal (but see Orekhov [1987] 1999:187). But, as Savinetsky (2002; Savinetsky et al. 2004) has shown, Chukotkan faunal assemblages contain abundant bird remains representing multiple families of colonial seabirds and waterfowl, as well as terrestrial species such as snowy owl and willow ptarmigan. Despite these shortcomings, these early works appear to be accurate with regard to the high density of settlements along the coast and the degree of reliance upon marine mammals. Results of the Bering Archeological Expedition, which mapped and sampled coastal sites in the early 1990s, indicate that settlements were systematically located less than 25 km away from each other in the early centuries AD, with adjustments made for rock outcroppings, sheer cliffs, and other landforms (Gusev et al. 1999). According to Gusev and his colleagues (1999), distance between settlements consistently declined through time, until a new pattern emerged in the 1300s, when distance took second priority to defensive position.

In the northeast part of the peninsula, small settlements become visible in the first two centuries AD. Gerlach and Mason (1992), who have compiled and calibrated dates for the coast, give a mean date for the Old Bering Sea (OBS) culture in Chukotka as c. 1460 BP. Many of the sites discussed below have OBS components, assuming relative chronological assignments are correct. However, once occupation of the coast intensified and maritime adaptations were firmly established, by the mid-first millennium, locations on headlands with access to fresh water were consistently used well into the proto and early historic periods. Along the northeastern coast, facing the Chukchi Sea, are several settlements with walrus remains. From the northwest to southeast, they include Cape Vankarem, with walrus and polar bear skulls; Seshan, which includes a cemetery component and what Dikov ([1977] 2003:176, 188) suggested are the remains of a ritual structure associated with walrus hunting; Inchoun, a village site located close to a walrus haul-out (Dikov [1977] 2003:165); and Uelen.

The sites around Cape Dezhnev, including Uelen, are relatively well known in comparison to other sites in Chukotka. Also known as East Cape, Cape Dezhnev is located in the northeastern corner of the peninsula marking the point where Bering Strait waters meet the Chukchi Sea. Uelen is a large site associated with the Old Bering Sea culture and known primarily for its cemetery and for evidence of whaling. To the south, the cemetery and settlement site of Ekven has yielded evidence for what Mason (1998) has suggested was a polity capable of generating a surplus and supporting a comparatively large population. The whaling evidence at both sites has been the focus of archaeological attention, but walrus remains are also present, presumably from earlier occupations.

Moving south from Dezhnev, Rudenko identified walrus at Yandygay in association with semi-subterranean houses; on Arakamchchen Island, Cape Chaplino, and at Kivak (Kiwak; Rudenko [1947] 1961:51–52, 53, 57), directly across the Strait of Anadyr from St. Lawrence Island. Moving northwest along the south side of the peninsula, Rudenko ([1947] 1961:71) reported a large settlement at Sireniki, the eastern part of which contained walrus. This site also contained a number of walrus tusk and bone artifacts, including the omnipresent bird bolas made from walrus teeth, as well as "amulets" made of walrus bone and wrapped or suspended with baleen. Today, Sireniki is a large village that continues to take sea mammals, although bowhead and gray whales now dominate. Farther along the coast, at Nunligran, are more cliffs full of nesting alcids and a small bay where the prehistoric village of Unanan is situated. Rudenko ([1947] 1961:89–94) found abundant evidence of walrus and ringed seal at the site; excavations in 2007, jointly directed by Sergey Gusev and Daniel Odess, recovered multiple walrus skulls, postcrania of phocid seals, and significant numbers of bird remains, likely alcid, which were identified by the author in the field.

WALRUS AS A DIETARY RESOURCE AND RAW MATERIAL

Studies of the role of walrus in prehistoric North Pacific subsistence are hampered by a number of factors, none of which are unique to the region. A similar situation prevails in Thule subsistence studies in the eastern Arctic (Savelle and McCartney 1988). Due to permafrost, materials excavated around the Bering and Chukchi seas are often in an excellent state of preservation. Wooden timbers, baleen containers, and skeletal remains are routinely recovered *in toto*. But early excavators, such as Henry Collins, who worked at complex stratified sites on St. Lawrence Island, made only the most cursory observations of faunal remains. Massive amounts of bone, like the two-meter-high pile produced (and photographed) by excavators at Kukulik in 1934–1935, were simply discarded (Geist and Rainey 1936). Though Geist and Rainey (1936:15) describe and illustrate some artifacts of walrus bone recovered from Kukulik, the 21st-century zooarchaeologist gathers from their work only that walrus meat was a major part of the diet, in addition to seal and whale. At the multicomponent site of Gambell, walrus and seal occur "in abundance" (Collins 1932:114), while at Nukleet at Cape Denbigh, walrus bone appears "throughout" the midden (Giddings 1964:37). Even the painstaking efforts of Dumond (1998:53, 2000:132) to reconstruct and report on 1930s excavations at Wales, Alaska, and the Hillside site on St. Lawrence were unsuccessful with regard to faunal remains. In the case of Bandi's excavations at Mayughaaq (Miyowagh) and at the site of Kitngipalak, on the west coast of St. Lawrence, the bulk of the fauna remains unanalyzed and unpublished (Bandi and Blumer 2002:50), though exploratory work by the author in 2010 indicates that the Kitngipalak fauna includes walrus, phocid seal, bearded seal, canids, sea urchin, fish, and murre (a seabird of the genus *Uria*).

Attempts to reconstruct subsistence on the basis of early-20th-century excavations are, therefore, a difficult undertaking. Even so, the available data is remarkably uniform: intensive walrus use occurred on St. Lawrence Island, while along the coast of Chukotka, assemblages generally consist of walrus and ringed or bearded seals, followed by (presumably) lesser numbers of dog, reindeer or caribou, and seabirds. Whale makes an occasional appearance at the earlier sites, often in architectural or funerary contexts, and was likely scavenged originally. Though Darwent (2006) demonstrated that the "Old Whaling" locality at Cape Krusenstern was actually a site where intensive phocid sealing took place, few zooarchaeological assemblages exist on either side of the Bering Strait that could be subjected to similar reanalysis. Faunal remains are recovered primarily from middens or, as at Cape Krusenstern, from semi-subterranean houses, which were excavated both as complete units and in portions (e.g., Geist and Rainey 1936:62).

While fauna from subsistence contexts went unrecorded in the early decades of Bering Sea archaeology, excavators tended to save artifacts made of walrus bone and tusk and noted when features employed whale or walrus bone as structural components. As a result, an impressive list of walrus products can be compiled, which I have categorized for the sake of simplicity as architectural materials, transport-related, hunting implements, tools, or ornaments.

The size and strength of walrus bones made certain skeletal elements ideal for architectural use. The walls of a cache at Miyowagh on St. Lawrence were lined with walrus bones (Collins 1937:73), while ribs were used to line a Birnirk burial at Tigara on the northwest coast of Alaska (Larsen and Rainey 1948:167). Tusks and baculi braced the walls of a house entranceway, prefiguring the later use of whale for the same purpose elsewhere in the Bering Sea region (Collins 1937:75). The multipurpose walrus baculum, when sharpened, was used as a bone peg in a structure at Kukulik (Geist and Rainey 1936:62). The same element was observed securing floorboards on the Chukchi Peninsula

(Geist and Rainey 1936:58–64; Lee and Reinhardt 2003:132). Walrus skins and intestines were used as construction materials and skylights. At Miyowagh, where preservation was excellent, Collins recovered a walrus hide above a set of house timbers (Collins 1937:37).

Walrus was also used for a number of transport-related purposes. In the recent past, the walrus stomach served as material for sails and the hide of female walruses for the covers of *angyapiget* (large open boats comparable to *umiat*) on St. Lawrence Island (Braund 1988). On both St. Lawrence and the Chukchi Coast, paired walrus tusks, sometimes flattened on one side, served as sled runners (Dumond 1998:64; Rudenko [1947] 1961:65). Such objects also appear in burial contexts. Dikov's ([1974] 2002:16–17, 20) excavations at Chini cemetery—dated to Old Bering Sea—recovered sleds with tusk runners that may have been used as platforms for food offerings. A Punuk context at St. Lawrence yielded a walrus baculum serving as a sled crossbar (Collins 1937:158, 231). Dumond (1998:65) documented a similar object in the materials Giddings excavated at the Hillside site in 1939.

The list of uses of walrus tusks in hunting implements is extensive, and includes harpoon heads, foreshafts and sockets, counterweights, and wound plugs. Bird bolas made of walrus teeth, discussed above, have been recovered in great numbers from Chukotka (Rudenko [1947] 1961) and St. Lawrence—Collins (1937:228) reported 61 from Ievoghiyoq. Mainland Alaska sites have yielded fewer examples; Bockstoce (1979:59) reports a single find from a Birnirk context at Cape Nome, and Stanford (1976) recorded several at Walakpa. At the proto-historic site of Tigara, Larsen and Rainey (1948:176) recovered a walrus baculum that they suggested had been used as a harpoon socket.

As with hunting implements, ivory was the favored material for many tools, employed for root picks, ice picks, and mattocks, which occur in both burial and domestic contexts (see multiple examples in Dikov [1974] 2002; Larsen and Rainey 1948; Rudenko [1947] 1961). In addition to tusk, walrus rib was used for "marlin spikes" (Larsen and Rainey 1948:87) and as a scraper for hideworking (Collins 1937:163). Collins (1937:175) documented a single, novel use of a pair of fused walrus teeth as a drill rest at Miyowagh.

Walrus scapula shovels have been described by several ethnohistorians (e.g., Bogoras 1904–1909:178). Similar tools have been recovered archaeologically from Miyowagh, Nunagiak, Birnirk, Kukulik, Ipiutak, Okvik, and Sireniki (Collins 1937:161, plate 150:166, 235; Ford 1959:143, figure 123; Geist and Rainey 1936:105; Larsen and Rainey 1948:87; Rainey 1941:512–513; Rudenko [1947] 1961:114). In each case, the edges of the scapula were trimmed or smoothed, four or six holes were created, and the scapula was hafted. Collins (1937:161) dated his Miyowagh example to an Old Bering Sea context. Geist and Rainey (1936:105) recovered six examples from what they believed to be a house occupied until 1879–1880. The walrus scapula shovel is thus one of the more conservative and enduring tool types in the Bering Sea region, persisting in its basic form for almost 2000 years.

A final use of walrus bone is as a raw material for personal ornaments. Rudenko ([1947] 1961:87) identified a perforated piece of walrus tusk at Sireniki (Sirhenik), which he interpreted as a pendant designed for suspension. At the same site, also recovered from a Punuk context, were a walrus phalanx and an astragalus, which Rudenko ([1947] 1961:87) suggested were amulets. Collins (1937:244) even reported finding a pair of snow goggles at Seklowaghyaget on St. Lawrence Island that were made from a walrus baculum.

To summarize, walrus bones were used for tools ranging from scapula shovels to sled crossbars; tusks were perforated for use as ornaments; blubber served as a fuel; and teeth were used as trade items. In northern Alaska during times of famine, the walrus hide was sometimes consumed (Murdoch [1892] 1988:61), a practice that likely occurred prehistorically as well. In contrast to seal and whale, both of which functioned primarily as subsistence items and, in the latter case, as a source of building material, every part of the walrus was intensively utilized at least as early as Old Bering Sea/Okvik times. As suggested above, the actual dietary contribution of walrus meat and organs may be consistently underestimated for the simple reason that the bones and tusks, rather than showing up in middens and other food residue discard contexts, were converted into tools and ornaments and employed for architectural purposes. These walrus artifacts, when recognized or recovered, often are considered separately from food remains. They remain analytically distinct, resulting in skewed estimates of the dietary contribution of walrus. Even assuming that we are underestimating the role of walrus in Bering Sea Eskimo diets, available data indicate that no other animal served so many needs for so long a time in the North Pacific.

Such intensive use required large numbers of animals and their efficient harvest. Although walrus hauled out on beach shingles may have been an early target of coastal hunters, female and young walrus using sea ice were likely preferred for a number of reasons. Hunting of walrus on sea ice was energetically and organizationally distinct from land-based hunting. Completely different technologies and strategies, discussed below, were required for success. Until the transition to whaling, walrusing demanded greater cooperation and skilled leadership than any other subsistence pursuit in the North Pacific.

WALRUS HUNTING STRATEGIES

Prehistoric Bering Sea hunting technology was dominated by toggling harpoons, parts of which have been used to construct regional chronologies. Sealing, walrusing, and whaling employed broadly similar projectile technologies. The size of ivory and lithic projectile points is generally the basis for determining whether the implement was used for small seals or the largest of sea mammals. The drag float, essential for walrusing on sea ice, appeared relatively early in the cultural sequence (i.e., Old Bering Sea/Okvik) in Chukotka and St. Lawrence Island, reaching the Alaska mainland much later, during the Birnirk period, AD 700 to 1000 (Sheehan 1985:138). In combination, the harpoon and the drag float made walrusing a viable subsistence strategy when pursued cooperatively. Because male and female walruses use different habitats, prehistoric hunters could preferentially take female walruses and calves when working cooperatively. In the winter and spring, walrus spend most of their time foraging in the leads and polynyas amongst broken ice. Female walruses, in particular, favor ice floes, where they rest and sleep with calves born the preceding spring (Fay 1985). In contrast, male walruses haul out in the hundreds and even thousands at many predictable and well-known places along the coasts of St. Lawrence Island, Chukotka, and Alaska. They generally prefer isolated rocky or sandy shingles (Fay 1985). Male walruses can be taken by hunters on foot alone or, more likely, in groups of two or three individuals. They can be butchered on land and the portions transported directly to the settlement site. Weather and ice conditions play comparatively small roles in the exploitation of male versus female walruses.

Given the time, energy, and labor requirements necessary to locate, approach, harvest, butcher, and transport females on sea ice versus males on coastal shingles, the preference for females and calves as documented ethnographically (Ellanna and Sherrod 1983:48) at first appears counterintuitive. However, Eskimo of the 19th and early 20th centuries indicated that the skin of female walruses was more suitable for boat covers because it was thinner, more durable and waterproof, and easier to work than the tough, scarred, and warty skins of male walruses. The taste of the meat of female walruses and calves has historically been preferred to that of males by some St. Lawrence Islanders, an admittedly subjective preference. However, the tusks of female walruses are preferred because they have fewer cracks, an objective consideration in the production of hunting implements and tools (Braund 1988:64; Brooks

1953:510). The skin of calves was favored for specific lines and lashings and was distinguished linguistically by the Siberian Yupiit of Chaplino from lashings of seal or bearded seal (Menovščikov 1968:436–437).

A single St. Lawrence Island boat of the 19th and early 20th centuries required two to three female walrus hides to cover it, plus the skins of two calves for lashings. Boats required new covers every 2 to 3 years, more often if the skin cover was damaged by ice or did not dry properly between uses (Braund 1988:23–24, 51). Presumably, the hide of female walruses was also preferred as roofing material. If these preferences for the meat, tusks, and skins of female walrus and their calves existed among prehistoric Eskimo, their harvest would dictate a cooperative hunting strategy. Because of the use of ice floe habitats by female walrus with young, prehistoric hunters would face ecological conditions and labor requirements that differed significantly from those pertaining to the harvest of male walrus.

ICE CONDITIONS

Accessing female walrus on spring ice floes required *umiat*. Although a single kayaker would be able to get to walrus amidst broken ice, he would not be able to return with his catch if the lead closed up and he had to drag both kayak and walruses over the ice. An *umiaq* crew would be capable of carrying both watercraft and a female walrus weighing as much as 1100 kg—perhaps as many as four walruses at a time—across the ice to leads in broken ice conditions. Such ice conditions have not occurred as far south as the Aleutian Islands in recent memory, making the Aleut use of the kayak much more practical than it would have been farther north (but see Crockford and Frederick 2007 and this volume for evidence of sea ice expansion during the Neoglacial).

Kayaks were also used at Nunivak, which is located at the southern edge of the winter pack today. Given Nunivak's location, kayak hunting was most likely a better strategy than a crewed *umiaq*, which required at least four or five men to haul the boat (Braund 1988:16–17; Hughes 1960:105). In addition to carrying the *umiaq* across the ice, both unloaded and later loaded with walrus, crew numbers were dependent upon the labor required to haul one or more walrus out of the water, butcher them in a timely manner, and return to shore fully loaded, even in difficult weather. The spring ice conditions, when walrus were most concentrated along their migration routes, made hunting by *umiaq* the most appropriate strategy. Later in the season, when walrus were more dispersed and ice conditions had changed, Bering Sea hunters occasionally pursued walrus by kayak (Murdoch [1892] 1988:328), but the key to the spring harvest was a communal hunt by walrus crews and their captains.

LABOR AND ORGANIZATIONAL REQUIREMENTS

Efficient walrusing required organization, timing, and the ability to mobilize and integrate members of the crew both on land and at sea. Menovščikov (1968:435–438) documented 15 different expressions used by crews to identify a walrus, its age, its position in the water, and the direction in which it was swimming. A single young walrus, for example, could be distinguished linguistically using a minimal number of terms from a walrus herd swimming toward shore. Likewise, a 1-year-old walrus was identified by a different term than that used for a walrus with small tusks. An additional 20 terms referred to direction relative to the *umiaq*. Demonstrative pronouns such as *samna*, meaning "that one on the southern side," were used to provide the designated striker with a precise orientation as rapidly and accurately as possible. Terms for the *umiaq* itself were similarly elaborated (Menovščikov 1968). This specialized vocabulary, shared by the crew and its captain, demonstrates the high level of cooperation and integration that could be achieved by a closely knit kin group tied to an accomplished leader.

The "traditional" role of whaling captain, or *umialik*, was characterized by wealth and numerous kinsmen, concepts that Burch (1975: 209–210, 212n; see also Spencer 1972) suggests were virtually synonymous for early historic Inupiat. The energetics and time constraints involved in walrusing required a set of skills and organizational focus very similar to that of successful whaling. At many locations in the Bering Strait and the Strait of Anadyr, walrus populations are concentrated for short periods of time during migration. On King Island and the Diomedes, walrus were reportedly available for a maximum of 4 weeks sometime between April and June, assuming that modern migration patterns approximate prehistoric ones (Burns 1965; Ellanna and Sherrod 1983). Depending on weather and ice conditions, this time could be considerably shortened, as when storms made navigation of leads in sea ice too risky.

To take greatest advantage of this short migration window, walrusing crews needed to be efficient, well organized, and able to work in concert. These requirements were achieved through kinship and crew membership, as well as through the leadership and managerial abilities of the walrus captain as evidenced in economic, social, and ritual activities. The importance of kin relations with regard to whaling has been well-documented (e.g., Cassell 1988; Ellanna 1988; Ellanna and Sherrod 1983; Spencer 1972), with close consanguines forming the nucleus of a crew in the 19th and early 20th centuries. These kinsmen spent time together throughout the year, sharing households and community structures, manufacturing and maintaining equipment, and celebrating feasts and rituals.

Ellanna (1988) has argued that the limiting factor in marine mammal subsistence adaptations was too few hunters, an observation made by Spencer in the 1970s. Spencer (1972:116) suggested that the minimum whaling crew in northern Alaska was six men. Ellanna reported that for 20th-century walrusing, Gambell skin boat crews were composed of seven or eight men, but that more men were needed at Wales, King Island, and the Diomedes due to larger boat sizes (Ellanna 1988). During ethnographic work at Gambell in the mid-1950s, Hughes observed that winter walrus hunting required a minimum of four men just to haul the boat across the ice when no leads were available (Hughes 1960:105).

In some cases, it was necessary to supplement the whaling crew, and by implication, the walrus crew, with those who were neither consanguines nor affines. Trading partnerships that extended beyond the tightly knit and bounded kin groups could potentially supply the need for additional crew members. Co-marriage was another way to establish fictive kinship and integrate crew members, even when individuals came from different societies (*sensu*, Burch 1980). Burch (1972:29) has noted that co-marriage partners could spend an entire year with partners from another region. This situation would provide the opportunity to integrate an outsider male into the crew and solidify regional alliances.

In sum, walrus behavior plus ice and ocean conditions specific to the North Pacific mandated the use of *umiat*. The need for a large, skilled crew and strong leadership derived from the parameters set by *umiaq* size in combination with sea ice conditions and the weight and habitat preferences of female walrus and their calves. Early *umiat* were likely much smaller than the motorized monsters found in the Bering and Chukchi seas today; their crews were probably as small as was safe and practicable. But as hunting successes accrued through the years, the successful *umialik* and his wife became lodestones for kinsmen, who sought crew membership, a reliable food supply, and the intangible advantages of association. As such kinsmen and their families began to settle near successful *umialiit*, the number of houses and, eventually, community structures began to increase. In this way, the socioeconomic organizational focus upon walrusing set the stage for larger populations and settlements as well as the more labor-demanding enterprise of whaling. The difference between these two pursuits is therefore in scale, rather than in structure.

DISCUSSION

Despite the centrality of walrus in subsistence and daily life, whaling has been considered the hallmark Eskimo adaptation to the arctic environment. Ethnographic and archaeological literature on the North Pacific has focused on whaling as the key subsistence activity in the lives of both modern and ancient Eskimos and their predecessors at least since Larsen and Rainey's (1948) *Ipiutak and the Arctic Whale Hunting Culture* (e.g., Cassell 1988; Chelenov and Krupnik 1984; Freeman et al. 1992; Krupnik 1987, 1993a; McCartney 1995a, 1995b, 2003; Sheehan 1985; Whitridge 1999; Worl 1980). In part, this focus is due to the phenomena believed to be related to whaling, including settlement of the High Arctic and major developments in socio-political organization (Krupnik 1993b). Among the modern cultural groups inhabiting the Arctic, whaling is not only a subsistence activity but also an important marker of ethnic identity, and so has received abundant attention from ethnographers and archaeologists (e.g., Brewster 2004; Freeman et al. 1998; Krupnik 1987).

The earliest date for indigenous whaling in the North Pacific remains a contentious issue, plagued by poor temporal control within sites, marine reservoir effects, and lack of standardization in radiocarbon dates (Dumond and Griffin 2002; Gerlach and Mason 1992; Khassanov and Savinetsky 2006; Mason 1998). Depending on the researcher, whaling was developed by Old Bering Sea, Birnirk, Punuk, or Thule peoples (Mason and Gerlach 1995:2). Complicating the picture is the recent report that inhabitants of a Korean site were whaling thousands of years earlier than people presumably began whaling in the North Pacific (Lee and Robineau 2004). The "Old Whaling" site at Cape Krusenstern, Alaska, argued by archaeologist Louis Giddings to represent the earliest settlement of whalers in the western Arctic, introduced an outlier into the Alaska data that has confounded archaeologists for nearly five decades (Giddings 1967; Giddings and Anderson 1986) and will doubtless continue to do so despite Darwent's (2006; see also Mason and Gerlach 1995 for a discussion of Giddings' evidence) conclusion that the inhabitants of beach ridge 53 at Cape Krusenstern were sealing, rather than whaling.

Those who favor whaling as a prime mover in the Arctic view the transition from seals and walrus to whales as a pivotal development in social organization, the origins of a surplus-based political economy, and the emergence of leaders in the form of *umialiit*, or boat captains. They argue that the whaling captain as a leader in both subsistence and ritual, in combination with increasing population, endemic warfare, and appropriate technology, led to larger settlements. Members of these settlements participated in whale hunts that united group members in new ways. Rather than the supposedly less complex kin- and marriage-based organization that structured much of North Pacific prehistory, whaling initiated innovative social forms that persisted through the early historic period (Cassell 1988; Sheehan 1985).

Dates and cultural associations for the whaling adaptation are significant in North Pacific prehistory because, in addition to social complexity and leadership, whaling has been associated with increased raiding and technological developments such as the drag float, which helped prevent the loss of harpooned animals (Bockstoce 1976; Mason 1998). Nevertheless, all the cultural features required for successful whaling were present earlier, when walrus and seals were the primary resources. This argument may be settled if zooarchaeologists can demonstrate that female walruses were preferentially exploited prehistorically, as they were in the ethnographic past—a point I return to below.

Cassell (1988:95) has argued that Inupiaq winter village sites, such as Point Barrow and Point Hope, were intentionally located near whale migration routes. Yet, walrus have historically passed these same sites on their spring migrations. Pacific walrus are known to have ventured eastward through the Arctic Ocean as far as Barter Island and the village of Kaktovik. Given that settlement at Alaska sites appears to predate whaling activities, walrus exploitation, in combination with sealing, provides a better explanation for their locations on the coast and proximity to migration routes and major walrus haul-outs. Although the present distribution of walrus may not be a completely accurate indication of past haul-out locations, anecdotal evidence suggests that walruses consistently return to the same location year after year in the absence of disturbance by humans or major changes in sea ice conditions or availability of benthic resources.

To my knowledge, only Linda Ellanna (1988: 77; Ellanna and Sherrod 1983) has suggested that walrus hunting played a central role in the emergence of a more complex system of maritime production. Perhaps due to her involvement with the Eskimo Walrus Commission, Ellanna emphasized that *both* walrusing and whaling produced social forms that departed from standard models of hunter-gatherer demography and social organization. Distinctive cultural features that appeared on the North Pacific coast include decreased mobility and semi-permanent settlements, larger populations relative to inland sites, and a conservative social system structured around the *umialik* and his crew (Ellanna 1988), which Burch (1980) has demonstrated was composed of local family members in the early 1800s. A "local" family was large and bilaterally extended, included up to four generations, and occupied two or more adjacent dwellings (Burch 1980:262), a definition that can reasonably be extended into Bering Sea pre- and protohistory.

In order to hunt walrus efficiently in the North Pacific during spring migration, skin boats (sing. Inupiaq *umiaq*; St. Lawrence Yupik *angyak* = boat, *angyapik* = skin or "real" boat) were essential. Kayaks, which were used for whaling by solitary hunters in the Aleutian Islands (Black 1987; Holland 1992; Laughlin 1980), were unsuitable. The Yukon-Kuskokwim Delta appears to be a transitional region between the use of kayaks to the south and the northern use of *angyapiget* and *umiat*. Walrus were hunted from both Nunivak (Lantis 1984) and Nelson islands, but were far less important to the diet than seal or herring (Fienup-Riordan 1983). The ice conditions at the mouth of the delta vary from year to year, and in some cases the delta never freezes, making spring hunting by kayaks (*qayat* on Nelson Island) a viable alternative to *umiat*, as hunters did not face the necessity of dragging watercraft and harvested sea mammals across ice floes in search of leads.

In Chukotka, the beginning of the walrus season was heralded by the arrival of snipe, which meant that near-shore hunting would soon be possible (Menovščikov 1968:434–435). Later, when ice floes retreated from the shore, hunters would follow walrus in *umiat*. Ethnohistoric and ethnographic evidence indicates that the preferred watercraft for whaling and walrusing in the North Pacific over the last 250 years has been some form of a large, open skin boat covered with walrus or bearded seal hides. Along the coasts of Chukotka and St. Lawrence Island, ice floes persist much longer than on the Yukon-Kuskokwim Delta, and female walruses flock to the floes with their calves. *Umiat*, and thus multiperson walrusing crews, were necessary to this more northern subsistence strategy for the reasons discussed above—habitat preferences of walrus, sea ice conditions in the region, and the labor and organizational requirements for hunting, harvesting, and transporting walrus across ice floes.

Walrus hunting, rather than whaling, provided the impetus for the development of social complexity in the North Pacific. Whaling then developed on a walrus-hunting foundation of sophisticated technology and complex social organization that privileged the role of the boat captain. The reasons for the general shift from walrusing to whaling are unknown, but factors critical to human exploitation of walrus include predictability of timing and location of migration routes and haul-outs, location and abundance of invertebrate prey, and extent and distribution of sea ice. Human hunters were familiar with when and where walrus passed specific points along the coast. Familiarity with sea ice patterns, animal behavior, and boating technology, in addition to complex social organization, enabled hunters at some sites, such as the complex of settlements in the northwest corner of St. Lawrence Island, to subsist almost exclusively on walrus at certain times of the year. Disruption of migration patterns due to changes in temperature or current flow, for example, would create changes in resource abundance that increased subsistence insecurity, potentially prompting a shift to whaling.

Data presented above indicate that Eskimo settlements in the early centuries of the first millennium AD were located at key sites on walrus spring migration routes. Site density appears to be closely related to the availability of seals, walrus, and possibly seabirds. During periods of increasing density, pressure from the inhabitants of other sites, as well as stress on the available marine resources, may have prompted prehistoric hunters to intensify their activities, with concomitant changes in social organization and possibly a refocus on cetacean prey. Late-20th-century scholars have noted the relationship between settlement location and bowhead migration routes (e.g., Worl 1980:306), but have not linked location to walrus migration. Earlier observers, however, often made this connection. Bogoras (1931:476), for example, noted that on the Chukchi Peninsula, settlements were established on outer capes or on the windy sides of islands "where there is always a chance for the sudden pursuit of . . . a big walrus". Rudenko ([1947] 1961:178) made the same link, suggesting that walrus distribution would help archaeologists clarify the settlement patterns of sea mammal hunters.

Based on ethnohistoric evidence, faunal data, and the distribution and time depth of sites with walrus remains around the coasts of the Bering and Chukchi seas, I argue that cooperative walrusing preceded whaling. However, sufficient osteological data does not yet exist to demonstrate that females were preferentially exploited over males. Though highly suggestive, the evidence of settlement size and location, and zooarchaeological and artifactual remains must be augmented in order to demonstrate conclusively that walrusing, rather than whaling, was the prime mover in the development of social complexity in the North Pacific. One solution to this problem comes with the application of osteometry—bone measurements—to the question of whether male or female walrus remains dominate prehistoric sites.

Studies of Atlantic walruses (*Odobenus rosmarus rosmarus*) have successfully employed osteometric data to distinguish between the sexes, but similar work remains to be done on the Pacific subspecies (Garlich-Miller and Stewart 1998; Weckerly 1998). There are few individuals of known sex and complete crania and postcrania in museum collections, but results of a preliminary study by the author of Pacific walrus of known sex at the American Museum of Natural History are encouraging.

Although the sample is too small for generalization, significant differences were identified in key cranial and mandibular dimensions, unsurprising given the degree of sexual dimorphism in modern walrus. With a larger sample of walrus of known sex, parameters for males and females can be established and applied in the field. Given the logistical constraints of archaeological work in the Arctic, it is frequently impossible to transport faunal assemblages to laboratories and curation facilities. The sheer size and weight of walrus remains makes their transport and curation expensive, time-consuming, and ultimately prohibitive. DNA analysis, though becoming increasingly common in zooarchaeological research, is more costly and less efficient than the field recording of a set of osteological measurements. In contrast to DNA work, bone measurements can be taken accurately by students, volunteers, and other nonprofessionals with little experience in the study of faunal remains. Finally, the ability to sex walrus as excavations proceed enables investigators to alter excavation depth and location, modify recovery procedures, and make more efficient use of the limited time available in the 2-month arctic summer. Development of osteometric parameters for

Pacific walrus, the subject of ongoing research, will eventually enable investigators to use proportions of male and female walrus in an assemblage as proxy indicators for the hunting strategies used to harvest them.

OF WHALES AND WALRUSES: CONCLUSIONS

To accurately characterize the subsistence adaptations of North Pacific Eskimos, archaeologists must be committed to examining all of the evidence, even when it has the potential to destabilize long-held assumptions about when and how people of the Bering Strait region became whalers. The ubiquity of ivory artifacts has, paradoxically, made walrus as a source of food and other raw materials virtually invisible. The spectacle of prehistoric whalebone monuments, paired with an archaeological imagination fuelled by written accounts, photographs, and film footage of whalers taking a bowhead weighing 100,000 kilos and measuring nearly 20 meters makes it easy to dismiss the contributions of walrus, seabirds, and other taxa to the diet. The continued fetishization of the whale at the expense of the walrus will only limit our understanding of the cultural dynamics of a critical period in arctic prehistory, rather than further our knowledge of the various ways humans adapted to the land- and waterscapes of the North Pacific.

This exploration of walrus exploitation has proposed the idea that communal walrusing under the direction of an *umialik*-like leader may have provided the foundation for the development of social complexity in the North Pacific. Complex social organization, traditionally associated with whaling, may have actually been initially driven by the occurrence of seasonal, predictable, and clumped resources. Such resources, available during a short period of time, were most efficiently harvested under the direction of a highly organized, relationally integrated *umialik* and his crew. The acquisition of surplus, the status and notoriety of a successful *umialik*, and the assurance of provisions during lean years would have attracted kin and their affines to a successful settlement and encouraged the establishment and maintenance of fictive kin relationships. Burch (2006) has demonstrated in detail how this process occurred among "traditional" Inupiat, while Ellanna (1988; Ellanna and Sherrod 1983) and Hughes (1960) have documented more recent manifestations of similar organizational systems.

In addition, this chapter has demonstrated how essential the walrus and its products were to inhabitants of the coasts of the Bering and Chukchi seas and provided an example of how humans engaged with and were in turn affected by sea mammal migration routes and habitat preferences. The walrus, which aroused such disgust among some early observers (e.g., Elliott 1881), provided a subsistence foundation, building materials, fuel, trade goods, and ivory for innumerable tools, ornaments, and hunting implements. The walrus has given its name to islands in English (i.e., the Walrus Islands in the Bering Sea), as well as to settlements on St. Lawrence Island (i.e., Ievoghiyoq = "place of the walrus") (Collins 1937:33). A clear archaeological record of intensive use of the walrus dates to at least the first centuries AD and throughout the Old Bering Sea/Okvik period. A walrus-based subsistence strategy in the region likely originated centuries earlier, but is difficult to discern archaeologically. Future research may clarify the chronology of this adaptation and the subsistence transition to whaling.

Evidenced by historical ecology studies elsewhere (e.g., Braje et al. 2007; Heckenberger et al. 2003, 2007), the human impact on prey animal populations can be significant in terms of their distribution, diversity, and size. One possible explanation for the transition to whaling in the Bering Sea region is a decline in the number and availability of walrus as prey. Walrus have a relatively long gestation period (10 to 11 months) and are able to breed successfully only when they reach 10 (females) or 15 (males) years of age (Fay 1985). A hunting focus on reproductive-age females and calves alone, or more likely in combination with other factors (such as changes in sea ice, availability of benthic prey, and shifts in water mass patterns or composition) could have resulted in local declines of walrus herds. Whaling may have emerged as an alternative strategy that only became viable when walrus were depleted (see Mason 1998 for a discussion of other factors). If this is the case, the earliest whaling in the western Arctic may be found at sites where humans were most dependent upon walrus. Although the evidence does not yet exist to support this argument, it remains a viable hypothesis for explaining one of most significant shifts in subsistence economies in the Arctic.

In sum, subsistence adaptations to North Pacific marine environments are more complex than the spectacular finds of whalebone structures and monuments would suggest. Whaling was almost certainly linked to walrusing, either as a response to walrus population declines or as the next step in a process of economic intensification. The link between the heroic Eskimo whaler, wielding his harpoon in death-defying contest with the great bowhead, and the rise of social complexity and leadership must be uncoupled, or minimally reconsidered, in our reconstructions of North Pacific prehistory. Krupnik (1987; Chelenov and Krupnik 1984) has demonstrated that bowhead was not the only whale being exploited by the Eskimo of Chukotka. To this we must add several other species whose contribution to Eskimo subsistence and lifeways have hitherto been largely discounted in favor of more charismatic cetaceans. First among these taxa is the walrus, which provided subsistence and material resources for successful adaptations on St. Lawrence Island, the Chukchi Peninsula, and parts of coastal northwest Alaska.

As work progresses in the North Pacific, our view of marine mammals as the mainstay of subsistence will shift to include the contributions of waterfowl, seabirds, and their eggs. Recent research by Moss and Bowers (2007) has demonstrated heavy reliance upon ducks, geese, and murres as early as AD 700 at Deering, on the north coast of the Seward Peninsula. Given that spring migrations of many birds overlap with some of the leanest times of the year for human inhabitants, birds must have formed a major, if seasonal, resource. As discussed above, seabird distribution in the western Bering Sea, like that of walruses, is heavily dependent upon nutrients provided by the Anadyr water mass.

As predicted by Mason and Gerlach (1995:5), the zooarchaeological data for walrus exploitation is comparatively "impoverished" along the western coast of Alaska, with the notable exceptions of Wales, at the tip of Seward Peninsula, and Point Hope, where Larsen and Rainey (1948:25) observed that "[e]normous herds of walrus" haul out in the spring as they follow the pack ice north. St. Lawrence Island, a center of walrus exploitation prehistorically and into the early 20th century, is located in the middle of the migratory path of walrus coming from the south. The depth of the Bering–Chukchi continental shelf and the route of nutrient-rich Anadyr water create an ideal habitat for the benthic invertebrates upon which walrus rely.

These observations have obvious implications for reconstruction of human subsistence patterns; but settlement pattern evidence also indicates that at least some marine mammal migration routes and preferred haul-outs have significant time depth and stability. By implication, some of the hydrodynamic processes in the Bering Sea are broadly similar to prehistoric patterns. A better understanding of these factors is critical to conservation efforts to preserve the Pacific walrus, which is rapidly losing sea ice habitat. During the late summer and fall of 2007, biologists observed that walruses were spending more time at haul-outs in greater numbers than previously observed. Loss of sea ice means that walrus are spending less time dispersed on ice floes and more time congregating, and overcrowding, familiar haul-outs. Early estimates suggest that as many as four thousand walruses died from being crushed on the Russian side of the Bering Strait (Joling 2007), a pattern that will become more common as sea ice retreats and floes become smaller and more dispersed.

The natural extension of the research discussed here is the development of a comprehensive atlas of human settlement, walrus haul-outs and migration routes, and breeding colonies of alcids, sea ducks, and gulls, taking into consideration variables affecting sea ice, currents, temperature, and salinity. From the perspective of historical ecology, the land- and waterscapes of the North Pacific represent dynamic systems in which humans were fully engaged. It is not possible to fully understand human settlement and use of the North Pacific ecosystem without tracking reliance upon sea mammals back to the physical and biological variables that underpinned them. While coastal foragers had no idea of the origin or chemical composition of the water masses that sustained them, they were nevertheless affected by them and alert to major alterations in their patterns.

In many ways, those now living along the coasts of the Bering Sea are experiencing changes and being forced to adapt in ways that parallel those of prehistoric inhabitants. Alaska and Chukotka Natives are observing fluctuations in the availability of sea mammal and bird resources; they are experiencing coastal erosion and are altering their settlement locations; they are noticing changes in the extent and distribution of sea ice; and they are being forced, as were their predecessors, to adapt by making changes in subsistence practices and social organization. Although the processes are similar, no matter whether they occurred at a small village dependent upon walrus over 1000 years ago, or at a comparatively modern village at the far reaches of the Russian Federation, Natives of the North Pacific face changes on a scale never experienced by their ancestors. Historical ecology can tell us how people dealt with change in the past, but only by applying its insights to the present can we prepare for the massive changes to come.

ACKNOWLEDGMENTS

Financial support for this research was provided in part by Alaska EPSCoR (NSF-0701898) and the state of Alaska. Archaeological investigation at Unanan in 2007 was conducted under the direction of Daniel Odess and Sergey Gusev, with funding by the National Science Foundation (NSF-0714062) and the Shared Beringian Heritage Program of the National Park Service. For facilitating collections research, I thank Eileen Westwig of the Department of Mammalogy, American Museum of Natural History. At the University of Alaska Museum of the North, my research was made possible by Jeff Rasic, James Whitney, Chris Houlette, Scott Shirar, and Fawn Carter (Department of Archaeology); Link Olson and Brandy Jacobsen (Department of Mammalogy); and Kevin Winker (Department of Ornithology). I thank Tiger Burch, Christyann Darwent and Daniel Odess for discussion and input on many of the points raised in this chapter. Thanks also to Todd Braje and Torrey Rick for the invitation to contribute this chapter, and for their comments, which have greatly improved it.

REFERENCES CITED

Ackerman, R.E.

1988 Settlements and Sea Mammal Hunting in the Bering–Chukchi Sea Region. *Arctic Anthropology* 25(1):52–79.

Bandi, H. G., and R. Blumer

2002 Investigations by Swiss Archaeologists on St. Lawrence Island, Alaska. In *Archaeology in the Bering Strait Region: Research on Two Continents*, edited by D. E. Dumond and R. L. Bland, pp. 25–59. Department of Anthropology, University of Oregon, Eugene, OR.

Black, L. T.

1987 Whaling in the Aleutians. *Etudes/Inuit/Studies* 11(2):7–50.

Bockstoce, J. R.

1976 On the Development of Whaling in the Western Thule Culture. *Folk: Journal of the Danish Ethnographic Society* (Copenhagen) 18:41–46.

1979 *The Archaeology of Cape Nome, Alaska*. University Museum, Philadelphia.

Bogoras, W.

1904–1909 *The Chukchee* XI. E.J. Brill, Leiden.

1931 Elements of the Culture of the Circumpolar Zone. In *Annual Report of the Board of Regents of the Smithsonian Institution, 1930*, pp. 465–482. U.S. Government Printing Office, Washington, D.C.

Braje, T. J., D. J. Kennett, J. M. Erlandson, and B. J. Culleton

2007 Human Impacts on Nearshore Shellfish Taxa: A 7000-Year Record from Santa Rosa Island, California. *American Antiquity* 72(4): 735–756.

Braund, S. R.
1988 *The Skin Boats of Saint Lawrence Island, Alaska.* University of Washington Press, Seattle.
Brewster, K. (editor)
2004 *The Whales, They Give Themselves: Conversations with Harry Brower, Sr.* University of Alaska Press, Fairbanks.
Brooks, J. W.
1953 The Pacific Walrus and Its Importance to the Eskimo Economy. *Transactions of the North American Wildlife Conference* 18:503–511.
Burch, E. S.
1972 Alliance and Conflict: Inter-Regional Relations in North Alaska. In *Alliance in Eskimo Society,* edited by L. Guemple, pp. 17–39. University of Washington Press, Seattle.
1975 *Eskimo Kinsmen: Changing Family Relationships in Northwest Alaska.* West Publishing Co., St. Paul, MN.
1980 Traditional Eskimo Societies in Northwest Alaska. In *Alaska Native Culture and History,* edited by Y. Kotani and W. B. Workman, pp. 253–304. National Museum of Ethnology, Osaka.
2006 *Social Life in Northwest Alaska.* University of Alaska Press, Fairbanks.
Burns, J. J.
1965 *The Walrus in Alaska: Its Ecology and Management.* Alaska Department of Fish and Game, Juneau, AK.
Cassell, M. S.
1988 "Farmers of the Northern Ice": Relations of Production in the Traditional North Alaskan Inupiat Whale Hunt. In *Research in Economic Anthropology,* edited by B. L. Isaac, pp. 89–116. vol. 10. JAI Press, Greenwich, CT.
Chelenov, M. A., and I. I. Krupnik
1984 Whale Alley: A Site on the Chukchi Peninsula, Siberia. *Expedition* (Philadelphia) 26(2): 6–15.
Collins, H. B.
1932 Prehistoric Eskimo Culture on St. Lawrence Island. *Geographical Review* 22(1):107–119.
1937 *Archeology of St. Lawrence Island* Smithsonian Institution, Washington, D.C.
Cook, J., Captain
1784 *A Voyage to the Pacific Ocean Undertaken by Command of His Majesty for Making Discoveries in the Northern Hemisphere,* vol. III. John Stockdale, Scatcherd and Whitaker, John Fielding, and John Hardy, London.
Crockford, S. J., and S. G. Frederick
2007 Sea Ice Expansion in the Bering Sea during the Neoglacial: Evidence from Archaeozoology. *The Holocene* 17(6):699–706.
Darwent, C.
2006 Reassessing the Old Whaling Locality at Cape Krusenstern, Alaska. In *Dynamics of Northern Societies,* edited by J. Arneborg and B. Grønnow, pp. 95–101. National Museum of Denmark, Copenhagen.
Dikov, N. N.
[1974] 2002 *The Chini Cemetery: A History of Sea Mammal Hunters in Bering Strait.* Translated by R. L. Bland. Shared Beringian Heritage Program, National Park Service, Anchorage, AK.
[1977] 2003 *Archaeological Sites of Kamchatka, Chukotka, and the Upper Kolyma.* Translated by R. L. Bland. Shared Beringian Heritage Program, National Park Service, Anchorage, AK.
Dumond, D. E.
1998 *The Hillside Site St. Lawrence Island Alaska: An Examination of Collections from the 1930s.* Department of Anthropology, University of Oregon, Eugene, OR.
2000 *Henry B. Collins at Wales, Alaska 1936.* Department of Anthropology, University of Oregon, Eugene, OR.
Dumond, D. E., and D. G. Griffin
2002 Measurements of the Marine Reservoir Effect on Radiocarbon Ages in the Eastern Bering Sea. *Arctic* 55(1):77–86.
Ellanna, L. J.
1988 Demography and Social Organization as Factors in Subsistence Production in Four Eskimo Communities. In *Research in Economic Anthropology,* edited by B. L. Isaac, pp. 73–87. vol. 10. JAI Press, Greenwich, CT.
Ellanna, L. J., and G. K. Sherrod
1983 *The Hunting Ecology of the "Arctic Whaling" Populations of Bering Strait 1650 to 1981.* Eskimo Walrus Commission Nome and Alaska Department of Fish and Game, Juneau.
Elliott, H. W.
1881 *The Seal-Islands of Alaska.* Government Printing Office, Washington, D.C.
Elphick, C. S., and G. L. Hunt
1993 Variations in the Distributions of Marine Birds with Water Mass in the Northern Bering Sea. *The Condor* 95(1):33–44.
Fay, F. H.
1982 *Ecology and Biology of the Pacific Walrus,* Odobenus rosmarus divergens *Illiger.* U.S. Fish and Wildlife Service, Washington, D.C.
1985 *Odobenus rosmarus. Mammalian Species* 238: 1–7.
Fienup-Riordan, A.
1983 *The Nelson Island Eskimo: Social Structure and Ritual Distribution.* Alaska Pacific University Press, Anchorage, AK.

Ford, J. A.
1959 *Eskimo Prehistory in the Vicinity of Point Barrow, Alaska* Anthropological Papers of the American Museum of Natural History, vol. 47, pt. 1. American Museum of Natural History, New York.
Freeman, M. M. R., L. Bogoslovskaya, R. A. Caulfield, I. Egede, I. I. Krupnik, and M. G. Stevenson
1998 *Inuit, Whaling, and Sustainability*. Altamira Press, Walnut Creek, CA.
Freeman, M. M. R., E. E. Wein, and D. E. Keith (editors)
1992 *Recovering Rights: Bowhead Whales and Inuvialuit Subsistence in the Western Canadian Arctic*. Canadian Circumpolar Institute, University of Alberta, Edmonton, AB.
Garlich-Miller, J. L., and R. E. A. Stewart
1998 Growth and Sexual Dimorphism of Atlantic Walruses (*Odobenus rosmarus rosmarus*) in Foxe Basin, Northwest Territories, Canada. *Marine Mammal Science* 14(4):803–818.
Geist, O. W., and F. G. Rainey
1936 *Archaeological Excavations at Kukulik St. Lawrence Island, Alaska*. U.S. Government Printing Office, Washington, D.C.
Gerasimov, D. V., E. Y. Giria, V. V. Pitul'ko, and A. N. Tikhonov
2006 New Materials for the Interpretation of the Chertov Ovrag Site on Wrangel Island. In *Archaeology in Northeast Asia: On the Pathway to Bering Strait*, edited by D. E. Dumond and R. L. Bland, pp. 203–206. Translated by R. L. Bland. Department of Anthropology, University of Oregon, Eugene, OR.
Gerlach, S. C., and O. K. Mason
1992 Calibrated Radiocarbon Dates and Cultural Interaction in the Western Arctic. *Arctic Anthropology* 29(1):54–81.
Giddings, J. L.
1964 *The Archaeology of Cape Denbigh*. Brown University Press, Providence, RI.
1967 *Ancient Men of the Arctic*. Knopf, New York.
Giddings, J. L., and D. D. Anderson
1986 *Beach Ridge Archeology of Cape Krusenstern: Eskimo and Pre-Eskimo Settlements around Kotzebue Sound, Alaska*. National Park Service, Washington, D.C.
Glassow, M. A.
2005 Prehistoric Dolphin Hunting on Santa Cruz Island, California. In *The Exploitation and Cultural Importance of Sea Mammals*, edited by G. G. Monks, pp. 107–120. Oxbow Books, Oxford.
Grebmeier, J. M., L. W. Cooper, H. M. Feder, and B. I. Sirenko
2006 Ecosystem Dynamics of the Pacific-Influenced Northern Bering and Chukchi Seas in the Amerasian Arctic. *Progress in Oceanography* 71:331–361.
Gusev, S. V., A. V. Zagoroulko, and A. V. Porotov
1999 Sea Mammal Hunters of Chukotka, Bering Strait: Recent Archaeological Results and Problems. *World Archaeology* 30(3):354–369.
Hall, E. S., and L. Fullerton (editors)
1990 *The Utqiagvik Excavations*, 3 vols. North Slope Borough Commission on Inupiat History, Language and Culture, Barrow, AK.
Harritt, R. K.
1995 The Development and Spread of the Whale Hunting Complex in Bering Strait: Retrospective and Prospects. In *Hunting the Largest Animals: Native Whaling in the Western Arctic and Subarctic*, edited by A. P. McCartney, pp. 33–50. Canadian Circumpolar Institute, University of Alberta, Edmonton, AB.
2004 A Preliminary Reevaluation of the Punuk–Thule Interface at Wales, Alaska. *Arctic Anthropology* 41(2):163–176.
Heckenberger, M. J., A. Kuikuro, U. T. Kuikuro, J. C. Russell, M. Schmidt, C. Fausto, and B. Franchetto
2003 Amazonia 1492: Pristine Forest or Cultural Parkland? *Science* 301:1710–1714.
Heckenberger, M. J., J. C. Russell, J. R. Toney, and M. J. Schmidt
2007 The Legacy of Cultural Landscapes in the Brazilian Amazon: Implications for Biodiversity. *Philosophical Transactions of the Royal Society B* 362:197–208.
Holland, K. M.
1992 In the Wake of Prehistoric North Pacific Sea Mammal Hunters. *Arctic Anthropology* 29(2): 63–72.
Hughes, C. C.
1960 *An Eskimo Village in the Modern World*. Cornell University Press, Ithaca, NY.
Joling, D.
2007 Climate Change Blamed as Thousands of Walruses Die in Stampedes. *The Seattle Times*, 14 December 2007.
Kennett, D. J., and J. P. Kennett
2000 Competitive and Cooperative Responses to Climatic Instability in Coastal Southern California. *American Antiquity* 65(2):379–395.
Khassanov, B. F., and A. B. Savinetsky
2006 On the Marine Reservoir Effect in the Northern Bering Sea. In *Archaeology in Northeast Asia: On the Pathway to Bering Strait*, edited by D. E. Dumond and R. L. Bland, pp. 193–202. Translated by R. L. Bland. Department of Anthropology, University of Oregon, Eugene, OR.

Kirch, P. V., and T. L. Hunt (editors)
1997 *Historical Ecology in the Pacific Islands: Prehistoric Environmental and Landscape Change.* Yale University Press, New Haven.
Krupnik, I. I.
1987 The Bowhead vs. the Gray Whale in Chukotkan Aboriginal Whaling. *Arctic* 40(1):16–32.
1993a *Arctic Adaptations: Native Whalers and Reindeer Herders of Northern Eurasia.* University Press of New England, Hanover, NH.
1993b Prehistoric Eskimo Whaling in the Arctic: Slaughter of Calves or Fortuitous Ecology? *Arctic Anthropology* 30(1):1–12.
Lantis, M.
1984 Nunivak Eskimo. In *Handbook of North American Indians, Arctic*, edited by D. Damas, pp. 209–223. vol. 5. Smithsonian Institution Press, Washington, D.C.
Larsen, H., and F. G. Rainey
1948 *Ipiutak and the Arctic Whale Hunting Culture.* American Museum of Natural History, New York.
Laughlin, W. S.
1980 *Aleuts: Survivors of the Bering Land Bridge.* Holt, Rinehart and Winston, New York.
Lee, M., and G. A. Reinhardt
2003 *Eskimo Architecture: Dwelling and Structure in the Early Historic Period.* University of Alaska Press, Fairbanks.
Lee, S. M., and D. Robineau
2004 Les cétacés des gravures rupestres néolithiques de Bangu-dae (Corée du Sud) et les débuts de la chasse à la baleine dans le Pacifique nord-ouest. *L'Anthropologie* 108(1): 137–151.
Mason, O. K.
1998 The Contest Between the Ipiutak, Old Bering Sea, and Birnirk Polities and the Origin of Whaling during the First Millennium A.D. along Bering Strait. *Journal of Anthropological Archaeology* 17:240–325.
Mason, O. K., and S. C. Gerlach
1995 The Archaeological Imagination, Zooarchaeological Data, the Origins of Whaling in the Western Arctic, and "Old Whaling" and Choris Cultures. In *Hunting the Largest Animals: Native Whaling in the Western Arctic and Subarctic*, edited by A. P. McCartney, pp. 1–31. Canadian Circumpolar Institute, University of Alberta, Edmonton, AB.
McCartney, A. P. (editor)
1995a *Hunting the Largest Animals: Native Whaling in the Western Arctic and Subarctic.* Canadian Circumpolar Institute, University of Alberta, Edmonton, AB.
1995b Whale Size Selection by Precontact Hunters of the North American Western Arctic and Subarctic. In *Hunting the Largest Animals: Native Whaling in the Western Arctic and Subarctic*, edited by A. P. McCartney, pp. 83–108. Canadian Circumpolar Institute, University of Alberta, Edmonton, AB.
2003 *Indigenous Ways to the Present: Native Whaling in the Western Arctic.* Canadian Circumpolar Institute, University of Alberta, Edmonton, AB.
McEwan, C., C. Barreto, and E. Neves (editors)
2001 *Unknown Amazon: Culture in Nature in Ancient Brazil.* British Museum Press, London.
Menovščikov, G. A.
1968 Popular Conceptions, Religious Beliefs and Rites of the Asiatic Eskimos. In *Popular Beliefs and Folklore Tradition in Siberia*, edited by V. Diószegi, pp. 433–449. Mouton & Co., The Hague.
Moss, M. L., and P. M. Bowers
2007 Migratory Bird Harvest in Northwestern Alaska: A Zooarchaeological Analysis of Ipiutak and Thule Occupations from the Deering Archaeological District. *Arctic Anthropology* 44(1):37–50.
Murdoch, J.
[1892] 1988 *Ethnological Results of the Point Barrow Expedition.* Smithsonian Institution Press, Washington, D.C.
Nelson, R. K.
1969 *Hunters of the Northern Ice.* University of Chicago Press, Chicago.
Orekhov, A. A.
[1987] 1999 *An Early Culture of the Northwest Bering Sea.* Translated by R. L. Bland. Shared Beringian Heritage Program, National Park Service, Anchorage.
Rainey, F. G.
1941 *Eskimo Prehistory: The Okvik Site on the Punuk Islands.* American Museum of Natural History, New York.
Rudenko, S. I.
[1947] 1961 *The Ancient Culture of the Bering Sea and the Eskimo Problem.* Arctic Institute of North America and University of Toronto Press, Toronto.
Savelle, J. M., and A. P. McCartney
1988 Geographical and Temporal Variation in Thule Eskimo Subsistence Economies: A Model. In *Research in Economic Anthropology*, edited by B. L. Isaac, pp. 21–72. vol. 10. JAI Press, Greenwich, CT.
Savinetsky, A. B.
2002 Mammals and Birds Harvested by Early Eskimos of Bering Strait. In *Archaeology in the*

Bering Strait Region: Research on Two Continents, edited by D. E. Dumond and R. L. Bland, pp. 274–305. Translated by R. L. Bland. Department of Anthropology, University of Oregon, Eugene, OR.

Savinetsky, A. B., N. K. Kiseleva, and B. F. Khassanov

2004 Dynamics of Sea Mammal and Bird Populations of the Bering Sea Region of the Last Several Millennia. *Palaeogeography, Palaeoclimatology, Palaeoecology* 209:335–352.

Sheehan, G. W.

1985 Whaling as an Organizing Focus in Northwestern Alaskan Eskimo Society. In *Prehistoric Hunter-Gatherers: The Emergence of Cultural Complexity*, edited by T. D. Price and J. A. Brown, pp. 123–154. Academic Press, Inc., Orlando.

1995 Whaling Surplus, Trade, War, and the Integrating of Prehistoric Northern and Northwestern Alaskan Economies, A.D. 1200–1826. In *Hunting the Largest Animals: Native Whaling in the Western Arctic and Subarctic*, edited by A. P. McCartney, pp. 185–206. Canadian Circumpolar Institute, University of Alberta, Edmonton, AB.

Spencer, R. F.

1972 The Social Composition of the North Alaskan Whaling Crew. In *Alliance in Eskimo Society*, edited by L. Guemple, pp. 110–120. University of Washington Press, Seattle.

Springer, A. M., and C. P. McRoy

1993 The Paradox of Pelagic Food Webs in the Northern Bering Sea, III: Patterns of Primary Production. *Continental Shelf Research* 13(5–6): 575–599.

Springer, A. M., C. P. McRoy, and M. V. Flint

1996 The Bering Sea Green Belt: Shelf-Edge Processes and Ecosystem Production. *Fisheries Oceanography* 5(3–4):205–223.

Stanford, D. J.

1976 *The Walakpa Site, Alaska: Its Place in the Birnirk and Thule Cultures*. Smithsonian Institution Press, Washington, D.C.

Steadman, D. W.

1995 Prehistoric Extinctions of Pacific Island Birds: Biodiversity Meets Zooarchaeology. *Science* 267(5201):1123–1131.

Walker, P. L.

1989 Cranial Injuries as Evidence of Violence in Prehistoric Southern California. *American Journal of Physical Anthropology* 80(3):313–323.

Weckerly, F. W.

1998 Sexual-Size Dimorphism: Influence of Mass and Mating Systems in the Most Dimorphic Mammals. *Journal of Mammalogy* 79(1):33–52.

Whitridge, P.

1999 The Prehistory of Inuit and Yupik Whale Use. *Revista de Arqueológica Americana* (Mexico City) 16:99–154.

Worl, R.

1980 The North Slope Inupiat Whaling Complex. In *Alaska Native Culture and History*, edited by Y. Kotani and W. B. Workman, pp. 305–320. National Museum of Ethnology, Osaka.

4

Neoglacial Sea Ice and Life History Flexibility in Ringed and Fur Seals

Susan J. Crockford and S. Gay Frederick

TWO INTERESTING LIFE history questions about North Pacific pinnipeds beg for zooarchaeological input. One is the observation that the ringed seal, *Phoca hispida*, an Arctic species that also inhabits the Bering Sea, appears to have two morphologically distinct ecotypes: (1) a large, territorial form that gives birth and breeds on the immobile, shorefast ice (aka "fast ice") that forms along terrestrial shorelines; and (2) a smaller, early maturing form that lives and breeds offshore, within the mobile Arctic pack ice (aka "sea ice"). The existence of two distinctly sized ringed seal ecotypes, initially brought to the attention of biologists by Inuit hunters (e.g., Brendan Kelly, University of Alaska SE, Juneau, pers. comm.), has since been confirmed by scientific research directed specifically at this topic (Fedoseev 1975; Finley et al. 1983) and corroborated by other studies on general ringed seal biology (Born et al. 2004; Davis et al. 2008; Ferguson et al. 2000; Ferguson 2006; Wiig et al. 1999). At issue, however, is whether offshore-breeding ringed seals constitute a significant portion of the global population or a very minor one.

Certainly, distinct ecological morphotypes that differ in size (an example of so-called resource polymorphism) are well documented in a number of other vertebrate species. Mammalian examples include Antarctic killer whales (*Orcinas orca*; Pitman and Ensor 2003; Pitman 2004; Pitman et al. 2007), minke whales (*Balaenoptera acutorostrata*; Arnold et al. 1987; Reeves et al. 2002, Best 1985), bottlenosed dolphins (*Tursiops truncata*; Charleton et al. 2006; Tezanos-Pinto et al. 2008; Morin et al. 2006), African elephants (*Loxodonta africana*; Debruyne et al. 2003; Debruyne 2005), and bison (*Bison bison*; Geist 1990, 1998; Wilson 1996). Numerous other examples are known among birds, reptiles, and amphibians (Skúlason and Smith 1995; Smith and Skúlason 1996) and may include one or more differences in morphology, behavior, life history characteristics, and color. Such ecological plasticity has been shown to be critical to evolutionary survival, especially in unstable environments (Keeley et al. 2007; Leimar et al. 2006; Smith and Skúlason 1996). Therefore, skeletal size variation within a population that is discernable from zooarchaeological remains, and which correlates to distinct ecological habitats, may be an indicator of important life history flexibility and ecological plasticity. If a significant portion of the global population of ringed seals lives and breeds exclusively in offshore pack ice, and is characterized by its small size, should we not expect noticeably small ringed seal remains to be strongly represented in zooarchaeological assemblages from sites occupied by maritime-adapted hunters who routinely exploited pack ice habitat throughout the Arctic?

Another phenomenon of interest is the evidence we now have from archaeological sites along the west coast of North America that northern fur seals (NFS), which currently migrate annually to breeding rookeries in the Bering Sea (Gentry 1998), formerly established rookeries as far south as California that were attended by nonmigratory animals (Crockford et al. 2002; Etnier 2002; Moss et al. 2006: Newsome et al. 2007). None of these satellite rookeries are older than about 4500 BP (all dates are radiocarbon years before present, uncalibrated, unless stated otherwise), and several appear to have survived into the late historic period (Crockford et al. 2002; Lyman 1988, 1989; Gifford-Gonzalez et al. 2005). While we have evidence that maternal nursing regimes were also different among nonmigratory fur seals (Burton et al. 2001; Newsome et al. 2007), no one has yet advanced a reason for such a major shift in life history strategy. Modern records confirm that this species is remarkably consistent in its habits and has been for more than a century: the peak date of birth on the Pribilof Islands has been virtually unchanged in over 100 years (Bigg 1990; Elliot 1887; Gentry 1998) and NFS are noted for their remarkable migratory drive and natal site fidelity (e.g., Baker et al. 1995; Baker 2007). What would prompt a change to this apparently rigid life history pattern?

We propose that comprehensive analysis of prehistoric marine mammal remains from a site in the eastern Aleutians, occupied at the height of the Middle Holocene Neoglacial, may assist in answering the above questions about ringed seals and fur seals because both species inhabit the Bering Sea portion of the North Pacific (Figure 4.1) during critical periods of their life history. The Neoglacial was a period of cold climate that lasted from approximately 4700 to 2500 years ago in most regions (e.g., Calkin et al. 2001; Dyke and Savelle 2001; Mason and Barber 2003; Razjigaeva et al. 2004). Any associated increases in seasonal pack ice in the Bering Sea during that time would have affected all marine mammals utilizing the region, as pack ice conditions virtually define this ecosystem: the extent of seasonal pack ice each year is governed by wind moving ice south from the Arctic through the Bering Strait (Rigor and Wallace 2004). During most of the 20th century, the pack ice front reached its southern-most position sometime in April (about as far as the Pribilof Islands, rarely beyond) and receded quickly thereafter (Grebmeier et al. 2006; Overland and Stabeno 2004). In other words, the Pribilofs have always been in a strategically vulnerable position in relation to seasonal Bering Sea pack ice.

While most resident pinniped species in the Bering Sea utilize pack ice as a mobile platform for giving birth, mating, and moulting (so-called pagophilic, or ice-obligate, taxa), a few temperate species utilize ice-free terrestrial beaches in the southern region for such purposes. Pagophilic species include walrus (*Odobenus rosmarus*), bearded seal (*Erignathus barbatus*), ringed seal (*Phoca hispida*), spotted seal (*Phoca largha*), and ribbon seal (*Phoca faciata*), while temperate species include the northern fur seal (*Callorhinus ursinus*), Steller sea lion (*Eumatopias jubata*), and harbor seal (*Phoca vitulina*). For the purpose of this discussion, we assume that habitat types utilized by these species for giving birth and mating were not appreciably different during the Neoglacial than they are today.

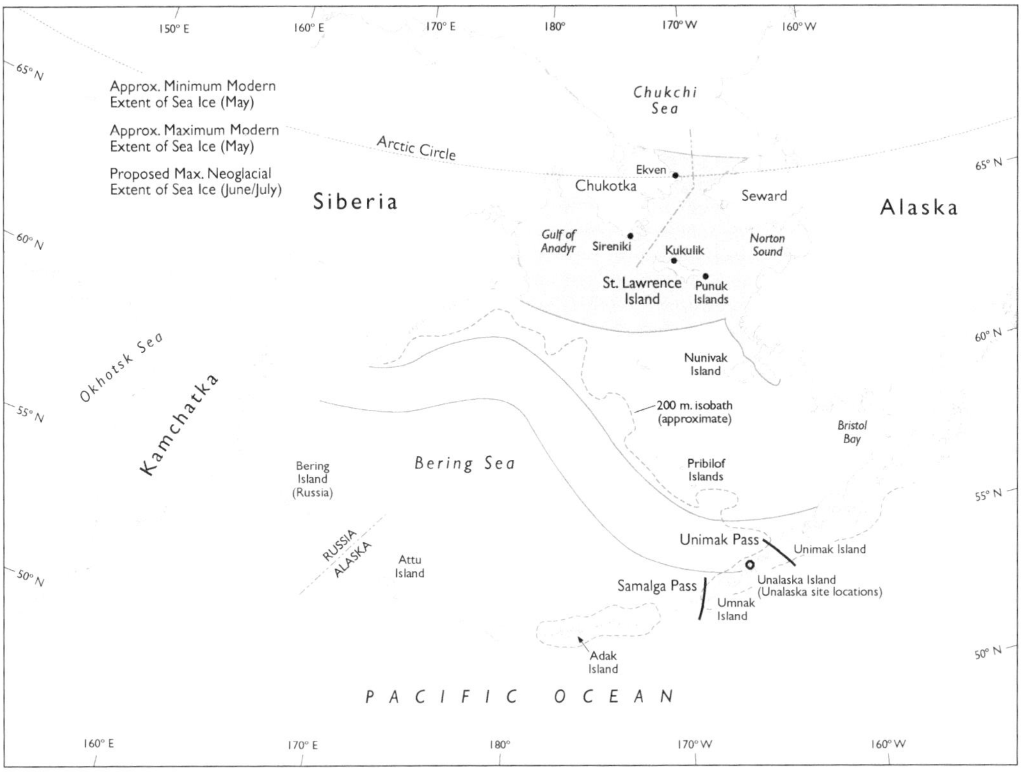

FIGURE 4.1. Modern minimum/maximum extent of spring pack ice (May) for the central and eastern Bering Sea (from http://www.beringclimate.noaa.gov/essays_mcnutt.html), with proposed maximum Neoglacial pack ice extent (June/July). Amaknak Bridge, UNL-050, is located on a small island off Unalaska, circled (adapted from Crockford and Frederick 2007).

Here we provide evidence that Neoglacial sea ice expansion (Figure 4.1) pushed Bering Sea populations of pack ice–breeding ringed and bearded seals south as far as the eastern Aleutians and kept them there until early summer, making these Arctic-adapted species easily accessible to ancient Aleut hunters. Extensive pack ice development would also have made the Pribilof Islands unsuitable as early summer pupping grounds for fur seals, forcing them to establish rookeries away from ice-covered waters and icy winds. These conclusions are based on a comprehensive analysis of skeletal remains recovered from an archaeological site off Unalaska Island in the eastern Aleutians (Figure 4.2) that was occupied at the height of the Neoglacial period, ca. 3500 to 2500 BP. Attention to taxonomic identity of postcranial phocid seal remains and estimation of ontogenic age for all taxa was instrumental in generating unique proxy evidence that pack ice extent in the southern Bering Sea changed markedly during the Neoglacial (Crockford and Frederick 2007), which sheds significant new light on the origins of Thule culture (Crockford 2008) and on important aspects of ringed seal and fur seal life history.

AMAKNAK BRIDGE ZOOARCHAEOLOGY

REGIONAL SETTING

The eastern Aleutians have been occupied by people for at least 9000 years (Knecht and Davis, 2001). The Amaknak Bridge site (UNL50) is located on Amaknak Island in Unalaska Bay near the seaport of Dutch Harbor and the city of Unalaska (Figure 4.2), on an old marine terrace about 2 meters above present sea level. Archaeologists Rick Knecht, Museum of the Aleutians, and Rick Davis, Bryn Mawr College,

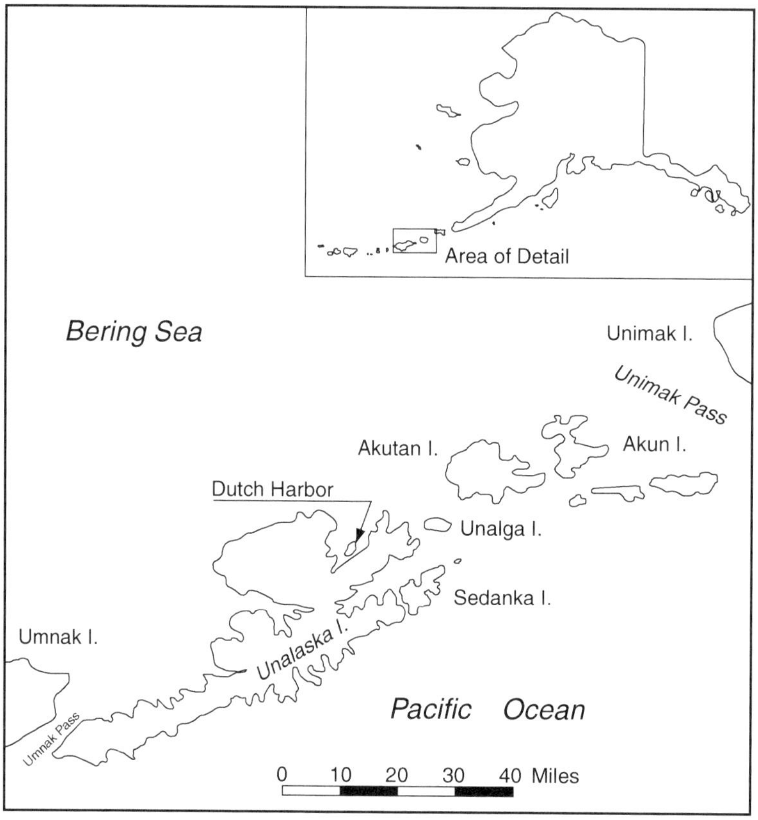

FIGURE 4.2. Location of the fishing port of Dutch Harbor on Unalaska Island, the general locale of the Amaknak Bridge site, near the modern city of Unalaska (from Knecht and Davis 2004).

conducted salvage excavations at the site over the summer of 2003 (Knecht and Davis 2004, 2008), exposing several semi-subterranean dwellings and associated shell middens containing well-preserved animal bone.

ZOOARCHAEOLOGICAL ANALYSIS METHODS

A representative subsample of the bird, fish, and mammal remains was analyzed, with a total of 42,359 specimens identified to family level or better (5947 of these are mammal). Basic quantification is by Number of Identified Specimens (NISP). Age-at-death for young juveniles (<1 yr) and sex determinations for pinnipeds were based on comparison to modern specimens of known age and sex (listed in Appendices 4.1 and 4.2), as well as published references (Etnier 2002, 2004; Storå 2000). This approach vastly increases the amount of information discernable, even for phocid seals (e.g., Hodgetts 2005; Storå and Lõugas 2005). General age estimates for animals > 1 yr are based on relative size and epiphyseal fusion: unfused epiphyses and smaller than adult size are "juvenile," fused or partially fused epiphyses with suture lines evident, close or equal to adult size, are "subadult," while those with fused epiphyses and no suture lines evident are "adult."

Age-at-death determinations were crucial to our interpretation of this assemblage, especially for the two dominant species (NFS and ringed seal). We spent extra effort determining species identity for as much of the small phocid seal component of the sample as possible and for ringed seals, this also entailed an assessment of putative ecotypes. These topics are discussed in more detail below.

AGE-AT-DEATH DETERMINATION, NFS

Age of juvenile NFS remains was estimated using four comparative fur seal skeletons from animals with approximate known ages (see Appendix 4.2 for descriptions; see Crockford et al. 2002 for measurements of selected elements). These specimens compare favorably (estimated age vs. size) with measurements reported by Etnier (2002, 2004) for a much larger sample. We also had a subadult female (epiphyses unfused, ca. 2–3 years old) and a fully adult male (epiphyses fully fused, >8 years) available for comparison.

As an additional check against our subjective age estimation method, all measurable specimens of mandible, humerus, radius, ulna, femur, and tibia that had been judged subjectively to be young juvenile or younger (based on comparison to modern specimens of known age) were selected for comparison to the large sample of modern juvenile fur seal measurements reported by Etnier (2002, 2004). All Amaknak Bridge NFS that were assessed subjectively as either "young juvenile" (newly weaned, ca. 4–6 months) or "newborn" (unweaned, less than 4 months) fell within or below the range of measurements of modern animals estimated by Etnier to be between 1 and 10 months of age (Crockford et al. 2004), suggesting that the additional step of measuring archaeological bones did not increase the level of accuracy in age estimates. This is almost certainly due to the fact that few truly known aged animals exist in any collection (i.e., animals tagged the day of birth). As is true for most specimens of wild species, all NFS specimens are assigned the mean pupping date for the species as their "birth date" and age is then back-calculated from the date of death or date of collection.

AGE-AT-DEATH DETERMINATION, RINGED SEALS

As most of the prehistoric ringed seal remains from the Amaknak Bridge site were immature, we attempted to estimate an ontogenic age for as many specimens as possible. Jan Storå (2000) has provided valuable data on fusion times and sequences for postcranial elements of ringed seal, based on comparative specimens collected in the Baltic Sea. In total, he examined 97 complete skeletons of ringed seal and 29 incomplete ones. Unfortunately, only two specimens represent the critical newborn period (<2 months, "skeletal age 1") and only one represents the next developmental stage ("skeletal age 2"). As a consequence, the fusion dates he gives for the youngest ages are based on a very small sample and given the bias in collection of ringed seals discussed above, these almost certainly represent the large, shorefast ecotype. However, the data complement information gathered from comparative specimens of harbor seal in the University of Victoria collection and ringed seal material we examined from other institutions (see Appendix 4.1). Storå's study identifies several useful developmental events for estimating the age of young ringed seals, which we summarize below. He notes that the sequence of fusion is the same in all species, although the precise timing of individual fusion events may vary somewhat.

- Fusion of vertebral bodies to the arches begins at about 3 months of age and is complete by 6 to 8 months (Figure 4.3a, b). Fusion starts with the cervicals and is followed by the lumbars; the thoracics fuse last. A line of fusion between centrum and arch may be visible, especially in the thoracics, until early in the second year of life. Thus, completely unfused vertebral bodies and arches definitely represent animals less than 3 months of age.
- Fusion of the distal epiphysis of metacarpus I usually starts at 2 months of age.
- Fusion of the distal epiphysis of metatarsus I does not usually start until 3 months of age.
- Fusion of the acetabulum (pelvis) usually begins at about 3 months of age (Figure 4.3c).
- Fusion of the proximal epiphysis of the humerus (head to greater tubercle components) usually begins at about 4 months of age.

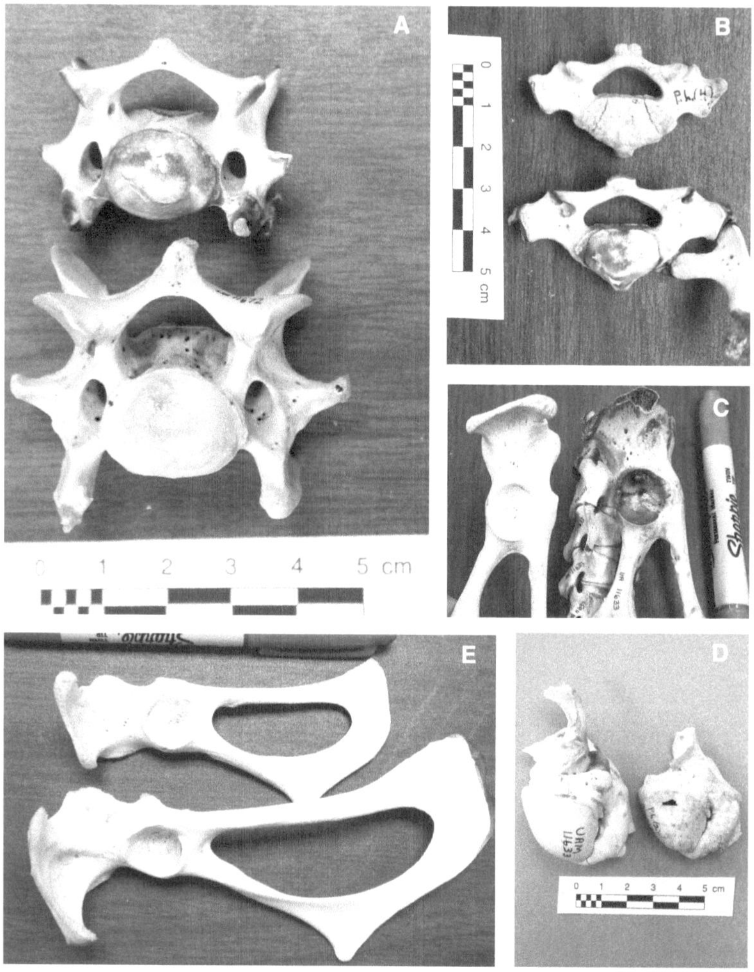

FIGURE 4.3. Ringed seals, ecotype size and development. (A) Top to bottom, cervical vertebrae (C6) of juvenile (shorefast ecotype), UAM 11633 (8–12 wks), vs. adult female (pack ice ecotype), UAM 11672. (B) Top to bottom, thoracic vertebrae (T1) of Amaknak Bridge juvenile (putative pack ice ecotype), AB 201438, vs. juvenile (shorefast ecotype), UAM 11633 (8–12 wks). (C) Left to right, innominates of Amaknak Bridge juvenile (putative pack ice ecotype), AB200133, vs. juvenile (shorefast ecotype), UAM 11633 (8–12 wks); see text for discussion. (D) Left to right, auditora bullae, juvenile (shorefast ecotype), UAM 11633 (8–12 wks), vs. Amaknak Bridge juvenile (putative pack ice ecotype), AB 200121. (E) Top to bottom, innominates of Amaknak Bridge juvenile (putative pack ice ecotype), AB200133, vs. adult female (pack ice ecotype), UAM 11672. See Appendix 4.1 for notes on modern specimens (UAM numbers).

- Fusion of the supraglenoid tubercle of the scapula does not usually begin until about 5 months of age.

TAXONOMY OF ARCTIC PHOCIDS

North Pacific phocids have distinct life history characteristics that may prove invaluable to the interpretation of a zooarchaeological assemblage, and thus unless a species identity can be determined for as much of the phocid component of a sample as possible, there may be a significant loss of valuable information. However, it has been customary for those working on North Atlantic seals, for which there is a relatively long history of analysis, to be very conservative regarding taxonomic identifications because of real and/or perceived difficulties in distinguishing among smaller species. This tradition extends back at least as far as Jeppe Møhl's work on prehistoric seal remains from Greenland over 30 years ago (Møhl 1979) and has been adopted by many eastern North American faunal analysts (e.g., Woollett et al. 2000). While certain identification problems certainly exist for phocid seal remains, as discussed below, we contend that limiting identification of phocid species to a few skeletal elements (e.g., auditory bullae, bony palate, maxillae, mandibles, and humeri only, as per Woollett et al. 2000) may result in an unnecessary reduction of sample size and, potentially, a significant loss of valuable information.

We believe that a more rational approach to small phocid bone identification is needed for North Pacific taxa. If humeri of small seals are judged to be diagnostic to species, there is no valid reason to expect that other postcranial elements will not be equally distinctive. For example, Møhl's (1979) analysis of material from Nugarsuk in northwest Greenland included ulnae as well as humeri (with illustrations of both, showing diagnostic features). We suggest that the question of whether a particular element from closely related species is diagnostic is often more a question of available time, effort, and access to adequate comparative material, rather than lack of distinctive taxonomic differences.

A good example of this phenomenon is provided by the closely related species pair of harbor seal and spotted seal, which are very similar in external appearance. Spotted seals are truly pagophilic: they give birth primarily on ice and are associated with pack ice for much of the year. They are found only in the North Pacific (mostly in the Bering Sea and Okhotsk Sea) and their distribution overlaps with that of harbor seals over a narrow zone in the south only, where hybridization is known to occur. Although a complete biological assessment, published decades ago, led Shaughnessy and Fay (1977) to conclude that the spotted seal is a distinct and separate species, the taxonomy was not changed to reflect this until it was confirmed by mitochondrial DNA analysis (O'Corry-Crowe and Westlake 1997; Mizuno et al. 2003). As a consequence, the old taxonomy (which considers the spotted seal a subspecies of *P. vitulina*) predominates in pre-21st-century literature for virtually all biological and archaeological studies of the region.

Shaughnessy and Fay (1977) described cranial and mandibular features between these two species, including the shape of the auditory bullae, the glenoid fossa of the skull, and the alignment of the premolar teeth (which are straight in spotted seal and oblique in harbor seal, Figure 4.4a). Since traditional taxonomy seldom considers the postcranial skeleton, it is not surprising that only cranial and mandibular features are mentioned in Shaughnessy and Fay's comparison. However, our analysis suggests that differences in the shape of other skeletal elements also separate these two species, particularly the articular surfaces of all major limb elements (femur, tibia, radius, and ulna) (Figure 4.5). Although beyond the scope of this paper, a comprehensive comparison of the postcranial skeleton of these two species would definitely be useful.

In contrast, the bearded seal, the largest, most ancient, and most genetically distinct of the arctic phocid seals (Arnason et al. 2006), has very distinctively shaped skeletal elements that are virtually impossible to confuse with any other species, even for juvenile remains. The very large size of newborns, which are slightly longer than newborn walrus (Burns 1970; Reeves et al. 2002), make distinguishing even neonatal bearded seal remains rather straightforward. Similarly, in our opinion, virtually all ribbon seal skeletal elements are highly diagnostic for all age classes (Figure 4.5), including carpals and tarsals of young juveniles. The ribbon seal is another strictly North Pacific pagophilic species with a very distinctive pelage (Burns 1970; Reeves et al. 2002)—its uniquely banded hair coat seems to echo the degree of skeletal difference from other small seals in the

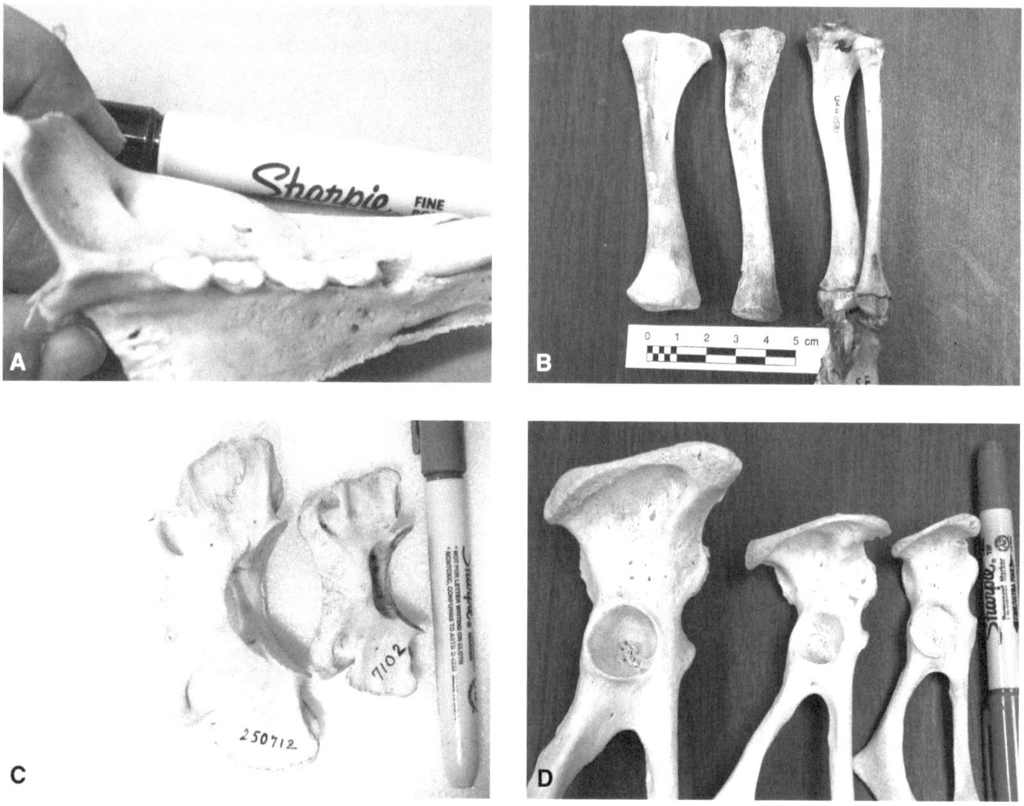

FIGURE 4.4. (A) Upper left maxillary tooth row of spotted seal, UAM 11675, showing the straight tooth alignment characteristic of this species (cf. diagonal alignment in the closely related harbor seal). (B) Species differences in shape for the tibia of some juvenile seals, left to right: spotted seal, UAM16605 (2–4 wks); Amaknak Bridge ringed seal, AB 200152; ringed seal, UAM 11581 (6–8 wks). (C) Species differences in shape and size for the cervical vertebra # 1 (atlas), left to right: adult harbor seal, NMNH 250712, vs. adult ringed seal, NMNH 7102. (D) Species differences in shape and size for the innominate: adult male spotted seal, UAM 11675; adult female ringed seal (pack ice ecotype), UAM 11672; Amaknak Bridge ringed seal, AB200133. See Appendix 4.1 for notes on modern specimens (UAM/NMNH numbers).

region. While the ribbon seal is genetically distinct from all other North Pacific seals (Arnason et al. 2006), it is quite closely related to the Atlantic harp seal (*Phoca groenlandicus*): together, ribbon and harp seals comprise the second-most-ancient lineage of Arctic phocids.

We also found ringed seal skeletal elements to be highly diagnostic in size and shape. These skeletal differences are almost certainly a reflection of their evolutionary antiquity: of all Arctic phoicids, only the bearded/hooded seal and ribbon/harp seal groups are older (Arnason et al. 2006; Harington 2008). While there is some size overlap between large ringed seals and small harbor or spotted seals, there are significant morphological shape differences in most postcranial skeletal elements, irrespective of size or age (Figures 4.4 and 4.5).

RINGED SEAL ECOTYPES

All prehistoric remains of adult ringed seal we examined from the Amaknak Bridge site were significantly smaller than museum specimens available for comparison that had been collected from nearshore locations with shorefast ice, as were most juvenile remains of similar developmental stage (Figure 4.3). Ringed seals in the Bering Sea inhabit both pack ice (large sheets or chunks of sea ice that drift with the currents and wind) and shorefast ice (immobile ice attached to land, aka "fast ice"). While it is often stated that animals found in the pack ice

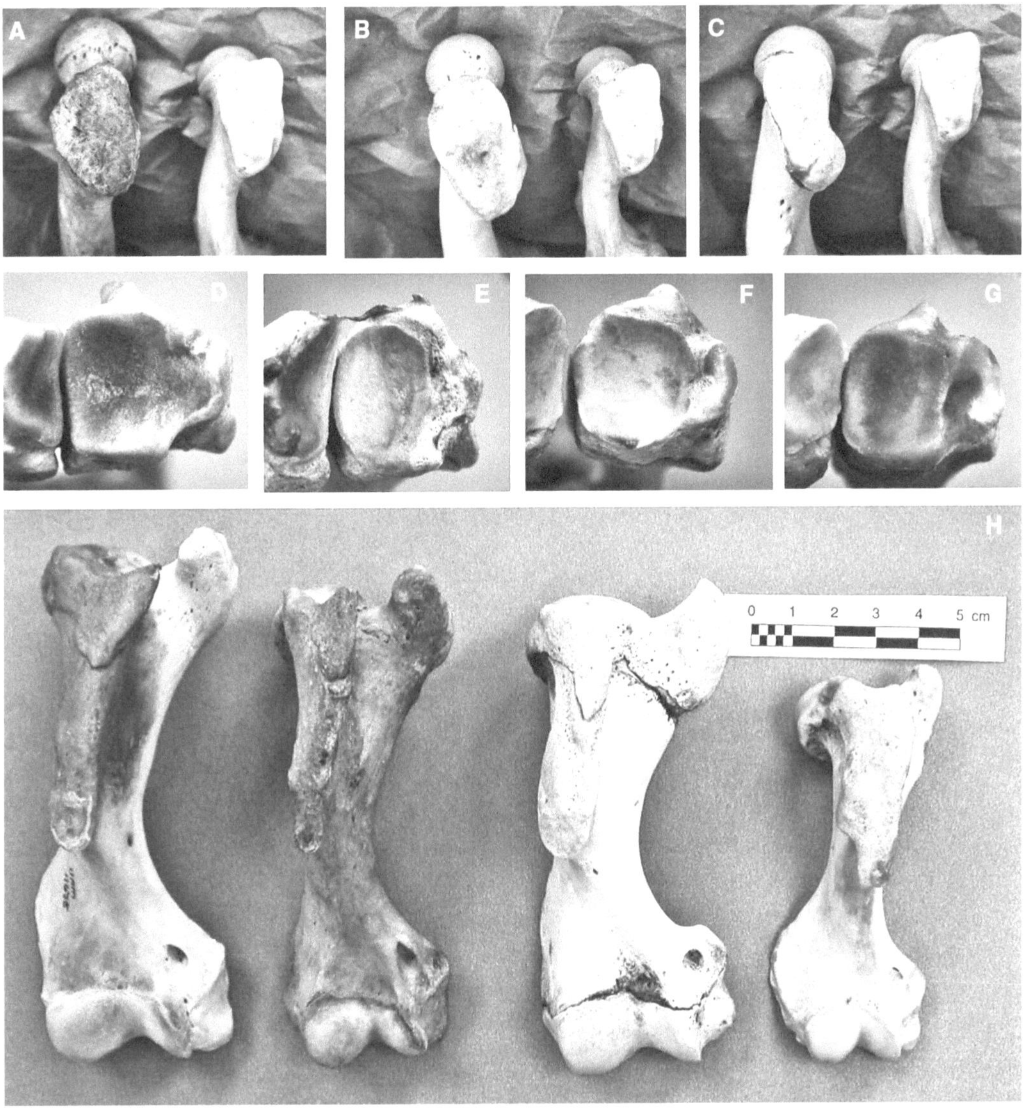

FIGURE 4.5. Top row, shape difference between adult ringed seal and other species in proximal femur shape, left to right: (A) harbor seal, UVic 86/85, vs. ringed seal, UAM 11672; (B) spotted seal, UAM 11675, vs. ringed seal, UAM 11672; (C) ribbon seal, UAM 11469, vs. ringed seal, UAM 11672. Middle row, shape difference between species of phocid seals (adult), distal tibia/fibula: (D) ringed seal, UAM 11672; (E) ribbon seal, UAM 11469; (F) spotted seal, UAM 11675; (G) harbor seal, UVic 86/85. Bottom row, shape difference between species of phocid seals (adult), L. humerus: (H) left to right, spotted seal, UAM 11675, harbor seal, UVic 86/85, ribbon seal, UAM 11469, ringed seal, UAM 11672. See Appendix 4.1 for notes on modern specimens (UAM/UVic numbers).

are groups of nonbreeding subadults that have been excluded from preferred territories in shorefast ice by older, dominant animals (Hammill and Smith 1991; Holst et al. 1999; Krafft et al. 2006, 2007; Laidre et al. 2008; Smith and Hammill 1981; Smith et al. 1991; Stirling and Øritsland 1995), it is has been clear for some time that the pack ice is not restricted to immature animals. Evidence of ringed seal breeding in the pack ice comes from the Bering Sea (Burns 1970), Sea of Okhotsk, and Chukchi Sea (Fedoseev 1975), Davis Strait of the central Canadian Arctic (Finley et al. 1983), and the Barents Sea (Wiig et al. 1999).

Fedoseev (1975) concluded from his preliminary study in the North Pacific that pack

TABLE 4.1

Subset of measurements from adult male ringed seals (≥8 yr), fast ice vs. pack ice specimens

Body measurements in cm; cranial measurements in mm.

MEASUREMENT	N	FAST ICE MEAN (±SD)	N	PACK ICE MEAN (±SD)	P
Body length	5	133.5±4,58	9	125.1±3.89	** <0.01
Zygomatic width	4	112.3±1.45	6	104.5±3.21	** <0.01
Mastoid width	4	112.0±1.98	7	106.1±2.07	** <0.01
Width of occipital condyles	4	58.7±1.35	8	56.2±1.85	* <0.05
Length of tympanic bullae	5	38.9±2.16	9	36.6±1.57	* <0.05
Length of lower tooth row	5	50.5±1.65	8	47.9±1.21	** <0.01
Length of upper tooth row	5	57.2±2.71	9	53.9±1.29	** <0.01
Length of lower jaw	5	114.6±3.56	8	107.6±3.67	** <0.01

NOTE: From Finley et al. 1983, table 3. Specimens archived in the National Museum, Ottawa, catalogue numbers NMC 45419-45469. Highly significant values for *p* are marked **.

ice–breeding seals were smaller at maturity as well as at birth and that they matured faster than shorefast ice–breeding seals. Finley and colleagues provided morphometric evidence from Baffin Bay in support of Fedoseev's work, of which we present a subsample here (see Table 4.1): Baffin Bay pack ice seals were not only smaller but matured about a year earlier than shorefast-breeding seals. In conclusion, Finley et al. (1983:171) suggested that the age structure of the pack ice population they analyzed "approximates that of a normal breeding population."

Despite the evidence that two distinct morphotypes indeed exist—apparently the consequence of the population dividing itself among two distinct ecological habitats that are consistently available across the Arctic—ringed seals are characterized as giving birth invariably within shorefast ice habitats exclusively (e.g., Frost et al. 2004; Hammill and Smith 1991; Holst et al. 1999, 2001; Kingsley et al. 1985; Laidre et al. 2008; Lydersen and Gjertz 1986; Reeves et al. 2002; Smith and Hammill 1981; Smith et al. 1991; Stirling and Øritsland 1995). Where pack ice–breeding is acknowledged to occur, it is largely dismissed as an insignificant phenomenon (e.g., Burns 1970; Kraftt et al. 2007; Nowak 2003). However, given that offshore pack ice throughout the Arctic remains virtually unsurveyed for all life forms, including polar bears and ringed seals (e.g., Aars et al. 2006), such a conclusion must be considered premature.

Some biologists have concluded that a substantial population of ringed seals must inhabit the offshore Arctic pack ice, based in part on estimates of the number of seals required to support known populations of polar bears, their primary predator (e.g., Stirling and Øritsland 1995; Wiig et al. 1999; Derocher et al. 2002; Stirling 2002). Finley et al. (1983:171), based on a survey of Baffin Bay in the central Canadian Arctic, concluded that "the pack ice supports a much larger population of seals than does the bordering area of fast ice" and that these animals inhabit ice that lies well offshore. This offshore habitat is probably beyond the reach of biologists' survey aircraft, which seldom manage to penetrate more than about 100 km north of the concentrated pack ice edge (e.g., Aars et al. 2009; Frost et al. 2004). As for polar bears, this limited survey effort has resulted in uncertainty regarding the global population size for ringed seal, which is currently estimated at about 7 million (Davis et al. 2008; Wiig et al. 1999; Reeves et al. 2002).

While the primary productivity of pack ice over deep Arctic waters is presumed to be uniformly low and, thus, unsuitable habitat for seals and other marine mammals (e.g., Fischbach et al. 2009), this assumption is based on very few actual measurements collected over huge distances and time periods (Aars et al. 2008). Ringed seals eat primarily young (1–2-year-old) polar cod (*Boreogadus saida*) that live under the ice, as well as the amphipods and small copepods (shrimp-like invertebrates) that polar cod themselves eat. In shallow waters, ringed seals eat other types of fish in addition to older, benthic-dwelling polar cod (e.g., Born et al. 2004; Finley et al. 1983; Labansen et al. 2007). However, both polar cod and their prey have been found to live under ice of all types, including multiyear and first-year drifting pack ice regardless of the ocean depth (Lønne and Gulliksen 1989). In other words, it is apparent that young polar cod do not require ice that is positioned over shallow, continental shelf waters, and, therefore, neither do ringed seals nor the polar bears that prey upon them. If offshore pack ice over deep water is productive enough for polar cod, it should be suitable for ringed seals and polar bears also. This perspective is important because it vastly expands what biologists consider suitable habitat for a number of arctic marine mammals and explains the ability of this ecosystem to sustain these species over millions of years despite dramatic shifts in ice extent and persistence (e.g., Harington 2008).

The most recent evidence in support of these conclusions comes from a study of genetic diversity in the ringed seal, which showed virtually no regional subdivision across the Arctic (Davis et al. 2008). The authors, in attempting to explain these results, suggest that preferred landfast ice breeding locations for mature ringed seals may be more limited than previously thought and propose that competition for landfast ice breeding territories may drive many mature seals offshore to breed in the pack ice. This suggests to us an explanation for how and why ringed seal ecotypes develop and persist. If strongly territorial, large bodied individuals of both sexes out-compete weakly territorial, smaller-bodied seals for access to limited coastal breeding habitats, smaller-bodied and more rapidly-maturing animals will end up breeding in the pack ice (see discussions of body size vs. relative growth rates in Crockford 2006). Certainly, within landfast ice habitats, ringed seals of both sexes set up territories during the breeding season (Smith and Hammill 1981; Kraftt et al. 2007), and both reportedly show territorial aggression during this time (Smith and Hammill 1981; Stirling 1977, cited in Lydersen and Gjertz 1986). In other species, a strong correlation has been found between body size and degree of territorial aggression, especially in males that stake out breeding territories; in some taxa, these traits appear to be positively linked to variations in plasma testosterone levels (e.g., Sinervo et al. 2000). Thus, strongly territorial behavior and an associated larger body size that allows highly territorial animals to successfully outcompete smaller-bodied individuals for preferred breeding sites may drive continual recruitment of small-bodied seals into the offshore pack ice environment, leading to the development of two ecotypes that are morphological distinct but genetically indistinguishable.

Moreover, a related issue that has important implications for zooarchaeology is the high probability that the ringed seals that breed on coastal shorefast ice are the source of virtually all skeletal specimens of this species held at museums and other institutions because of their increased accessibility. This means that comparative skeletons available in museums may vary widely in size and many collections are likely to be heavily biased toward specimens of the large form. For example, Storå (2000:205) reports "a number of individuals with a very small body size" in his age estimation study of ringed seals collected from the Baltic Sea, although he gives no possible explanation for this phenomenon. Thus, for example, a collecting bias may account for the fact that virtually all ringed seal specimens that we examined in the National Museum of Natural History (NMNH) collection, Smithsonian Institution, were large enough to overlap in size with small harbor seals. As a consequence, making note of specimen collection location is especially important for this species, although caution is needed: curators may not always distinguish pack ice from shorefast ice collection locations in their records, in part because this information is not routinely provided (see Appendix 4.1 for an example). Landfast breeding populations of ringed seals are virtually the only ones examined by biologists studying pupping and breeding phenomena (e.g., Davis et al. 2008; Holst et al. 1999; Krafft et al. 2006, 2007; Smith and Hammill 1981; Stirling and Øritsland 1995). Although modern Inuit hunters in many areas, such as west Greenland (Finley et al. 1983), and perhaps the Bering and Chukchi Seas as well, probably access pack ice populations during annual harvests, this information seems not to make its way back to research biologists or museum curators.

ZOOARCHAEOLOGICAL ANALYSIS RESULTS

PAGOPHILIC SPECIES

Substantial numbers of bearded and ringed seal remains in the assemblage indicate that spring pack ice reached a more southerly position during the Neoglacial than it does today and persisted until summer. Today, bearded seals in the Bering Sea passively migrate south with the advancing pack ice in spring (Simpkins et al. 2003). This moving ice platform takes animals at the leading edge of the pack quite far south. Females haul out on the ice to give birth from about mid-March to early April (Burns 1970), and pups are nursed for 24 days or less. Adult bearded seals mate after pups are weaned and continue to haul out on the ice during an extended molting period that can last several months, during which time the receding ice pack usually carries them north. Because juvenile and subadult animals molt in the early spring while mature mammals are giving birth, all age classes of bearded seals may be available when the pack ice is at its full extent.

Similarly, ringed seals in the Bering Sea also move south with the expanding pack ice in the spring (Burns 1970; Simpkins et al. 2003). They haul out to give birth from March through May, with a peak in mid-April (Fedoseev 1975). Most animals continue to spend extensive periods of time on the ice between May and June, when all but young of the year haul out to molt. Young juveniles, including newly weaned young, stay in the water at the ice edge to feed while older animals molt. In the fall, all age classes are usually found together, feeding in open water south of the pack ice or in open polynyas (Born et al. 2004; Holst et al. 1999; Wiig et al. 1999).

BEARDED SEAL REMAINS

Bearded seal remains from the Amaknak Bridge site represent 4% of the total identified sample (264/5947), and while this may seem an insignificant amount, it is substantially more than has been reported from archaeological sites in the Canadian Arctic Archipelago and Greenland. For example, for the mammalian remains (NISP 26,642) reported by Møhl (1979) for the Thule site of Nugarsuk in west Greenland (with species totals adjusted to account for the stated bias in identification effort for small phocids, as per Davis 2001), relative frequencies break down thus: 87% ringed seal, 9% harp seal, 1% bearded seal, 0.4% harbor seal, 0.1% walrus, and 3% other species (such as polar bear, caribou, whale, etc.). Similarly, in a comparison of the marine mammal remains from 61 arctic sites compiled by Maribeth Murray (2008, table 1), only three contained more than 1% bearded seal: *Walakpa* (Barrow, AK, ca. 1500–1100 BP) had 1.5% bearded seal vs. 59.5% *Phoca* sp. (undetermined small seals); *Coldsite* (Devon Island, Canada, ca. 4000 BP) had 4.3% bearded seal vs. 81.5% ringed seal; *Skraeling*

(Ellesmere Island, Canada, ca. 900 BP) had 2.5% bearded seal vs. 66.4% ringed seal.

The only faunal assemblages we found reported in the literature that had frequencies of bearded seal exceeding 4% of the total NISP are reported by Woollett et al. (2000), from three small Thule and historic Inuit sites in Frobisher Bay (south Baffin Island, Canada) that date from the 16th to the early 20th century. NISPs ranged from 110 to 133,and all pinniped species totals were based on counts of auditory bullae, boney palate, maxillae, mandibles, and humeri only. In these sites, bearded seal made up 6–8% of the sample, ringed seal 69–82%, harp seal 2–6%, harbor seal 5–7%, and walrus 4–12%.

Therefore, we conclude that even low percentages of bearded seal remains in an archaeological assemblage are reliable indicators of pack ice habitat. In addition, as bearded seals haul out on pack ice to give birth and nurse their young in early spring, the presence throughout the site assemblage of neonatal (0–3 weeks) and newly-weaned (ca. 4–8 weeks) individuals indicate that the ice edge habitat which bearded seals prefer must have been available close to the Amaknak site thoughout the spring (March–May). The bearded seal sample described here provides substantial evidence that at the height of the Neoglacial, pack ice was probably present off Unalaska Island well into summer on a regular basis.

RINGED SEAL REMAINS

In contrast to bearded seal, ringed seal remains comprised an astonishing 32% of the total NISP (1905/5947) recovered from the Amaknak Bridge site. The bulk of the ringed seal remains (90%) are estimated to have been about 2 to 6 months of age (Crockford et al. 2004), the age class that feeds at the edge of the pack ice during early summer. The very low frequency of adult and neonatal remains suggests little successful hunting for ringed seals took place during the birthing season, and the lack of subadult and older juvenile material argues against the suggestion that ringed seals were harvested from open water in the fall (where all age classes feed together).

The vast majority of the Amaknak Bridge juvenile remains are also small for their developmental stage, as compared to most juveniles in modern collections (Figure 4.3), which suggests they represent the pack ice ecotype described previously. If pack ice ecotypes are smaller at birth than shorefast ice ecotypes and have faster early growth rates, suture fusion in pack ice juveniles should be further advanced at all ages than shorefast ecotypes. The fact that most of the Amaknak Bridge juveniles are only slightly larger but better developed than the 8-to-12-week-old shorefast ecotype available for comparison (UAM 11633, see Appendix 4.1) supports our interpretation that the Amaknak Bridge ringed seal sample is dominated by pack ice ecotype juveniles aged 3 to 4 months of age (range, 2 to 6 months). The few adult remains recovered are also smaller than adults of modern animals collected from shorefast ice locations, similar in size to the adult female pack ice individual available (UAM 11672, see Appendix 4.1) and typical specimens of Baikal seal (*Phoca sibirica*), the smallest of all phocid seals. Lack of suitable modern samples of known-aged ringed seals (both adult and young juveniles), taken from known pack ice vs. shorefast ice locations, precluded a more detailed analysis. We contend, however, that since Amaknak Bridge bearded seals must represent animals harvested from the edge of late spring/early summer pack ice habitat, ringed seals were very likely harvested at the same time and thus likely represent the small pack ice ecotype.

We believe that the strong representation in this sample of ringed seals 2 to 6 months of age provides additional evidence that at the height of the Neoglacial, pack ice must have been present off Unalaska Island and persisted well into summer on a regular basis over several centuries, a phenomenon that is unprecedented historically. The small size of the ringed seal skeletal elements, compared to those from comparative specimens of similar developmental stage taken from shorefast ice

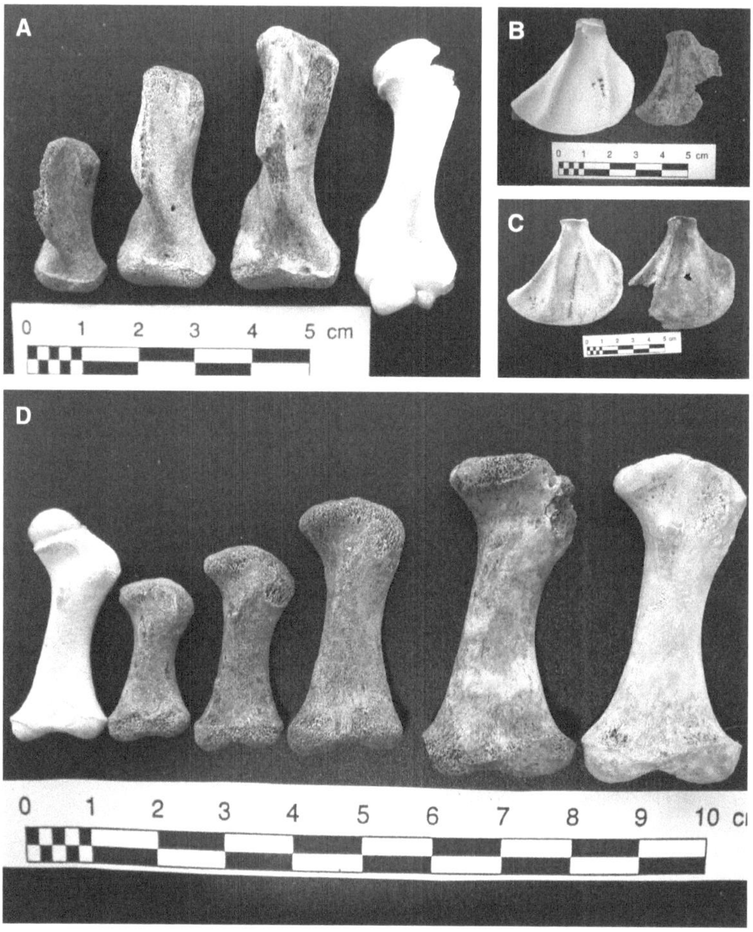

FIGURE 4.6. Modern northern fur seal (UVic and RBCM numbers) vs. specimens from La Push, WA (LAP numbers), with estimated or known ages. (A) Humerus, L to R: LAP 224 (fetal); LAP 225 (0–2 wks); LAP 221 92-4 wks); RBCM F80-11 (2 wks). (B) scapula, L to R, RBCM F80-11 (2 wks) vs. LAP 219 (0–2wks). (C) scapula, L to R, UVic 90/140 (6–8 wks) vs. LAP 220 (8–12 wks). (D) femur, L to R, RBCM F80-11 (2 wks); LAP 058 (fetal); LAP 217 (0–2 wks); LAP 215 (2–4 wks); LAP 121 (4–6 wks); UVic 90/140 (6–8 wks). Note that specimens numbered 058 and 121 are from the controlled excavation (undisturbed and dated), the others are from undated disturbed deposits.

habitats, suggests very strongly that ancient Aleut hunters were harvesting the same pack ice ecotype described by modern Inuit hunters and scientific researchers.

TEMPERATE SPECIES: NORTHERN FUR SEAL REMAINS

Substantial numbers of NFS remains were recovered from the Amaknak Bridge site (40% of the total NISP, 2377/5947). Almost half of all NFS remains that could be aged (46%) were subjectively estimated as representing unweaned pups (Figure 4.6), while 28% were judged to be adult or close to adult sized, with epiphyses fused (the line of epiphyseal fusion being absent or just visible). Within this oldest category (NISP 529), 71% (NISP 373) were potentially breeding age adults (adult/subadult), and 84% of these were male, recognizable due to the extreme sexual dimorphism characteristic of this species. Other age classes were poorly represented.

The presence of all age groups and both sexes of NFS, including newborn and still-nursing infants, provides strong evidence for the presence of pupping grounds in the site vicinity (Crockford et al. 2002; Etnier 2002; Newsome et al. 2007). The presence of unweaned newborns in particular confirms that hunting of NFS occurred, at least part of the time, on breeding rookeries. Today, the primary Bering Sea rookeries are on the Pribolof Islands (Gentry 1998), about 400 km to the north of Unalaska. This is almost certainly farther than residents of the eastern Aleutians could have traveled regularly to hunt. The nearby Aleutian site of Chaluka reported by Etnier (2002) suggests that a fairly substantial rookery must have existed close to the site at the time it was occupied, either on the island of Unalaska itself or somewhere nearby. The presence of a well-established local rookery that no longer exists and was not recorded during the 18th-century commercial sealing boom (Kenyon and Wilke 1953) indicates that the distribution pattern of NFS was very different during the prehistoric period than it is today. We contend that the same expansion of Bering Sea pack ice that brought arctic seals to the doorstep of Amaknak Bridge hunters also made the Pribilofs inaccessible to fur seals during their early summer pupping season (June–July) or rendered them too inclement (due to ice-associated winds) for survival of fur seal newborns (Trites and Antonelis 1994).

ESTABLISHMENT OF NFS ROOKERIES ALONG THE NORTHWEST COAST AND ALEUTIAN ISLANDS

EVOLUTIONARY HISTORY OF NFS ROOKERIES

NFS have the longest evolutionary history of all otariids and have existed in the North Pacific as a discrete lineage for about 25 million years (Arnason et al. 2006). Although it is often assumed that the Pribilof Islands have always been used by NFS as a breeding rookery, during the Last Glacial Maximum of the Pleistocene (and at least once before then), the Pribilofs were part of the Bering Land Bridge and thus inaccessible to fur seals from at least 25,000 to 11,500 BP (Berta and Sumich 1999; England and Furze 2008; Guthrie 2004). Consequently, even if NFS did not use these islands prior to the onset of the Holocene, the contemporary use of the Pribilofs suggests that fur seal life history strategies are sufficiently flexible to adjust to fairly major disruptions in their migratory habits and preferred breeding environment. The fact that nonmigratory NFS colonies established in southern CA in 1968 are still flourishing today is proof that such life history flexibility still exists (Gentry 1998; Nowak 2003; Peterson et al. 1968; Testa 2007).

The Pribilof Islands would have been exceptionally suitable as rookery habitat for fur seals during the Holocene, since the islands were never occupied by aboriginal people and were too far away from the nearest land masses (at least 400 km) for aboriginal hunters to access from elsewhere (Figure 4.1). Such a refuge from prehistoric hunting pressure makes it extremely doubtful that fur seals failed to utilize the Pribilofs before the Neoglacial period, given that aboriginal populations dependent on marine mammal hunting were already well established in the surrounding region at that time (Knecht and Davis 2001). During the Neoglacial, however, increased winds in the Bering Sea appear to have forced spring pack ice south to the eastern Aleutians, where it persisted well into summer (June/July). The Pribilof Islands almost certainly became more and more unacceptable as a pupping location for NFS from about 4700 BP onward. Although we cannot say for certain that breeding colonies of fur seals existed on the Pribilofs prior to the Neoglacial, we do know that rookeries were established there as soon as Neoglacial conditions ameliorated at about 2700 to 2500 BP, as natural accumulations of fur seal bones dated to that period have been recovered (Etnier et al. 2005).

EVIDENCE OF NONMIGRATORY NFS ROOKERIES

We suggest that the inaccessibility of the Pribilofs from May through July (due to Neoglacial

TABLE 4.2
NFS Measurements from La Push, WA (45CA23)

FAUNAL #	ELEMENT	SIDE	ESTIMATED AGE	GL (MM)
LAP23T206	Femur	R	ca. 24–36 wks	49.7
LAP23T025	Femur	L	ca. 24–36 wks	50.0
LAP23T204	Femur	L	ca. 24–36 wks	50.3
LAP23T205	Femur	R	ca. 24–36 wks	50.4
LAP23T134	Femur	L	ca. 24–36 wks	53.2
LAP23T095	Femur	R	ca.24–36 wks	54.9
LAP23T211	Femur	R	ca. 8–12 wks	44.4
LAP23T057	Femur	L	ca. 8–12 wks	46.1
LAP23T005	Femur	R	ca. 8–12 wks	44.0
LAP23T209	Femur	R	ca. 8–12 wks	44.5
LAP23T003	Femur	L	ca. 8–12 wks	45.3
LAP23T026	Femur	L	ca. 6–8 wks	35.8
LAP23T002	Femur	L	ca. 6–8 wks	40.2
LAP23T001	Femur	L	ca. 6–8 wks	41.1
LAP23T207	Femur	R	ca. 6–8 wks	41.8
LAP23T210	Femur	R	ca. 6–8 wks	41.8
LAP23T004	Femur	R	ca. 6–8 wks	42.1
LAP23T024	Femur	R	ca. 6–8 wks	42.3
LAP23T212	Femur	R	ca. 6–8 wks	42.4
LAP23T121	Femur	L	ca. 6–8 wks	42.7
LAP23T213	Femur	R	ca. 6–8 wks	43.4
LAP23T008	Femur	R	ca. 2–6 wks	28.9
LAP23T022	Femur	R	ca. 2–6 wks	29.5
LAP23T007	Femur	R	ca. 2–6 wks	30.5
LAP23T023	Femur	R	ca. 2–6 wks	31.4
LAP23T216	Femur	L	ca. 2–6 wks	32.7
LAP23T010	Femur	L	ca. 2–6 wks	32.7
LAP23T009	Femur	L	ca. 2–6 wks	33.1
LAP23T215	Femur	R	ca. 2–6 wks	33.4
LAP23T218	Femur	R	<0–2 wks	24.2
LAP23T011	Femur	L	<0–2 wks	26.5
LAP23T006	Femur	R	<0–2 wks	26.8

TABLE 4.2 *(continued)*

FAUNAL #	ELEMENT	SIDE	ESTIMATED AGE	GL (MM)
LAP23T217	Femur	L	<0–2 wks	27.1
LAP23T058	Femur	R	<0 wks (fetal)	21.4
LAP23T177	Humerus	L	ca. 8–12 wks	59.2
LAP23T096	Humerus	L	ca. 6–8 wks	52.5
LAP23T222	Humerus	L	ca. 2–6 wks	39.0
LAP23T020	Humerus	L	ca. 2–6 wks	41.5
LAP23T221	Humerus	R	ca. 2–6 wks	42.4
LAP23T226	Humerus	L	<0–2 wks	31.3
LAP23T225	Humerus	R	<0–2 wks	35.8
LAP23T133	Humerus	R	<0–2 wks	37.2
LAP23T224	Humerus	R	<0 wks (fetal)	25.6
LAP23T191	Radius	R	ca. 8–12 wks	69.0
LAP23T229	Radius	R	ca. 6–8 wks	63.3
LAP23T021	Radius	R	ca. 2–6 wks	39.3
LAP23T228	Radius	R	ca. 2–6 wks	43.1
LAP23T013	Radius	R	ca. 2–6 wks	43.6
LAP23T012	Radius	R	<0–2 wks	35.7
LAP23T227	Radius	L	<0–2 wks	38.7
LAP23T015	Radius	R	<0 wks (fetal)	23.5
LAP23T014	Radius	R	<0 wks (fetal)	24.7
LAP23T016	Radius	L	<0 wks (fetal)	30.2

NOTE: Specimens numbered T001–T191 were recovered during controlled excavation of intact and dated deposits, while those numbered T200–T228 were recovered during backdirt screening of disturbed deposits.

pack ice expansion) was likely the event that prompted fur seals to establish nonmigratory breeding colonies along the Aleutian Archipelago and the west coast of North America, as documented by numerous zooarchaeological studies (Crockford et al. 2002; Gifford-Gonzalez et al. 2005; Moss et al. 2006). While these published accounts are convincing, however, they are not the only examples that exist: many other recently-excavated sites on the Northwest Coast and throughout the Aleutians, especially those subject to small-scale excavations and described in unpublished accounts, continue to yield evidence of nonmigratory NFS rookeries.

We report here previously unpublished measurements (Table 4.2) and comparative photos (Figure 4.6) of some very young NFS long bones recovered from one such site, a salvage recovery on the west coast of the Olympic Peninsula, WA (Wessen 2006). This late prehistoric site was occupied during a narrow time period somewhere between ca. 600 to 1000 BP (two C^{14}

dates are reported: 880 ± 50 and 660 ± 50, from the top and bottom of excavated deposits, respectively). Recovery efforts at the site generated faunal remains from a 2×2-meter excavation in an intact portion of the site, where the mammalian component was found to be heavily dominated by NFS remains of all age classes, including adults and very young pups (93% of the total NISP 1130). A few additional NFS remains were recovered during screening of backdirt from a disturbed portion of the site. The youngest of these remains indisputably represent unweaned pups, which represent the strongest evidence available that local rookeries attended by nonmigratory NFS were being harvested by aboriginal hunters.

WHY AND WHEN NONMIGRATORY ROOKERIES WERE ESTABLISHED

While decimation of prehistoric Aleutian and Northwest Coast colonies can quite reasonably be attributed to a combination of aboriginal and European hunting practices (e.g., Gentry 1998; Newsome et al. 2007), no one before this time has suggested an explanation for why such rookeries would have been established in the first place during the Middle Holocene, nor why these colonies were composed of nonmigratory animals. We contend that pack ice expansion at the beginning of the Neoglacial encouraged some fur seals, especially those with relatively weak migratory instincts and/or natal site fidelity, to establish new rookeries throughout the eastern North Pacific. Many of these new rookeries were very successful and survived well into the historic period (e.g., Crockford et al. 2002). Individuals with strong migratory instincts and powerful natal site fidelity probably become relatively rare during the Neoglacial. Reinvasion of the Pribilof Islands could only have occurred after the climate ameliorated, when individuals with strong migratory instincts would again have had an evolutionary advantage, given their unique ability to utilize this seasonally available rookery location that is so well protected from human harvest. Therefore, the currently dominant life history pattern of NFS, where virtually all breeding and pupping take place on the Pribilofs in June and July, after which females and juveniles migrate south to feed over the winter, is one of at least two strategies available to this species.

CONCLUSIONS AND DISCUSSION

The zooarchaeological data from the Amaknak Bridge site provides evidence that for about 2000 years during the mid-Holocence, seasonal pack ice in the Bering Sea extended farther south and persisted longer, on a regular basis, than it does today. While significant climatic cooling during the Neoglacial is known to have occurred, this is the first evidence for such a dramatic increase in Bering Sea pack ice during this period (Crockford and Frederick 2007). The comprehensive skeletal analysis described here provides two excellent examples of how animal life history flexibility can be ascertained from zooarchaeolgical material: for one species, this flexibility was known about but not recognized as being especially important and the other species had a flexibility not previously appreciated before zooarchaeological analysis provided the evidence.

The first example, that two distinctly sized ecotypes of ringed seal exist, had been discussed by biologists for decades but was largely dismissed as a minor ecological and life history feature. However, a somewhat tentative conclusion that pack ice–breeding seals must make up a significant portion of the global population is now supported by phylogenetic evidence, and this has implications for both archaeologists and biologists. Combined with the knowledge that ancient Inuit hunters probably harvested large numbers of ringed seals from the southern edge of the Arctic pack ice, we suggest that assemblages of ringed seals from other prehistoric Arctic sites may also contain large numbers of small-sized, pack ice seals and perhaps should be re-analyzed (e.g., Woollett et al. 2000, who remarked on small seals from the Labrador component of their site material). Biologically speaking, having two life history strategies—one that utilizes shorefast ice and another that takes advantage of mobile pack ice—has almost certainly given ringed seals the evolutionary flexibility they have needed to survive in the Arctic over millions of years; such flexibility certainly needs to be factored into modern ecological studies and population estimates.

Our second example challenges the notion that current distributions and life history strategies of NFS in the North Pacific have been in place throughout the Holocene (e.g., Reeves et al. 2002). If we accept the assumption that the Pribilof Islands were utilized by NFS as summer rookeries in the Early Holocene, then Neoglacial sea ice expansion explains the post-Middle Holocene dates for the establishment of local rookeries throughout the Aleutian Archipelago and the Northwest Coast. Although archaeological sites older than ca. 5000 BP are admittedly rare in the eastern North Pacific, no sites that predate ca. 4700 BP have yet been reported that show evidence of nonmigratory NFS rookeries (Gifford-Gonzales et al. 2005). If the Pribilofs did not host NFS rookeries during the Early Holocene, then the current life history pattern of this species, where 99% of the population gives birth in the Bering Sea to offspring that are weaned by 4 months of age, was established less than 200 years ago when the last well-established rookeries of nonmigratory animals were wiped out. If so, a pattern may have existed for the 2300 years prior (and at other times in their history), whereby the entire NFS population was composed of both migratory and nonmigratory subgroups. Unfortunately, we simply do not have enough evidence at this time to determine which of these options is correct.

Both of these examples remind us that many species are more flexible and adaptable in their life history strategies than conventionally thought and that zooarchaeological data can be an invaluable addition to contemporary biological and ecological studies that attempt to explain such differences. Since life history flexibility may be what has allowed both ringed seals and NFS to persist over millions of years and countless shifts in environmental conditions, evidence of life history flexibility ascertainable from the zooarchaeological record is as important to biologists as it is to archaeologists.

ACKNOWLEDGEMENTS

We are indebted to the Ounalashka Corporation and Qawalangin Tribe of Unalaska, the Museum of the Aleutians and Richard Knecht and Richard Davis for the opportunity to work on this material. In addition, we thank managers and curators of the following institutions for making their skeletal collections available: Department of Anthropology, University of Victoria, Canada; Royal British Columbia Museum, Victoria; Canada; Smithsonian Institution, Washington, DC; University of Alaska Museum, Fairbanks; Burke Museum, Seattle, WA.

APPENDIX 4.1

Phocid Seal Comparative Skeletons Used for Estimating Age and Species Determination

SEAL	GENDER	DATA
P. vitulina		
UVic 85-8	-	(stillborn? < 1 wk)
UVic 97/42	-	(newborn, ca. 1 wk)
UVic 88/124	-	(newborn, ca. 1 wk)
UVic 86/85	-	(subadult)
UVic 84/48	-	(subadult)
P. largha		
UAM 16605	F	(unweaned, ca. 2–4 wks)[a]
NMML 196	F	skull (ca. 8–12 wks)[c]
BM 34221	M	(juvenile/subadult)[c]
UAM 11676	F	(adult, mother to 16605)[a]
UAM 11675	M	(adult? mate of 11676)[a]
NMML 402	M	skull (adult/subadult)[b]
P. hispida		
UAM 11581	F	(weaned, ca. 6–8wks)[a]
UAM 11633	M	(ca. 8–12 wks)[d]
NMNH 504210	-	(ca. 8–12 wks)[d]
NMNH 504214	-	(juvenile)[d]
UAM 11672	F	(adult/subadult)[a]

APPENDIX 4.1 *(continued)*

SEAL	GENDER	DATA
NMNH 7102	-	(adult)[d]
NMNH 305081	M	(adult/subadult)[d]
P. fasciata		
UAM 16619	M	(unweaned, ca. 3–4 wks)[a]
UAM 11529	M	(weaned, ca. 6–8 wks)[a]
UAM 11469	M	(subadult/adult)[a]
E. barbatus		
NMNH 16112	F	(fetus, at/near term)[c]
NMNH 550412	-	(weaned, ca. 6–8 wks)[c]
BM 34219	-	(juvenile)[c]
NMNH 16116	F	(adult/subadult)[c]
NMNH 550250	F	(adult/subadult)[c]
NMNH 550251	F	(adult/subadult)[c]

[a]Indicates collection known from pack ice habitat.
[b]Indicates ice habitat unknown.
[c]Indicates collection presumed to be from pack ice habitat.
[d]Indicates collection presumed to be from shorefast ice habitat.
INSTITUTIONAL DESIGNATORS: UVic, University of Victoria, Anthropology; UAM, University of Alaska Museum, Fairbanks; NMML, National Marine Mammal Laboratory, Seattle; NMNH, National Museum of Natural History, Washington DC; BM, Burke Museum, Seattle.
NOTE: Age estimates given for juveniles < 1 yr are based on a combination of collection notes, collection dates, and mean dates of birth for that taxa. General age estimates for animals > 1 yr are based on collection date and epiphyseal fusion, such that unfused epiphyses and smaller than adult size are "juvenile," fused or partially fused epiphyses with suture lines evident, close or equal to adult size is "subadult," while those with fused epiphyses and no suture lines evident are "adult." UAM 11581 (*P. hispida*) and UAM 11529 (*P. fasciata*) were collected in April 1963 by a research team working out of Gambell, St. Lawrence Island, within days of each other. Since the unweaned *P. fasciata* pup was almost certainly taken from pack ice, we presumed that the *P. hispida* specimen was also, especially as the occurrence of ringed seal "whitecoat" pups and breeding females in the pack ice were noted in the study (Burns 1970).

APPENDIX 4.2

Northern Fur Seal Comparative Skeletons Used for Estimating Age of Young Pups, All from Rookeries on the Pribilofs, Alaska

SEAL	GENDER	DATA
RBCM F80/11	-	(unweaned, ca. 0–2 weeks)
UVic 90/141	M	(unweaned, ca. 5–6 weeks)[a]
UVic 90/140	F	(unweaned, ca. 6–8 weeks)[a]
UVic 87/26	M	(weaned, 6–8 month)

[a] Sex designated here corrected from Crockford et al. 2002.
INSTITUTIONAL DESIGNATORS: UVic, University of Victoria, Anthropology; RBCM, Royal B.C. Museum, Victoria.

REFERENCES CITED

Aars, J., N. J. Lunn, and A. E. Derocher (editors)
2006 *Polar Bears: Proceedings of the 14th Working Meeting of the IUCN/SSC Polar Bear Specialist Group, 20–24 June 2005, Seattle, Washington, USA.* Occasional Paper of the IUCN Species Survival Commission 32. IUCN, Gland (Switzerland) and Cambridge (UK).

Aars, J., T. A. Marques, S. T. Buckland, M. Andersen, S. Belikov, A. Boltunov, and Ø. Wiig
2009 Estimating the Barents Sea Polar Bear Subpopulation Size. *Marine Mammal Science* 25: 35–52.

Arnason, U., A. Gullberg, A. Janke, M. Kullberg, N. Lehman, E. A. Petrov, and R. Väinölä
2006 Pinniped Phylogeny and a New Hypothesis for Their Origin and Dispersal. *Molecular Phylogenetics and Evolution* 41:345–354.

Arnold, P. W., H. Marsh, and G. Heinshon
1987 The Occurrence of Two Forms of Minke Whales in East Australian Waters with a Description of External Characters and Skeleton of the Diminutive or Dwarf Form. *Scientific Reports of the Whales Research Institute* 38:1–46.

Baker, J. D.
2007 Post-Weaning Migration of Northern Fur Seal (*Callorhinus ursinus*) Pups from the Pribilof Islands, Alaska. *Marine Ecology Progress Series* 341:243–255.

Baker, J. D., G. A. Antonelis, C. W. Fowler, and A. E. York
1995 Natal Site Fidelity in Northern Fur Seals, *Callorhinus ursinus. Animal Behavior* 50:237–247.

Berta, A., and J. L. Sumich
1999 *Marine Mammals: Evolutionary Biology.* Academic Press, San Diego.

Best, P. B.
1985 External Characters of Southern Minke Whales and the Existence of a Diminutive Form. *Scientific Reports of the Whales Research Institute* 36:1–33.

Bigg, M. A.
1990 *Migration of Northern Fur Seals (*Callorhinus ursinus*) Off Western North America.* Canadian Technical Report of Fisheries and Aquatic Sciences 1764. Fisheries and Oceans Canada, Nanaimo.

Born, E. W., J. Teilmann, M. Acquarone, and F. F. Riget
2004 Habitat Use of Ringed Seals (*Phoca hispida*) in the North Water Area (North Baffin Bay). *Arctic* 57(2):129–142.

Burns, J.
1970 Remarks on the Distribution and Natural History of Pagophilic Pinnipeds in the Bering and Chukchi Seas. *Journal of Mammalogy* 51(3): 445–454.

Burton, R. K., J. J. Snodgrass, D. Gifford-Gonzalez, T. Guilderson, T. Brown, and P. L. Koch
2001 Holocene Changes in the Ecology of Northern Fur Seals: Insights from Stable Isotopes and Archaeofauna. *Oecologia* 128:107–115.

Calkin, P. E., G. C. Wiles, and D. J. Barclay
2001 Holocene Coastal Glaciation of Alaska. *Quaternary Science Review* 20:449–461.

Charlton, K., A. C. Taylor, and S. W. McKechnie
2006 A Note on Divergent mtDNA Lineages of Bottlenose Dolphins from Coastal Waters of Southern Australia. *Journal of Cetacean Research Management* 8(2):173–179.

Crockford, S. J.
2006 *Rhythms of Life: Thyroid Hormone and the Origin of Species.* Victoria, Trafford.
2008 Be Careful What You Ask For: Archaeozoological Evidence of Mid-Holocene Climate Change in the Bering Sea and Implications for the Origins of Arctic Thule. In *Islands of Inquiry: Colonisation, seafaring and the archaeology of maritime landscapes,* edited by G. Clark, F. Leach and S. O'Connor, pp. 113–131. Terra Australis 29 ANU E Press, Canberra.

Crockford, S., and G. Frederick
2007 Sea Ice Expansion in the Bering Sea during the Neoglacial: Evidence from Archaeozoology. *The Holocene* 17:699–706.

Crockford, S. J., Frederick, G., and R. Wigen
2002 The Cape Flattery Fur Seal: An Extinct Species of *Callorhinus* in the Eastern North Pacific? *Canadian Journal of Archaeology* 26:152–174.

Crockford, S., G. Frederick, R. Wigen, and I. McKechnie
2004 *Final Report on the Analysis of the Vertebrate Fauna from Amaknak Bridge, Unalaska, AK, UNL050.* In *Amaknak Bridge Site Data Recovery Project Final Report,* by R. Knecht and R. Davis, section 3.5. Contract No. MGS-STP-BR-0310(S)/52930. Copies available from the Museum of the Aleutians, Unalaska, and Pacific Identifications Inc., Victoria.

Davis, B. L.
2001 Sea Mammal Hunting and the Neoglacial: An Archaeological Study of Environmental Change and Subsistence Technology at Margaret Bay, Unalaska. In *Recent Archaeology in the Aleut Zone of Alaska,* edited by D. E. Dumond, pp. 71–85. University of Oregon Anthropological Papers No. 58, University of Oregon, Eugene.

Davis, C. S., I. Stirling, C. Strobeck, and D. W. Coltman

2008 Population Structure of Ice-Breeding Seals. *Molecular Ecology* 17:3078–3094.
Debruyne, R.
2005 A Case Study of Apparent Conflict between Molecular Phylogenies: The Interrelationships of African Elephants. *Cladistics* 21:31–50.
Debruyne, R., A. Van Holt, V. Barriel, and P. Tassy
2003 Status of the So-Called African Pygmy Elephant (*Loxodonta pumilio* (NOACK 1906)): Phylogeny of Cytochrome *b* and Mitochondrial Control Region Sequences. *C. R. Biologies* 326: 687–697.
Derocher, A. E., Ø. Wiig, and M. Andersen
2002 Diet Composition of Polar Bears in Svalbard and the Western Barents Sea. *Polar Biology* 25: 448–452.
Dyke, A. S., and J. M. Savelle
2001 Holocene History of the Bering Sea Bowhead Whale (*Balaena mysticetus*) in its Beaufort Sea Summer grounds off Southwestern Victoria Island, Western Canadian Arctic. *Quaternary Research* 55:371–379.
Elliot, H. W.
1884–7 Fur Seals (Life History, Habits, Distribution, etc). In *Fisheries and Fishery Industries of the U.S., Sec. 1, Natural History of Useful Aquatic Animals (S.M. 47-1, V.6, No. 124, pt 1,2)*, edited by G. B. Goode, pp. 75–113. Library of Congress, Washington, D.C.
England, J. H., and M. F. A. Furze
2008 New Evidence from the Western Canadian Arctic Archipelago for the Resubmergence of Bering Strait. *Quaternary Research* 70: 60–67.
Etnier, M. A.
2002 *The Effect of Human Hunting on Northern Fur Seal (*Callorhinus ursinus*) Migration and Breeding Distribution in the Late Holocene*. Unpublished Ph.D. dissertation, Department of Anthropology, University of Washington, Seattle.
2004 Reevaluating Evidence of Density-Dependent Growth in Northern Fur Seals (*Callorhinus ursinus*) Based on Measurements of Archived Skeletal Specimens. *Canadian Journal of Fisheries and Aquatic Sciences* 61:1616–1626.
Etnier, M. A., M. Malavansky, and T. Remick
2005 Enhancing Science Education Through the Investigation of a 2500-Year-Old Fur Seal Rookery on St. George Island, Alaska. Paper presented at the 16th Biennial Conference on the Biology of Marine Mammals (Abstracts pp. 85), San Diego.
Fedoseev, G. A.
1975 Ecotypes of the Ringed Seal (*Pusa hispida* Schreber, 1777) and Their Reproductive Capabilities. In *Biology of the Seal*, edited by K. Ronald and A.W. Mansfield, pp. 156–160. Rapports et Proces-verbaux des Reunions, Conseil International Pour L'Exploration de la Mer 169.
Ferguson, S. H.
2006 The Influences of Environment, Mating Habitat, and Predation on Evolution of Pinniped Lactation Strategies. *Journal of Mammalian Evolution* 13:63–82
Ferguson, S. H., M. K. Taylor, and F. Messier
2000 Influence of sea ice dynamics on habitat selection by polar bears. *Ecology* 81:761–772.
Finley, K. J, G. W. Miller, R. A. Davis, and W. R. Koski
1983 A Distinctive Large Breeding Population of Ringed Seal (*Phoca hispida*) Inhabiting the Baffin Bay Pack Ice. *Arctic* 36:162–173.
Fischbach, A. S., S. C. Amstrup, and D. C. Douglas
2007 Landward and Eastward Shift of Alaskan Polar Bear Denning Associated with Recent Sea Ice Changes. *Polar Biology* 30:1395–1405.
Frost, D. J., L. F. Lowry, G. Pendleton, and H. R. Nute
2004 Factors Affecting the Observed Densities of Ringed Seals, *Phoca hispida*, in the Alaskan Beaufort Sea, 1996–99. *Arctic* 57:115–128.
Geist, V.
1990 Agriculture Versus Bison in Canada's Wood Buffalo National Park. *Conservation Biology* 4:345–346.
1998 *Deer of the World: Their Evolution, Behavior, and Ecology*. Stackpole Books, Mechanicsberg, PA.
Gentry, R. L.
1998 *Behavior and Ecology of the Northern Fur Seal*. Princeton University Press, Princeton.
Gifford-Gonzalez, D., S. D. Newsome, P. L. Koch, T. P. Guilderson, J. J. Snodgrass, and R. K. Burton
2005 Archaeofaunal Insights on Pinniped-Human Interactions in the Northeastern Pacific. In *The Exploitation and Cultural Importance of Sea Mammals*, edited by G. Monks, pp. 19–38. Oxbow Books, Oxford.
Grebmeier, J., J. E. Overland, S. E. Moore, E. V. Farley, E. C. Carmack, L. W. Cooper, K. E. Frey, J. H. Helle, F. A. McLaughlin, and S. L. McNutt
2006 A Major Ecosystem Shift in the Northern Bering Sea. *Science* 311:1461–1464.
Guthrie, R. D.
2004 Radiocarbon Evidence of Mid-Holocene Mammoths Stranded on an Alaskan Bering Sea Island. *Nature* 429:746–749.
Hammill, M. O., and T. G. Smith
1991 The Role of Predation in the Ecology of the Ringed Seal in Barrow Strait, Northwest Territories, Canada. *Marine Mammal Science* 7:123–135.

Harington, C. R.
2008 The Evolution of Arctic Marine Mammals. *Ecological Applications* 18 (Suppl.):S23–S40.

Hodgetts, L.
2005 Dorset Palaeoeskimo Harp Seal Exploitation at Phillip's Garden (EeBi-1), North-western Newfoundland. In *The Exploitation and Cultural Importance of Sea Mammals,* edited by G. Monks, pp. 63–76. Proceedings of the 9th Conference of the International Council for Archaeozoology, Durham. Oxbow Books, Oakville, CT.

Holst, M., I. Stirling, and W. Calvert
1999 Age Structure and Reproductive Rates of Ringed Seals (*Phoca hispida*) on the Northwest Coast of Hudson Bay in 1991 and 1992. *Marine Mammal Science* 15:1357–1364.

Holst, M., I. Stirling, and K. A. Hobson
2001 Diet of Ringed Seals (*Phoca hispida*) on the East and West Sides of the North Water Polynya, Northern Bafffin Bay. *Marine Mammal Science* 17:888–908.

Keeley, E. R., E. A. Parkinson, and E. B. Taylor
2007 The Origins of Ecotypic Variation of Rainbow Trout: A Test of Environment vs. Genetically Based Differences in Morphology. *Journal of Evolutionary Biology* 20:725–736.

Kenyon, K. W., and F. Wilke
1953 Migration of the Northern Fur Seal, *Callorhinus ursinus. Journal of Mammalogy* 34:86–98.

Kingsley, M. C. S., I. Stirling, and W. Calvert
1985 The Distribution and Abundance of Seals in the Canadian High Arctic, 1980–1082. *Canadian Journal of Fisheries and Aquatic Sciences* 42:1189–1210.

Knecht, R., and R. Davis
2001 A Prehistoric Sequence for the Eastern Aleutians. In *Recent Archaeology in the Aleut Zone of Alaska,* edited by D.E. Dumond, pp. 269–288. University of Oregon Anthropological Papers No. 58, University of Oregon, Eugene.
2004 *Amaknak Bridge Site Data Recovery Project Final Report.* Submitted to the Museum of the Aleutians, Unalaska and Alaska Department of Transportation, Anchorage, Contract No. MGS-STP-BR-0310(S)/52930. Copies available from the Museum of the Aleutians, Unalaska.
2008 The Amaknak Bridge Site: Cultural Change and the Neoglacial in the Eastern Aleutians. *Arctic Anthropology* 45:61–78.

Krafft, B. A., K. M. Kovacs, A. K. Frie, T. Haug, and C. Lydersen
2006 Growth and Population Parameters of Ringed Seals (*Pusa hispida*) from Svalbard, Norway, 2002–2004. *ICES Journal of Marine Science* 63:1136–1144.

Kraftt, B. A., K. M. Kovacs, and C. Lydersen
2007 Distribution of Sex and Age Groups of Ringed Seals *Pusa hispida* in the Fast-Ice Breeding Habitat of Kongsfjorden, Svalbard. *Marine Ecology Progress Series* 335:199–206.

Labansen, A. L., C. Haug, and K. M. Kovacs
2007 Spring Diet of Ringed Seals (*Phoca hispida*) from Northwestern Spitsbergen, Norway. *ICES Journal of Marine Science* 64:1246–1256.

Laidre, K. L., I. Stirling, L. F. Lowry, Ø. Wiig, M. P. Heide-Jørgensen, and S. H. Ferguson
2008 Quantifying the Sensitivity of Arctic Marine Mammals to Climate-Induced Habitat Change. *Ecological Applications* 18(2, Suppl.):S97–S125.

Leimar, O., P. Hammerstein, and T. J. M. Van Dooren
2006 A New Perspective on Developmental Plasticity and the Principles of Adaptive Morph Determination. *The American Naturalist* 167:367–376.

Lønne, O. J., and B. Gulliksen
1989 Size, Age and Diet of Polar Cod, *Boreogadus saida* (Lepechin 1773), in Ice Covered Waters. *Polar Biology* 9:187–191.

Lydersen, C., and I. Gjertz
1986 Studies of the Ringed Seal (*Phoca hispida* Schreber 1775) in its Breeding Habitat in Kongsfjorden, Svalbard. *Polar Research* 4:57–63.

Lyman, R. L.
1988 Zoogeography of Oregon Coast Marine Mammals: The Last 3000 years. *Marine Mammal Science* 4:247–264.
1989 Seal and Sea Lion Hunting: A Zooarchaeological Study from the Southern Northwest Coast of North America. *Journal of Anthropological Archaeology* 8:68–99.

Mason, O. K., and V. Barber
2003 A Paleo-Geographic Preface to the Origins of Whaling: Cold Is Better. In *Indigenous Ways to the Present: Native Whaling in the Western Arctic,* edited by A. P. McCartney, pp. 69–107. Canadian Circumpolar Institute Press, Edmonton.

Mizuno, A. W., M. Onuma, M. Takahashi, and N. Ohtaishi
2003 Population Genetic Structure of the Spotted Seal (*Phoca largha*) Along the Coast of Hokkaido, Based on Mitochondrial DNA Sequences. *Zoological Science* 20:783–788.

Møhl, J.
1979 Description and Analysis of the Bone Material from Nugarsuk: An Eskimo Settlement Representative of the Thule Culture in West Greenland. In *Thule Eskimo Culture: An Anthropological Retrospective,* edited by A. P. McCartney, pp. 380–394. Mercury Series, Archaeology Survey of Canada Paper No. 88. National Museum of Man, Ottawa.

Morin, P. A., R. G. Leduc, K. M. Robertson, N. M. Hedrick, W. F. Perrin, M. Etnier, P. Wade, and B. L.Taylor
2006 Genetic Analysis of Killer Whale (*Orcinus orca*) Historical Bone and Tooth Samples to Identify Western U.S. Ecotypes. *Marine Mammal Science* 22:897–909.
Moss, M. L., D. Y. Yang, S. D. Newsome, C. F. Speller, I. McKechnie, A. D. McMillan, R. J. Losey, and P. L. Koch
2006 Historical Ecology and Biogeography of North Pacific Pinnipeds: Isotopes and Ancient DNA from Three Archaeological Assemblages. *Journal of Island and Coastal Archaeology* 1:165–190.
Murray, M. S.
2008 Zooarchaeology and Arctic Marine Mammal Biogeography, Conservation and Management. *Ecological Applications* 18 (Suppl):S41–S55.
Newsome, S. D., M. A. Etnier, D. Gifford-Gonzalez, D. L. Phillips, M. van Tuinen, E. A. Hadly, D. P. Costa, D. J. Kennett, T. P. Guilderson, and P. L. Koch
2007 The Shifting Baseline of Northern Fur Seal Ecology in the Northeast Pacific Ocean. *Proceedings of the National Academy of Sciences USA* 104:9709–9714.
Nowak, R. M.
2003 *Walker's Marine Mammals of the World.* John's Hopkins University Press, Baltimore.
O'Corry-Crow, G. M., and R. L. Westlake
1997 Molecular Investigations of Spotted Seals (*Phoca largha*) and Harbor Seals (*P. vitulina*), and Their Relationship in Areas of Sympatry. In *Molecular Genetics of Marine Mammals*, edited by A.E. Dizon, S.J. Chivers, and W.F. Perrin, pp. 291–304. Society for Marine Mammalogy Special Publication 3, Lawrence, KS.
Overland, J. E., and P. J. Stabeno
2004 Is the Climate of the Bering Sea Warming and Affecting the Ecosystem? *EOS (Transactions of the American Geophysical Union)* 85(33):309–316.
Peterson, R. S., B. J. Le Boeuf, and R. L. Delong
1968 Fur Seals from the Bering Sea Breeding in California. *Nature* 219: 899–901.
Pitman, R. L.
2004 Good Whale Hunting. *Natural History* December 2003/January 2004 24–28.
Pitman, R. L., and P. Ensor
2003 Three Forms of Killer Whales (*Orcinus orca*) in Antarctic Waters. *Journal of Cetacean Research Management* 5:131–139.
Pitman, R. L., W. L. Perryman, D. LeRoi, and E. Eilers
2007 A Dwarf Form of Killer Whale in Antarctica. *Journal of Mammalogy* 88:43–48.
Razjigaeva, N. G., T. A. Grebennikova, L. A. Ganzey, L. M. Mokhova, and V. B. Bazarova
2004 The Role of Global and Local Factors in Determining the Middle to Late Holocene Environmental History of the South Kuril and Komandar Islands, Northwestern Pacific. *Palaeogeography, Palaeoclimatology, Palaeoecology* 209:313–333.
Reeves, R. R, B. S. Stewart, P. J. Clapham, and J. A. Powel
2002 *National Audubon Society's Guide to Marine Mammals of the World.* Alfred A. Knopf, New York.
Rigor, I. G., and J. M. Wallace
2004 Variations in the Age of Arctic Sea-Ice and Summer Sea-Ice Extent. *Geophysical Research Letters* 31:L09401.
Shaughnessy, P. D., and F. H. Fay.
1977 A Review of the Taxonomy and Nomenclature of North Pacific Harbour Seals. *Journal of Zoology (London)* 182:385–419.
Simpkins, M. A., L. M. Hiruki-Raring, G. Sheffield, J. M. Grebmeier, and J. L. Bengtson
2003 Habitat Selection by Ice-Associated Pinnipeds near St. Lawrence Island, Alaska in March 2001. *Polar Biology* 26:577–586.
Sinervo, B., Miles, D. B., Frankino, W. A., Klukowski, M., and D. F. DeNardo
2000 Testosterone, Endurance, and Darwinian Fitness: Natural and Sexual Selection on the Physiological Basis of Alternative Male Behaviors in Side-Blotched Lizards. *Hormones and Behavior* 38:222–233.
Skúlason, S., and T. B. Smith
1995 Resource Polymorphisms in Vertebrates. *Trends in Ecology and Evolution* 10:366–370.
Smith, T. B., and S. Skúlason
1996 Evolutionary Significance of Resource Polymorphisms in Fishes, Amphibians, and Birds. *Annual Review of Ecology and Systematics* 27:111–133.
Smith, T. G., and M. O. Hammill
1981 Ecology of the Ringed Seal, *Phoca hispida*, in Its Fast Ice Breeding Habitat. *Canadian Journal of Zoology* 59:966–981.
Smith, T. G., M. O. Hammill, and G. Taugbøl
1991 A Review of the Development, Behavioural and Physiological Adaptations of the Ringed Seal, *Phoca hispida*, to Life in the Arctic Winter. *Arctic* 44:124–131.
Stirling, I.
2002 Polar Bears and Seals in the Eastern Beaufort Sea and Amundsen Gulf: A Synthesis of Population Trends and Ecological Relationships over Three Decades. *Arctic* 55 (Suppl. 1):59–76.
Stirling, I., and N. A. Øritsland
1995 Relationships between Estimates of Ringed Seal (*Phoca hispida*) and Polar Bear (*Ursus mari-*

timus) Populations in the Canadian Arctic. *Canadian Journal of Fisheries and Aquatic Sciences* 52:2594–2612.

Storå, J.

2000 Skeletal Development in the Grey Seal *Halichoerus grypus*, the Ringed Seal *Phoca hispida botnica*, the Harbour Seal *Phoca vitulina vitulina*, and the Harp Seal *Phoca groenlandica:* Epiphyseal Fusion and Life History. *Archaeozoologia* XI:199–222.

Storå, J., and L. Lõugas

2005 Human Exploitation and History of Seals in the Baltic during the Late Holocene. In *The Exploitation and Cultural Importance of Sea Mammals*, edited by G. Monks, pp. 95–106. Proceedings of the 9th Conference of the International Council for Archaeozoology, Durham. Oxbow Books, Oakville, CT.

Testa, J. W. (editor)

2007 Fur Seal Investigations, 2004–2005. U.S. Department of Commerce, NOAA Technical Memorandum NMFS-AFSC-174. Springfield, VA.

Tezanos-Pinto, G., C. S. Baker, K. Russell, K. Martien, R. W. Baird, A. Hutt, G. Stone, A. A. Mignucci-Giannoni, S. Caballero, T. Endo, S. Lavery, M. Oremus, C. Olavarría, and C. Garrigue

2008 A Worldwide Perspective on the Population Structure and Genetic Diversity of Bottlenose Dolphins (*Tursiops truncatus*) in New Zealand. *Journal of Heredity* in press.

Trites, A. W., and G. A. Antonelis

1994 The Influence of Climatic Seasonality on the Life Cycle of the Pribilof Northern Fur Seal. *Marine Mammal Science* 10:311–324.

Wessen, G.

2006 Archaeological Activities Associated with the Construction of the Quileute Senior Center, (45CA23) La Push, Washington. Prepared for Quileute Housing Authority, La Push, WA.

Wiig, O., A. E. Derocher, and S. E. Belikov

1999 Ringed Seal (*Phoca hispida*) Breeding in the Drifting Pack Ice of the Barents Sea. *Marine Mammal Science* 15:595–598.

Wilson, M. C.

1996 Late Quaternary Vertebrates and the Opening of the Ice-Free Corridor, with Special Reference to the Genus *Bison*. *Quaternary International* 32:97–105.

Woollett, J. M., A. S. Henshaw, and C. P. Wake

2000 Palaeoecological Implications of Archaeological Seal Bone Assemblages: Case Studies from Labrador and Baffin Island. *Arctic* 53:395–413.

5

A 4500-Year Time-Series of Otariid Abundance on Sanak Island, Western Gulf of Alaska

Matthew W. Betts, Herbert D.G. Maschner, and Veronica Lech

ARCHAEOLOGICAL REMAINS REVEAL that Steller sea lions (*Eumetopias jubatus*) and northern fur seals (*Callorhinus ursinus*) were crucially important to the subsistence strategies of most prehistoric Aleut (e.g., Crockford et al. 2004; Tews 2005; Yesner 1981, 1988). This relationship developed over millennia and was mediated both by "natural" factors characteristic of all predator-prey systems, and by social factors that became entangled in the hunt and its related products.

The hunting of otariids in the western Gulf of Alaska (GOA) was fundamentally disrupted by the insertion of Aleuts in the Russian fur trade in the 18th century and the subsequent entry of Aleut into the "world system." It continued to be altered by the introduction of state and federal legislations governing the harvesting of otariids, both for commercial and subsistence-level usages. As noted by Etnier (2004:90–91), management policies currently attempt to strike a careful balance between threats to otariid populations (caused by recent climate change and human harvesting) and the perceived impact of otariids on commercial fish stocks. The harvesting methods and limits set by various government legislation are profoundly different from those that otariids in the GOA have adapted to over the last 6000 years.

Therefore, the modern era witnesses a profoundly different sort of relationship between humans and these animals, which has altered both the ancient predator-prey dynamic and the cultural aspects associated with their harvest. As the behavior of their predators (and predation rates) have fundamentally changed, so to have otariid populations themselves. For example, analysis of archaeological data suggests that modern northern fur seal ranges, breeding strategies, and feeding habits are drastically different than those of prehistoric populations (Newsome et al. 2007). Studies utilizing biological data collected during the past half-century (e.g., Kruse et al. 2001; Trites et al.

2007; Winship and Trites 2006) indicate that otariid populations in the Gulf of Alaska have decreased dramatically in size in the late 20th century, with proposed causes ranging from climate change, to overfishing of prey, to trophic cascades caused by the destruction of whale populations from commercial harvests.

As Newsome et al. (2007; see also Etnier 2004) have demonstrated, developing management policies for these species is hampered by the "shifting baselines syndrome" (Pauly 1995), where our understanding of sustainability and stock health can be biased by the limited time depth of biological and ecosystem data. Simply put, the shifting baselines syndrome suggests "modern" data series are inadequate, because they are derived from populations under pressure from "unnatural" modern harvesting, high climatic variability, and significant habitat change caused by industrial and commercial impacts on ecosystems.

It is therefore a *conditio sine qua non* that to develop adequate conservation management policies, we must incorporate a longer-term perspective that considers ancient otariid population variability and their adaptations to predators, prey, and environmental fluctuations prior to the modern era. Recent efforts have attempted to incorporate a longer-term perspective in the assessment of otariid stocks through the analysis of archeological records (e.g. Etnier 2002, 2004; Lyman 2003b; Newsome et al. 2007; Trites et al. 2007; this volume). This type of conservation zooarchaeology (see papers in Lyman and Cannon 2004a for an overview) is a burgeoning field in archaeofaunal studies. At this early stage of the discipline's development, the general goal is to produce the necessary pre-modern population data, while at the same time exploring the appropriate methods and theory essential to incorporate these new data into conservation research.

In this chapter, we assume these basic objectives, and investigate the history of changing otariid populations in the western Gulf of Alaska, while at the same time discussing the methodological and theoretical issues necessary to interpret these shifts within a conservation biology framework (for related discussion see Lyman 2006; Lyman and Cannon 2004b). Specifically, we establish a baseline record of shifting otariid abundance in the Western Gulf of Alaska, which we assess against changes in human behavior, otariid behavior, and climatic regimes. We also discuss the use of zooarchaeological quantification techniques and related foraging theory for interpreting variability in the frequencies of prehistoric animal populations.

This analysis is conducted on a high-resolution faunal record from Sanak Island, western Gulf of Alaska, which spans approximately 4500 years. We utilize the classification otariid, meaning belonging to the family Otariidae, throughout the chapter. We have chosen this coarse family-level taxon as a unit of analysis because we sometimes experienced problems distinguishing between juvenile and/or fragmentary Steller seal lion and northern fur seal (*Callorhinus ursinus*) remains. We chose to combine the two taxa in our analysis primarily to increase sample sizes (see discussion below), although we note that the vast majority of the remains are likely Steller sea lion.

THEORETICAL AND METHODOLOGICAL BACKGROUND

There are many different quantification techniques utilized by zooarchaeologists to measure the relative abundances of taxa in faunal assemblages. The literature on quantification procedures is vast (see Grayson 1984; Lyman 1994, 2008), and we will not attempt to develop an in-depth discussion here. Instead, we focus our analysis on a potentially useful tool in conservation zooarchaeology, the abundance index (AI). This measure is useful for tracking shifts in the relative abundance of prehistoric animal populations because it can be linked to foraging theory in a way that allows inferences to be drawn about the impacts of exploitation pressure, environmental change, and prey behavior (see discussion below) on their encounter rate (and absolute abundance). We also note the measure is believed to be relatively robust and generally resistant to taphonomic and collection biases, as long as these biases are systemic to all the assemblages being compared (Ugan and Bright 2001).

Abundance indices have seen increasing use among zooarchaeologists to track changes in the return and encounter rates of prey through time (e.g., Bayham 1979; Broughton 1994a, 1994b, 1997, 1999; Butler, 2000; Cannon, 2000; Grayson, 2001; Hildebrandt and Jones, 1992; Janetski, 1997; Lupo 2007; Lyman, 2003a, 2003b; Nagaoka 2001, 2002, 2006), and are beginning to penetrate into conservation zooarchaeology research (e.g. Darwent and Darwent 2004; Lyman 2003a, 2003b; Maschner et al. 2008). While the index is widely used among zooarchaeologists, its theoretical relationship to foraging theory is sometimes questioned by nonspecialists (as several peer reviews and symposia discussant overviews have made clear to us), and hence we feel it appropriate to provide an in-depth discussion of the technique and its associated theoretical relationship to foraging theory.

An abundance index (AI) is a simple measure, representing a normed ratio of a "highly ranked" taxon (A) to a "lower-ranked" taxon (B), measured as $AI = A/A + B$. When calculated, index values more than 0.5 indicate greater frequencies of the higher-ranked taxon than the lower-ranked taxon, while values less than 0.5 indicate greater frequencies of the lower-ranked taxon than the more highly ranked taxon. Here "rank" is typically based on body size, with the assumption that large-bodied animals contain more calories and therefore have higher return rates than smaller-bodied prey (e.g., Broughton 1994b, 1997, 1999; Grayson 2001).

The abundance index is often linked to changes in foraging efficiency and return rate via the prey choice, or diet breadth, model (McArthur and Pianka 1966). The prey choice model (see Stephens and Krebs 1986; Lupo 2007), assumes that foragers search for all prey simultaneously and take resources whenever they are encountered. Diet breadth in this situation is determined by ranking resources by the amount of calories they provide per unit of effort (typically time) and "by adding resources into their diet in rank order from highest to lowest until the return rate per unit of time is maximized" (Lupo 2007:147; see also MacArthur and Pianka 1966). According to the model, foragers determine prey rank by taking into account the natural abundance of taxa (their encounter rate), their body size (e.g., calories), their post-encounter pursuit costs, and processing costs. Determining the return rates of individual taxa is a difficult process, especially in prehistoric contexts (e.g. Betts 2000), so zooarchaeologists typically rank animals based on body size (Lyman 2003b:377), with the assumption that the largest animals will donate the largest amount of calories and therefore should be the most highly ranked in terms of the overall foraging strategy.

The prey choice model assumes that prey is taken in proportion to its natural abundance—the so-called proportional hunting hypothesis (see Betts 2008:139; Yesner 1981:150). When encounter rates for large-bodied prey decline, a reduction in the overall return rate will occur, assuming that the abundances and handling times of the low-ranked prey in the diet breadth are not fluctuating. The model predicts that a reduction in highly ranked prey will result in the increased procurement of lower-ranked taxa in the diet breadth independent of the abundance of the low-ranked prey. This occurs because as encounters with the large-bodied prey decrease, encounters with more lowly ranked prey must increase, assuming a constant foraging return is the goal (e.g., Lupo 2007:147).

Therefore, decreasing proportions of large-bodied prey and increasing proportions of small bodied prey in a faunal assemblage when considered in tandem should imply a reduction in (1) the encounter rate of large-bodied taxa, (2) a decline in return rate, and (3) a decline in overall foraging efficiency. This interpretation of declining AIs as reflecting a decline in the absolute abundance of a high-ranked taxa pervades the literature (e.g. Broughton, 1994b, 1997,

1999; Butler, 2000; Cannon, 2000; Grayson, 2001; Hildebrandt and Jones, 1992; Nagaoka 2002) and is often attributed to human exploitation pressure. However, Wolverton (2005) has also used AIs to investigate the effects of climate on changing abundance.

However, it is important to note that a decline in the proportion of a large-bodied taxon in a faunal assemblage may not necessarily imply a decline in the abundance of the living population of that taxon. Even when carefully calculated, a decline in an AI only implies a decline in encounter rate, or the probability that a high-ranked taxon will be encountered during a foraging excursion (Betts and Friesen 2006; Lyman 2003b). Declines in encounter rates, also known as resource depressions (e.g., Charnov et al. 1976), can result from a decline in the abundance of prey, but they also may be caused by changes in their behavior, the behavior of their predators, or changes in their preferred habitat. Charnov et al. (1976:247) characterize a suite of predator-prey interactions that can affect encounter rates, and thus foster a resource depression (see also Betts and Friesen 2006:74; Butler 2001:96, Lupo 2007:160).

An *exploitation depression* is perhaps the most direct predator-linked type of resource depression; these are caused when predators reduce the number of prey to a point where encounter rates decline. Exploitation depressions are often linked to increases in the size of predator populations, although in human systems this can occur when changes in technology substantially reduce handling, search, or pursuit times. A *microhabitat depression* can occur when prey change their geographic distribution. These changes in distribution may be a response to predation, or might be caused by natural factors such as climatic and geomorphological changes that affect migrations or breeding areas. An *ecological depression* can occur when mortality increases because of non-predation-related factors such as availability of food, or climatic factors that limit survivability, fecundity, or recruitment (e.g., temperature, ice cover, etc.). Finally, a *behavioral depression* can occur when prey actively alter encounter rates by learning or developing behaviors that reduce their interactions with predators. Such reductions in encounter rates typically occur when predator-naive populations develop new avoidance behaviors when introduced to new predators (or new predator behaviors).

In his analysis of shifting otariid abundances on the Oregon coast, Lyman (2003b) makes a relevant distinction between local populations and metapopulations and their potential effects on encounter rates (and abundance indices). Local populations can be described as those groups of animals encountered within a set foraging zone around an archaeological site, which are donating the bulk of faunal remains to the archaeological deposits being analyzed. Metapopulations are animal populations in the region surrounding the site made up of multiple local populations. Metapopulations continually colonize areas when local populations decline, or accept individuals when local populations decolonize (or increase in size). As demonstrated by Lyman (2003b), shifting AI values may reflect both local and metapopulation responses to human or natural stimuli. For example, a resource depression caused by behavioral responses to predation or microhabitat shifts is most likely to affect local populations, rather than metapopulations. Declining encounter rates associated with these processes will not necessarily be linked with declines in the abundance of the metapopulation (and could even occur if the metapopulation increased significantly in size). However, ecological depressions and exploitation depressions are likely to affect both local populations and metapopulations (the latter assuming constant recolonization of local areas by the metapopulation). In the analysis that follows, we consider both the local and the metapopulation in our interpretation of changing otariid abundances.

THE ARCHAEOFAUNAL SAMPLE

The faunal database presented here is derived from eight discrete shell midden contexts that

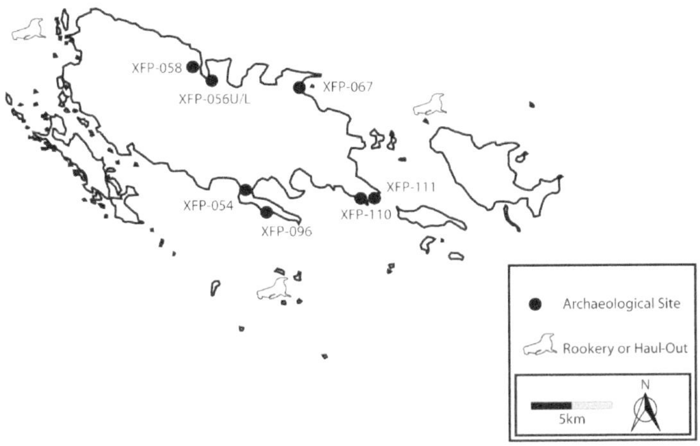

FIGURE 5.1. Sanak Island, indicating locations of sites and otariid rookeries/haul-outs.

span a period from ca. 4550 to 400 cal BP, with a relatively even temporal distribution (Figure 5.1). Excavations were conducted by trowel, and deposits were screened through 8 mm mesh, a screen gauge size we consider appropriate to specimen and element size for medium and large mammals in this region. Total sample sizes for all vertebrate remains range between 426 and 8339 NISP (Number of Identified Specimens). Dates utilized represent the mean calibrated date for each context or the pooled means derived from multiple assays; all dates were obtained from wood charcoal found in stratigraphic association with the mammal bone assemblages. The samples are excellently preserved, with relatively minimal weathering, burning, and post-depositional fragmentation (although predepositional fragmentation was high). Crucially, unlike many northern faunal assemblages, carnivore gnawing was exceptionally rare in all the faunal assemblages from Sanak Island, indicating that this taphonomic agent is not a significant factor affecting taxonomic frequencies.

We identified the marine mammal remains in this study by comparison to reference specimens loaned by the National Marine Mammal Laboratory (NMML), collections from Idaho State University and the Idaho Museum of Natural History, and collections housed at the Burke Museum. Regardless, in the case of juveniles or heavily fragmented bone, we sometimes ran into difficulty differentiating between Steller sea lion (*Eumetopias jubatus*) and northern fur seal (*Callorhinus ursinus*) remains. Although we note that adult northern fur seals appear to occur in very low frequencies in the middens, we have taken a cautionary approach by grouping all remains in the family Otariidae. Despite the limitations posed by the broad taxonomic classification, these data establish a baseline for understanding the relationship between otariids and the phocids and sea otters that were also harvested by peoples, a strategy used effectively by researchers in other regions (e.g., Broughton 1999; Lyman 2003a, 2003b).

We controlled for seasonality in the samples by choosing only semi-subterranean sites with middens in which analysis of migratory fauna indicated a cold season occupation. A 10 km foraging radius around all sites includes modern-day sea lion rookeries and haul-outs, which are located at Clubbing Rocks, Cherni Island, and Seal Rock, respectively (Figure 5.1). We tested whether the number of otariid remains in an assemblage (%NISP) were related to the distance of an archaeological site from the nearest haulout or rookery (distance measured in meters), but there was no statistical relationship ($r^2 = 0.0003$). Table 5.1 displays the faunal frequencies from Sanak Island.

ANALYSIS AND RESULTS

When calculating abundance indices, care must be taken in the selection of the "B" variable, or

TABLE 5.1

Mammalian Faunal Frequencies from Sanak Island

Number of Identified Specimens

SCIENTIFIC NAME	COMMON NAME	XFP-111	XFP-54	XFP-67	XFP-96	XFP-58	XFP-056 UPPER	XFP-056 LOWER	XFP-110
Rodentia	Rodentia			1	5	4			
Spermophilus parryii	Arctic ground squirrel	1							
Microtus sp.	Lemming/vole							3	
Cetacea	Cetacea	8	14	15	16	16	5	65	19
Canidae	Canidae					2			
Vulpes vulpes	Red fox			1	35	5		5	
Enhydra lutris	Sea otter	92	1	101	156	229	49	174	32
Otariidae	Otariidae	1	23	50	56	81	34	37	4
Callorhinus ursinus	Northern fur seal			7	10	14		8	3
Eumetopias jubatus	Steller sea lion		2	7	4	12	4	12	22
Odobenus rosmarus	Walrus		1						
Phocidae	Phocidae	35	8	89	92	176	111	106	17
Phoca sp.	Phoca sp.	9				8		37	20
Phoca vitulina	Harbor seal		1	4	2	3	2	1	3

Phoca hispida	Ringed seal	1							
Erignathus barbatus	Bearded seal			1					
Cervidae	Cervidae							1	
Rangifer tarandus	Caribou			2		2			
Total Identified Mammal		147	50	278	376	552	205	449	120
Unidentified Mammal		884	52	123	109	350	137	633	17
MEAN CALIBRATED DATE		2550 BC	1750 BC	590 BC	295 BC	80 BC	520 AD	1030 AD	1540 AD
Otariidae + Phocidae		46	34	157	164	294	151	201	69
Otariidae – Phocidae AI		0.021739	0.735294	0.407643	0.426829	0.363946	0.251656	0.283582	0.42029
Otariidae + Sea otter		93	26	165	226	336	87	231	61
Otariidae – Sea otter AI		0.010753	0.961538	0.387879	0.309735	0.318452	0.436782	0.246753	0.47541
Otter + Phocidae		137	10	171	250	371	162	318	72
Sea otter – Phocidae AI		0.671533	0.1	0.567251	0.624	0.579515	0.302469	0.54717	0.444444
Otariidae + Marine mammal		138	35	259	320	523	200	375	101
Otariidae – Marine mammal AI		0.007246	0.714286	0.247104	0.21875	0.204589	0.19	0.152	0.287129

lower-ranked taxon, to avoid the issue of "closed arrays" (e.g., Grayson, 1984; Lyman 2008), or the possibility that the bivariate proportional measure we are using is actually tracking shifts in the frequencies of the comparison taxa rather than shifts in the taxon of interest. Ugan and Bright (2001:1313), in their consideration of this effect, suggest that changes in the encounter rate (i.e., frequency) of B should not violate the assumptions of the prey model (i.e., AI values should still reflect real changes in the encounter rate of taxon A) so long as the changes in encounter rates for the low-ranked taxa change at rates slower than those of the high-ranked taxa. Furthermore, if we assume a goal of constant returns, any decline/increase in the higher-ranked taxon should, theoretically, be linked to a respective increase/decrease in the procurement (and thus archaeological abundance) of the lower-ranked taxon. Because of how abundance indices are calculated, these relationships would result in an amplification of the magnitude of changes in the index, but not a change in direction. Betts and Friesen (2006) argue that because of the ultimate uncertainty in understanding the mechanisms behind changes in the encounter rate of the low-ranked taxon and possible changes in encounter rate caused by cultural processes (such as intensification), it is impossible to ascertain which encounter rates (A or B) are actually changing. Their solution is to create multiple abundance indices across multiple taxa and to assess overall trends (Betts and Friesen 2006:71).

In this chapter, we utilize three separate abundance indices to assess shifts in the abundance of otariids. The first, an otariid–phocid (OP) index, calculated as OP = NISP Otariidae/ NISP Otariidae + NISP Phocidae, and second, an otariid–sea otter (OSO) index, calculated as OSO = NISP Otarridae/NISP Otariidae + NISP sea otter, track relationships between otariids and individual lower-ranked taxa (i.e., phocids and sea otters). The third index, an otariid–marine mammal (OMM) index, calculated as OMM = NISP Otariidae/NISP Otariidae + NISP sea otter + NISP Phocidae, monitors general shifts between otariids and all exploited marine mammals. Table 5.1 outlines the calculation of these indices and the species represented by the larger family-level classifications.

We chose the first two indices because they incorporated taxa that, although lower ranked than otariids, played very different roles in Aleut foraging strategies. Otariids, phocids, and sea otters are ubiquitous in Sanak middens, and hence were likely consistent members of the prehistoric diet breadth. However, their position and importance in terms of their return rate (caloric return based on searching, pursuit, and processing costs) was significantly different. Otariids were likely the highest-ranked noncetacean marine mammal in the vicinity of Sanak. They are very large pinnipeds (adults averaging 660 kg for Steller sea lion and 117.25 kg for northern fur seal), and the presence of local rookeries and haul-outs made them the most easily accessible taxa in local waters (while phocids are present in low densities in the nearshore areas around the island, their breeding colonies are located far offshore from Sanak Island; there are no known walrus rookeries or haul-outs in the vicinity of Sanak). Not only could a single otariid provide a vast quantity of meat and fat, their skins were crucial for manufacturing waterproof clothing and watercraft (Veniaminov 1984; Liapunova 1996). To put it simply, otariids were a keystone taxa for the inhabitants of Sanak Island.

Seals in the Phocidae family were also likely highly ranked taxa, although their smaller relative body size (average weight 78.125 kg for *Phoca* and *Pusa* genus seals) and small concentrations on local rookeries and haul-outs would substantially lower their average encounter rate. Lowest-ranked among the noncetacean marine mammals were sea otters, whose small size (average weight 27.7 kg), low-calorie meat, and low-fat anatomy (see discussion in Banfield 1974: 344–345), greatly reduces their return rate. Ethnographic observations provide some indication of their relative importance in Aleut diets, suggesting that their exploitation occurred only in times of desperation, a fact that likely corre-

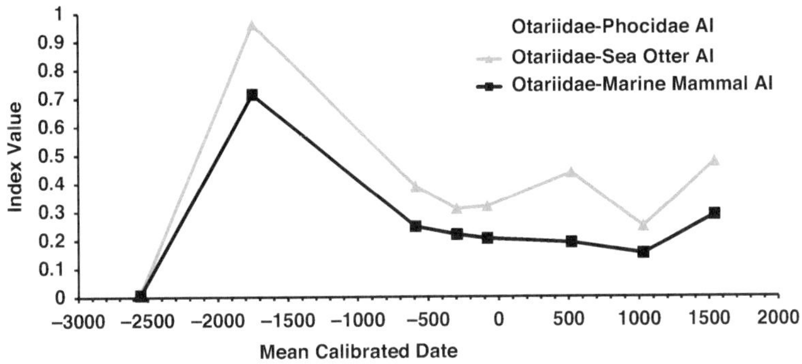

FIGURE 5.2. Abundance indices (AIs) for otariids in the Sanak Island assemblages.

lates with the low-calorie nature (e.g., low fat) and poor taste of their meat (Veniaminov 1984; see also Lech et al. this volume).

Figure 5.2 displays the multiple abundance indices we used to track frequencies of otariids in our sequence. The three indices correspond favorably (with one notable exception at AD 520), suggesting that all indices are tracking similar variations in the frequencies of otariid remains in the assemblages. In general, we see a drastic increase in otariid frequencies during the first millennium of occupation on Sanak, followed by a 500-year period of relative stasis. Frequencies then decline at AD 1030, but increase at AD 1540. The increase in abundance at AD 520 indicated by the OSO index is difficult to assess without additional data, and we speculate on the nature of this discordance later in the chapter.

Despite obvious variability, two significant declines in the abundance index are displayed by Figure 5.2. The first occurs between 1750 BC and 590 BC, and the second occurs between AD 520 and 1030. According to the prey choice model, these decreases should be interpreted as reductions in otariid encounter rates, and therefore resource depressions. Because otariids were likely the highest-ranked marine mammal (behind cetaceans) within the diet breadth of the inhabitants of Sanak, these two shifts likely represent a reduction in overall foraging efficiency. Concomitantly, the two very large increases in otariid abundance indices between 2550 and 1750 BC and AD 1030 and 1540 likely represent significant increases in otariid encounter rates, and therefore overall foraging efficiency. As outlined above, understanding the mechanisms responsible for the variability in otariid encounter rates must involve a consideration of both local populations and metapopulations, and their potential reactions to both predator behaviors and environmental factors.

VARIABILITY IN THE METAPOPULATION • *Climate and Human Demography*

To assess the relationship between palaeoclimatic changes and otariid abundances, we created a climatic model for the western Gulf of Alaska and Sanak Island. These climatic reconstructions (see summary in Table 5.2) are based on major changes in atmospheric and sea surface temperature, storm frequency and intensity, and productivity as reconstructed from a variety of sources (Calkin et al. 2001; D'Arrigo et al. 2005; Finney et al. 2002; Hu et al. 2001; Huesser et al. 1985, Jordan and Krumhardt 2003; Loso et al. 2006; Mann 2001; Mann and Hamilton 1995; Mason and Jordan 1993; Misarti 2007). From the perspective of the marine ecosystem, these perturbations may be referred to, very generally, as macro-level regime shifts, with their attendant relationships between productivity, ocean circulation, and effects on animal populations (e.g., Benson and Trites 2002; Polovina 2005). In general, cooler, more stormy regimes should be correlated with increased

TABLE 5.2

Palaeoclimatic Model for the Gulf of Alaska and Sanak Archipelago

PERIOD	DATE RANGE	APPROXIMATE CLIMATIC CONDITIONS
Early Holocene	7000–4200 BC	Cool and dry
	4200–2550 BC	Warm and wet
Neoglacial	2550–1200 BC	Cool and wet
	1200–100 BC	Cool and wet (increased storminess)
Pre-Medieval Warm Period	100 BC–AD 300	Warm
	AD 300–900	Cold
Medieval Warm Period	AD 900–1250	Warm and dry
Little Ice Age	AD 1250–1850	Cold and wet
Modern Era	AD 1850–Present	Very warm

SOURCE: Calkin et al. 2001; D'Arrigo et al. 2005; Finney et al. 2002; Hu et al. 2001; Huesser et al. 1985, Jordan and Krumhardt 2003; Loso et al. 2006; Mann 2001; Mann and Hamilton 1995; Mason and Jordan 1993; Misarti 2007.

oceanic productivity (Chavez et al. 2003:l), which will have positive impacts on top tier predators, such as otariids (e.g., Trites et al. 2007). Sustained warming should be linked to a decline in oceanic productivity, which should affect top-tier predators negatively.

Figure 5.3 represents the OP, OSO, and OSM indices projected against our climatic model. Here the light grey boxes represent periods of cooler and stormier palaeoenvironments, which were generally more productive, while the dark grey boxes represent warmer and generally less productive paleoenvironments. Interestingly, the climatic model is partially discordant—it correlates well with part of the sequence, but not all of it. For example, a relatively clear correlation occurs between our climate reconstruction and the post-Neoglacial sequence, that is, after ca. 100 BC. Specifically, we see a slight increase in otariid abundance in a cold period ca. AD 300 to 900 (if the OSO index is representative), a more abrupt decline in otariid abundance during the Medieval Warm Period ca. AD 900 to 1250, and a large increase in otariid abundance during the Little Ice Age ca. AD 1250 to 1850 (see Table 5.1 for general climatic patterns). We interpret this correspondence as otariid populations responding to climate and marine productivity shifts. The lack of correspondence between otariid shifts and the Neoglacial period suggest that other factors may be influencing otariid abundance during the beginning of the sequence, and we explore these possibilities below.

We believe the correlation between otariid abundance and climate in the more recent points in the sequence represents the response of the metapopulation to overall paleoclimatic conditions. During warmer and less productive climatic regimes, encounter rates for otariids declined, indicating resource depression. Because of the correlation with climate, we interpret these decreases as classic ecological depression (e.g., Charnov et al. 1976) of regional populations in the western Gulf of Alaska. In contrast, the increased otariid remains in middens during cold regimes can be interpreted as increased encounter rates, likely due to increases in the GOA otariid metapopulations as a result of increased oceanic productivity.

We can test these climate–otariid relationships by comparing the AI sequences against a model of changes in human populations. As top-tier predators, we expect that maritime

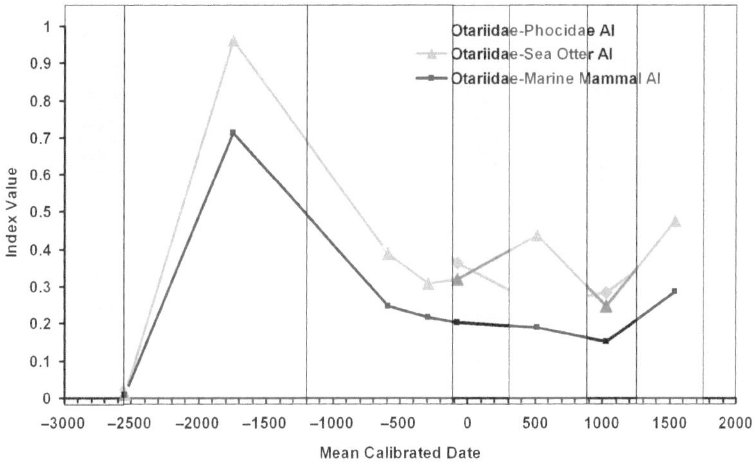

FIGURE 5.3. Otariid abundance compared to climatic model. Light gray boxes represent cold and productive oceanic environments, while dark gray boxes represent warm and less-productive environments.

hunter-gatherers should respond to shifts in oceanic productivity in a similar manner as otariids. Reconstructing human demographics from archaeological records is fraught with uncertainty, but changing village sizes, the number of houses per village, house sizes, and midden depths can provide relative proxies for population shifts (see Maschner et al. 2009a, for a detailed overview of the Sanak population reconstruction). Utilizing these indicators, the archaeological record indicates that after the initial colonization of Sanak at ca. 2550 BC (during the Neoglacial period) human populations were low and relatively stable. Three peaks in human population can be identified after this period, when the number of settlements, the size of villages, and midden depths increase relative to previous periods: ca. 200 BC to AD 100; ca. AD 300 to 1000; and ca. AD 1300 to 1740. The latter increase in population was extreme, by hunter-gatherer standards, as reflected in very large multiroom houses, multimeter midden depths, and very large villages. While the record suggests populations declined in the times between these population "peaks," our radiocarbon data indicates a continuous occupation on Sanak from ca. 2550 BC. Indeed, only one period, beginning ca. AD 1000 and lasting to approximately AD 1300, suggests a substantial depopulation event, as reflected in the low village numbers, small village sizes, and virtual lack of shell-middens (in fact, the data point used in the above sequence represents the only known shell midden on Sanak during this period). This evidence corresponds well with our understanding of the archaeological record on the western Alaska Peninsula, which could be described as a virtually depopulated landscape during this period (Maschner et al. 2009b).

Comparing our schematic population model (Figure 5.4) against otariid abundance produces interesting correlations between otariid abundance and human population levels on Sanak. While the trend during the Neoglacial is again murky, post-Neoglacial there is a positive correspondence between the number of humans and the number of otariids. An important corollary of this relationship is that it appears that humans were not overexploiting otariids in the past, at least on Sanak Island. If they were, declines in otariids should be associated with an increase in the number of predators (humans), which as described above, is the hallmark of a classic exploitation depression. In fact, the exact opposite is evident here; human and sea lion populations appear to respond to climate changes in exactly the same fashion, at least during the

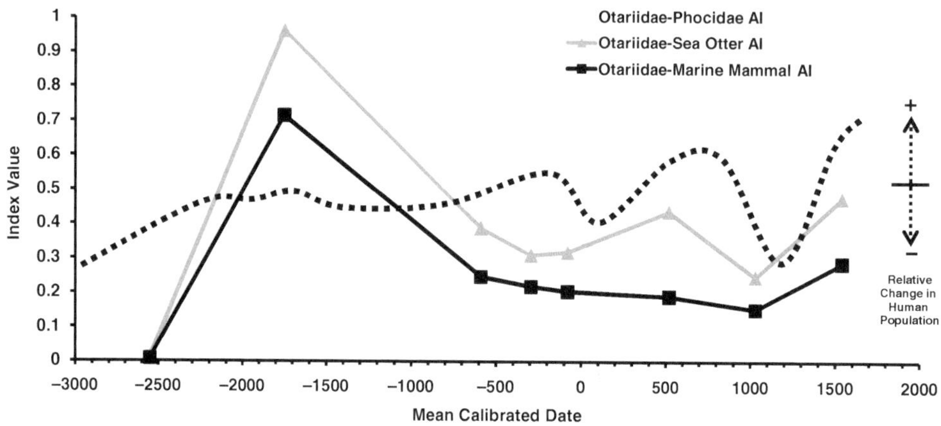

FIGURE 5.4. Otariid abundance compared to human population model (dashed line). Human population model is derived from analysis in Maschner et al. (2009a).

post-Neoglacial period. For example, during the Medieval Warm Period, both human populations and otariid abundance decline, while during the Little Ice Age, both human populations and otariid abundance increase. This strongly implies that changes in climate and oceanic productivity were a significant force in the population histories of large-bodied, top-tiered predators, which includes both humans and sea lions, in the western Gulf of Alaska.

PREDATOR/PREY BEHAVIOR • *Shifts in the Local Population*

Our analysis above implies that the declines and increases in otariid frequencies at ca. AD 1030 and 1540, respectively, were a response of the otariid metapopulation to environmental shifts—shifts that also appear to have affected human populations. We cannot identify a correlation between climatic variability and otariid abundance during the Neoglacial portion of the sequence, leading us to suspect that fluctuations in abundance may be a response of local otariid populations to human predation. One means to understand the fluctuations in the otariid abundance indices during the Neoglacial is to consider the changes in return/encounter rate that they imply from the perspective of Charnov et al.'s (1976) ecological model of resource depressions.

The first point in our sequence (Figure 5.2), dated to 2550 BC, was recovered from one of the oldest village sites on Sanak Island and represents the earliest known shell midden on the island. We know that coastlines were rebounding (isostatic rebound) during this period, as sites of similar age were once located on small islands that have since become part of Sanak Island as ocean-levels dropped (James Jordan 2008, personal communication). This is important because it suggests that some of the local rookeries and haul-outs near Sanak, which are currently located on low-lying rock outcrops and islands, may have been underwater at this early period. It is possible, therefore, that the very low otariid frequencies during this period represent an instance where local rookeries and haul-outs had not been established, because the land for them did not yet exist. If Sanak inhabitants had encountered a "predator-naïve" resident otariid population when they settled Sanak (ca. 3550 BC), encounter rates (and thus otariid abundance) should have been much higher than indicated in the 2550 BC assemblage.

The second point in the sequence is difficult to interpret, as it represents the period of highest relative otariid abundance in any of the middens. One possible explanation is that it represents an era where otariids first colonized the Sanak archipelago, representing the opposite of what Charnov et al. (1976) call a microhabitat depression. The relationship of this colonization to the abundance of otariid metapopulations remains unclear. We know from our climatic model (Figure 5.3) that oceanic productivity should have been high at this period, and therefore the size of the metapopulation might have been increasing. However, if this was a period where attractive locations for rookeries and haul-outs were first created due to a rebounding coastline, it might simply represent a colonization event rather than an increase in the size of the metapopulation. In any case, when otariids first established rookeries and haul-outs near Sanak, they were likely quickly exploited by Sanak inhabitants. Such a "predator-naïve" population, in the sense that it was adapting behaviorally to both a new environment and new predator pressures, would be much easier to exploit than a predator-wary population, and encounter rates could have been very high (Charnov et al. 1976:248). Whatever the case, we suggest that this increase in encounter rate may be due to a microhabitat shift in the local otariid population surrounding Sanak Island, caused by the establishment of rookeries and haulouts at locations recently created by lowering sea levels.

Such an explanation, which relies on prey behavior as a mechanism, may also account for the drastic decline in otariid abundance at 590 BC. This may be what Charnov et al. (1976) call a behavioral depression, where animal populations learned to avoid human predation, making them more difficult to hunt, and thus reducing return rates. Whether this was primarily a behavioral response or represented a slight relocation of otariids to new rookeries (therefore representing a microhabitat depression of sorts) is difficult to assess. We know that Steller sea lions and fur seals are wary of humans today, and that to some extent the location of their rookeries and haul-outs is mediated by human proximity. In the absence of any evidence for a demographic, climatic, or critical technological change in the archaeological record of Sanak, which may have influenced these trends, a behavioral depression is our preferred hypothesis.

We suspect that the actual process of learning to avoid human predation would have occurred much more rapidly than the period represented by our samples (ca. 1200 years). Regardless of when it occurred, the next ca. 1000 years of otariid abundances represented by our samples indicates a relatively stable encounter rate, a period of relative balance between local otariid populations and human exploitation—perhaps the closest thing to an equilibrium state that we can reconstruct from available archaeological data. Unfortunately, we have no idea what the absolute abundance of humans or sea lions was during this period, nor any absolute indication of the intensity of exploitation of these animals. What we can say is that our climatic and human demographic models, when compared to this portion of the sequence, suggest that otariid metapopulations were neither declining nor increasing during this time. If so, this may be a period in which human predation rates and otariid recruitment rates had reached a point of equilibrium. Lyman (2003b:385) makes a similar interpretation of a flat AI curve from a site on the Oregon coast. In that example, the period of stability also occurs after a period of increasing and then decreasing encounter rates (Lyman 2003b:figure 5.4). In both cases, it is significant that this "balanced" predator/prey relationship appears to occur only after a period where local otariid populations must have been adapting behaviorally to human predation. If the resource depressions indicated by these data are the result of a behavioral depression, then this implies that the following period of balanced predation rates and recruitment rates was only achieved by significant adaptation of otariid populations to human predation.

RELATIONSHIPS WITH SEA OTTER PROCUREMENT

From the framework of the prey choice (diet breadth) model, the fluctuation in the abundance of otariids, one of the most highly ranked taxa in the Aleut diet, would have had impacts on

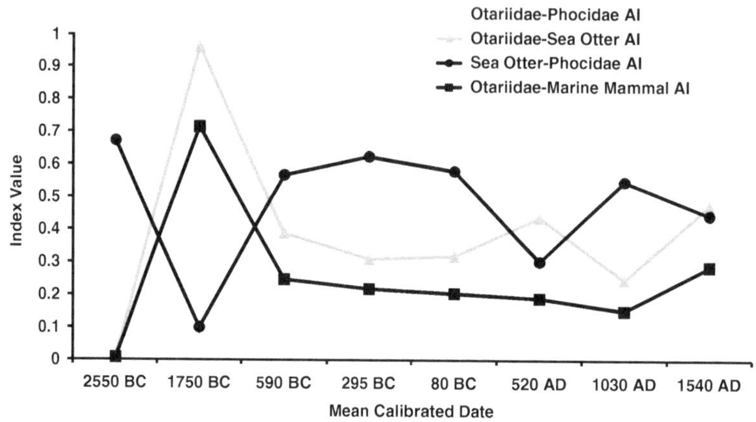

FIGURE 5.5. Otariid abundance compared to sea otter abundance.

the procurement of lower ranked taxa. In Figure 5.5 we compare our otariid data to an abundance index for sea otters (*Enhydra lutris*), calculated as SP = NISP sea otter/NISP Sea otter + NISP Phocidae. What is clear from the graph is that there is an inverse relationship between otariid abundance and sea otter abundance over the entire sequence. This sequence represents the impact that fluctuations in the abundance of otariids had on the procurement of sea otters from local populations surrounding Sanak Island. Because sea otters are significantly smaller bodied than otariids (see discussion above) and are lower ranked, we interpret this correspondence to reflect a shift in emphasis in marine mammal exploitation during times when otariid encounter rates declined. When otariid abundance was low, the procurement of sea otters increased to negotiate any energy deficits (assuming a goal of constant caloric return). This interpretation appears to be corroborated by evidence of a positive correlation between the frequency of cut marks on sea otter carcasses and their abundance in middens (Lech et al. this volume). The assumption here is that in times of stress, Aleut more intensively exploited animal carcasses to maximize energy return.

CONCLUSIONS

In the preceding analysis, we attempted to explain variability in otariid abundance over 4500 years on Sanak Island in relation to both climatic shifts and human predation. We believe the variability in the otariid abundance indices are best explained by the effect of climate and human predation on both local populations and metapopulations (e.g., Lyman 2003b). In the first part of the sequence, the wildly fluctuating abundance of otariids is likely related to the colonization by otariids and subsequent exploitation by humans of local rookeries and haul-outs, and the response of local otariid populations to human predation. Once both human and local otariid populations had adapted to each other, the sequence begins to reflect the response of otariid metapopulations to climatic fluctuations.

Our evidence suggests that humans and local populations of otariids developed a hard-fought natural predator/prey equilibrium over a period of about 2000 years (the actual process was likely much shorter, but our data cannot map it), after which both responded to climatic changes in the same manner. If our interpretation of the initial part of the sequence is correct, humans may be viewed as having a significant influence on the ecology of otariids for at least 4500 years on Sanak Island. If similar relationships developed in other areas of the western Gulf of Alaska where we have early evidence of otariid exploitation, then we might assume that the entire post-glacial history of otariid populations involved humans as a significant predator—put simply, the history of otariid populations cannot be understood without humans as a significant force.

The correspondence between otariid abundance and climate is an important outcome of this research. Specifically, during cold and productive climatic regimes, it appears that otariid abundance increased in Sanak middens, while otariid abundance decreased during warm and less productive oceanic climate regimes. The identification of otariid resource depressions during these warm periods is intriguing given the possible link between climatic warming and the decline of the western Steller sea lion in the modern era (for discussion, see Kruse et al. 2001). Given the above analysis, it seems clear that global warming will be a complicating factor in developing management policies for otariids in the western Gulf of Alaska.

Yet the revelation that ancient otariids developed significant behavioral responses to humans in the Western Gulf of Alaska (see Lyman 2003b and Etnier 2004 for a similar interpretation on the Washington and Oregon coasts) is also significant. What might have happened to Steller sea lions and northern fur seals when this significant exploitation pressure was removed? Furthermore, how are otariids responding to the shifts in harvesting patterns in the modern era? These questions are beyond the scope of this chapter, and they will require the integration of archaeological, historical, anthropological, and modern biological data to explore effectively. However, our present analysis both provides a premodern baseline that can aid in the development of modern management policies, while raising several issues that might frame research questions for conservation biologists.

ACKNOWLEDGMENTS

This research was supported by a grant from the National Science Foundation and support from Idaho State University and the Canadian Museum of Civilization. We gratefully acknowledge the support and access provided by the Sanak Corporation and the Pauloff Harbor Tribe. We would especially like to thank the many undergraduate and graduate students who helped with the laboratory analyses used in this paper, including Amber Tews and Julie Kramer. Finally, we would like to thank the book's editors, Todd Braje and Torbin Rick, for inviting us to participate in a wonderful session at the SAA, and for their good natured support through the editorial process for this volume.

REFERENCES CITED

Banfield, A.
1974 *The Mammals of Canada.* University of Toronto Press, Toronto

Bayham, F.
1979 Factors Influencing the Archaic Pattern of Animal Utilization. *Kiva* 44:219–235.

Benson, A. J., and A. W. Trites
2002 Ecological Effects of Regime Shifts in the Bering Sea and Eastern North Pacific. *Fish and Fisheries* 3:95–113.

Betts, M.
2000 *Opportunism, Optimality, and the Origin of Species: An Investigation of Thule Subsistence on Herschel Island, Yukon Territory.* Unpublished Master's thesis, Department of Anthropology, University of Toronto, Canada.
2008 *Subsistence and Culture in the Western Canadian Arctic: A Multicontextual Approach.* Mercury Series 169, Canadian Museum of Civilization, Gatineau

Betts, M., and T. M. Friesen
2006 Declining Foraging Returns from an Inexhaustible Resource? Abundance Indices and Beluga Whaling in the Western Canadian Arctic. *Journal of Anthropological Archaeology* 25:59–81.

Broughton, J.
1994a Late Holocene Resource Intensification in the Sacramento Valley, California: The Vertebrate Evidence. *Journal of Archaeological Science* 21:501–514.
1994b Declines in Mammalian Foraging Efficiency during the Late Holocene, San Francisco Bay, California. *Journal of Anthropological Archaeology* 13:371–401.
1997 Widening Diet Breadth, Declining Foraging Efficiency, and Prehistoric Harvest Pressure: Ichthyofaunal Evidence from the Emeryville Shellmound, California. *Antiquity* 71:845–862.
1999 *Resource Depression and Intensification during the Late Holocene, San Francisco Bay: Evidence from the Emeryville Shellmound Vertebrate Fauna.* University of California Anthropological Records, 32.

Butler, V.
2000 Resource Depression on the Northwest Coast of North America. *Antiquity* 74:649–661.

2001 Changing Fish Use on Magaia, Southern Cook Islands: Resource Depression and the Prey Choice Model. *International Journal of Osteoarchaeology* 11:88–100.

Calkin P. E., G. Wiles, and D. J. Barclay
2001 Holocene Coastal Glaciation of Alaska. *Quaternary Science Reviews* 20:449–461.

Cannon, M.
2000 Large Mammal Relative Abundance in Pithouse and Pueblo Period Archaeofaunas from Southwestern New Mexico: Resource Depression Among the Mimbres-Mogollon? *Journal of Anthropological Archaeology* 19:317–347.

Charnov, E., G. Orians, and K. Hyatt
1976 Ecological Implications of Resource Depression. *The American Naturalist* 110:247–259.

Chavez, F. P., J. Ryan, S. E. Lluch-Cota, and C. M. Ñiquen
2003 From Anchovies to Sardines and Back: Multidecadal Change in the Pacific Ocean. *Science* 10:217–221.

Crockford, S., G. Frederick, R. Wigen, and I. McKechnie
2004 *Analysis of the Vertebrate Fauna from Amaknak Bridge, Unalaska, AK, UNL050* (Final report). Report on File, Museum of the Aleutians, Unalaska, AK.

D'Arrigo, R., E. Mashig, D. Frank, R. Wilson., and G. Jacoby
2005 Temperature Variability over the Past Millennium Inferred from Northwestern Alaska Tree Rings. *Climate Dynamics* 24:227–236.

Darwent, C.M. and J. Darwent
2004 Where the Muskox Roamed: Biogeography of Tundra Muskox (*Ovibos moschatus*) in the Eastern Arctic. In *Zooarchaeology and Conservation Biology*, edited by R. L. Lyman and K. P. Cannon, pp. 61–87. University of Utah Press, Salt Lake City.

Etnier, M. A.
2002 *The Effects of Human Hunting on Northern Fur Seal (Callorhinus ursinus) Migration and Breeding Distributions in the Late Holocene.* Unpublished Ph.D. Dissertation, Department of Anthropology, University of Washington.
2004 The Potential of Zooarchaeological Data to Guide Pinniped Management Decisions in the Eastern North Pacific. In *Zooarchaeology and Conservation Biology*, edited by R. L. Lyman and K. P. Cannon, pp. 88–102. University of Utah Press, Salt Lake City.

Finney, B. P., I. Gregory-Eaves, M. S. V. Douglas, and J. P. Smol
2002 Fisheries Productivity in the Northeastern Pacific Ocean over the Past 2200 years. *Nature* 416:729–733.

Grayson, D. K.
1984 *Quantitative Zooarchaeology*. Academic Press, Orlando.
2001 Explaining the Development of Dietary Dominance by a Single Ungulate Taxon at Grotte XVI, Dordogne, France. *Journal of Archaeological Science* 28:115–125.

Hildebrandt, W., and T. Jones
1992 Evolution of Marine Mammal Hunting: A View from the California and Oregon Coasts. *Journal of anthropological Archaeology* 11:360–401.

Hu, F.S., E. Ito, T. A. Brown, B. B. Curry, and D. R. Engstrom
2001 Pronounced Climatic Variations in Alaska during the Last Two Millennia. *Proceedings of the National Academy of Sciences* 98(19): 10552–10556.

Huesser, C. J., L. E. Huesser, and D. M. Peteet
1985 Late-Quaternary Climate Change on the American North Pacific Coast. *Nature* 315:485–487

Janetski, J.
1997 Fremont Hunting and Resource Intensification in the Eastern Great Basin. *Journal of Archaeological Science* 24:1075–1088.

Jordan, J. W., and A. Krumhardt
2003 Postglacial Climate and Vegetation on the Western Alaska Peninsula. *Alaska Journal of Anthropology* 1:16–33

Kruse, G. H., M. Crow, E. Krygier, D. Lloyd, K. Pitcher, L. Rea, M. Ridgway, R. Small, J. Stinson, and K. Wynne
2001 *A Review of Proposed Fishery Management Actions and the Decline of Steller Sea Lions (Eumetopias jubatus) in Alaska: A Report by the Alaska Steller Sea Lion Restoration Team.* Regional Information Report No. 5J01-04. Alaska Department of Fish and Game, Division of Commercial Fisheries.

Loso, M. G., R. S. Anderson, S. P. Anderson, and P. J. Reimer
2006 A 1500-Year Record of Temperature and Glacial Response Inferred from Varved Iceberg Lake, Southcentral Alaska. *Quaternary Research* 66(1): 12–24.

Liapunova, R. G.
1996 *Essays on the Ethnography of the Aleuts: At the End of the Eighteenth and the First Half of the Nineteenth Century.* The Rasumson Library Historical Translation Series IX. University of Alaska Press, Fairbanks, AK.

Lupo, K. D.
2007 Evolutionary Foraging Models in Zooarchaeological Analysis: Recent Applications and Future Challenges. *Journal of Archaeological Research* 15:143–189.

Lyman, R. L.
1994 *Vertebrate Taphonomy*. Cambridge Manuals in Archaeology. Cambridge University Press, Cambridge.
2003a The Influence of Time Averaging and Space Averaging on the Application of Foraging Theory in Zooarchaeology. *Journal of Archaeological Science* 30:595–610.
2003b Pinniped Behavior, Foraging Theory, and the Depression of Metapopulations and Nondepression of a Local Population on the Southern Northwest Coast of North America. *Journal of Anthropological Archaeology* 22:376–388.
2006 Paleozoology in the Service of Conservation Biology. *Evolutionary Anthropology* 15:11–19.
2008 *Quantitative Paleozoology*. Cambridge University Press, New York.

Lyman, R. L., and K. P. Cannon (editors)
2004a *Zooarchaeology and Conservation Biology*. University of Utah Press, Salt Lake City.

Lyman, R. L., and K. P. Canon
2004b Applied Zooarchaeology, Because It Matters. In *Zooarchaeology and Conservation Biology*, edited by R. L. Lyman and K. P. Cannon, pp. 1–24. University of Utah Press, Salt Lake City.

MacArthur, R. H., and E. R. Pianka
1966 On Optimal Use of a Patchy Environment. *American Naturalist* 100:603–609.

Mann D. H.
2001 Climate during the Past Millennium. *Weather* 56:91–102.

Mann, D. H., and T. D. Hamilton
1995 Late Pleistocene and Holocene Paleoenvironments of the North Pacific Coast. *Quaternary Science Reviews* 14:449–471.

Maschner, H. D. G., M. Betts, K. Reedy-Maschner, and A. Trites
2008 A 4500-Year Time Series of Pacific Cod (*Gadus macrocephalus*): Archaeology, Regime Shifts, and Sustainable Fisheries. *Fishery Bulletin* 106(4): 386–394.

Maschner, H.D.G., M. Betts, J. Cornell, J. Dunne, B. Finney, N. Huntly, J. Jordan, A. King, N. Misarti, K. Reedy-Maschner, R. Russell, A. Tews, S. Wood, and B. Benson.
2009a An Introduction to the Biocomplexity of Sanak Island, Western Gulf of Alaska. *Pacific Science* 64(4): 673–708.

Maschner, H. D. G., B. Finney, J. Jordan, K. L. Reedy-Maschner, A. Tews, N. Misarti, G. Knudsen
2009b Did the North Pacific Ecosystem Collapse in AD 1250? In *The Northern World AD 900–1400*, edited by Herbert Maschner, Owen Mason, and Robert McGhee, pp. 33–57. University of Utah Press, Salt Lake City.

Mason, O. K., and J. W. Jordan
1993 Heightened North Pacific Storminess during Synchronous Late Holocene Erosion of Northwest Alaska Beach Ridges. *Quaternary Research* 40:55–69

Misarti, N
2007 *Six Thousand Years of Change in the Northeast Pacific: An Interdisciplinary View of Maritime Ecosystems*. Ph.D. Dissertation, University of Alaska, Fairbanks.

Nagaoka, L.
2001 Using Diversity Indices to Measure Changes in Prey Choice at the Shang River Mouth Site, Southern New Zealand. *International Journal of Osteoarchaeology* 11:101–111.
2002 Explaining Subsistence Change in Southern New Zealand Using Foraging Theory Models. *World Archaeology* 34:84–102.
2006 Prehistoric Seal Carcass Exploitation at the Shag Mouth site, New Zealand. *Journal of Archaeological Science* 33:1474–1482.

Newsome, S., M. Etnier, D. Gifford-Gonzalez, D. Phillips, M. van Tuinen, E. Hadly, D. Costa, D. Kennett, T. Guilderson, and P. Koch
2007 The Shifting Baseline of Northern Fur Seal Ecology in the Northeast Pacific Ocean. *Proceedings of the National Academy of Science* 104: 9709–9714

Pauly, D.
1995 Anecdotes on the Shifting Baseline Syndrome of Fisheries. *Trends in Ecology and Evolution* 10:430.

Polovina, J. J.
2005 Climate Variation, Regime Shifts, and Implications for Sustainable Fisheries. *Bulletin of Marine Science* 76:233–244.

Stephens, D., and J. Krebs
1986 *Foraging Theory*. Princeton University Press, Princeton.

Tews, A. M.
2005 *Prehistoric Eastern Aleut Subsistence: 5000 Years of Faunal Variations*. Unpublished master's thesis, Department of Anthropology, Idaho State University.

Trites, A. W., A. J. Miller, H. D. G. Maschner, M. A. Alexander, S. J. Bograd, J. A. Calder, A. Capotondi, K. O. Coyle, E. Di Lorenzo, B. P. Finney, E. J. Gregr, C. E. Grosch, S. R. Hare, G. L. Hunt, J. Jahncke, N. B. Kachel, H. J. Kim, C. Ladd, N. J. Mantua, C. Marzban, W. Maslowski, R. Mendelssohn, D. J. Neilson, S. R. Okkonen, J. E. Overland, K. L. Reedy-Maschner, T. C. Royer, F. B. Schwing, A. J. Wang, and J. X. L. Winship
2007 Bottom-Up Forcing and the Decline of Steller Sea Lions (*Eumetopias jubatus*) in Alaska:

Assessing the Ocean Climate Hypothesis. *Fisheries Oceanography* 16:46–67.
Ugan, A., and J. Bright
2001 Measuring Foraging Efficiency with Archaeological Faunas: The Relationship between Relative Abundance Indices and Foraging Returns. *Journal of Archaeological Science* 28:1309–1321.
Veniaminov, I.
1984 *Notes on the Islands of the Unalaska District* [1840]. L. T. Black and R. H. Goeghegan (trans.), R. A. Pierce (ed.). Alaska History, No. 27. Kingston: The Limestone Press.
Winship, A., and A. Trites
2006 Risk of Extirpation of Steller Sea Lions in the Gulf of Alaska and Aleutian Islands: A Population Viability Analysis Based on Alternative Hypotheses for Why Sea Lions Declined in Western Alaska. *Marine Mammal Science* 22:124–155.
Wolverton, S. R.
2005 The Effects of the Hypsithermal on Prehistoric Foraging Efficiency in Missouri. *American Antiquity* 70:91–106.
Yesner, D. R.
1981 Archaeological Applications of Optimal Foraging Theory: Harvest Strategies of Aleut Hunter-Gatherers. In *Hunter-Gatherer Foraging Strategies: Ethnographic and Archaeological Analyses,* edited by B. Winterhalder and E.A. Smith. pp.148–170. University of Chicago Press, Chicago.
1988 Effects of Prehistoric Human Exploitation on Aleutian Sea Mammal Populations. *Arctic Anthropology* 25:28–43.

6

An Analysis of Seal, Sea Lion, and Sea Otter Consumption Patterns on Sanak Island, Alaska

AN 1800-YEAR RECORD ON ALEUT CONSUMER BEHAVIOR

Veronica Leon, Matthew W. Betts, and Herbert D. G. Maschner

THE SANAK ISLAND BIOCOMPLEXITY project investigates the roles that humans played in North Pacific ecosystems during the Middle-to-Late Holocene. The underlying premise of the project is that prehistoric peoples, with traditional technology, were significant forces in marine ecosystems and, like all taxa, actively engineered environments to both their advantage—and potentially to their detriment (Maschner et al. 2008, n.d.). Excavations of shell middens on Sanak Island have generated a 4500-year time-series of marine mammal remains. An analysis of otariid and sea otter remains (Figure 6.1; Betts et al. this volume) suggests that the frequencies of Steller sea lions, northern fur seals, and sea otters fluctuated over time, likely in response to both natural resource depressions and the adaptation of local populations to human predation (see also Tews 2005).

Since body size plays an important role in carcass processing decisions, fluctuations in the abundance of large-bodied marine mammals must have had impacts on the butchery and transport of their carcasses. In general, large animals must be butchered to allow for easier transport, storage, and consumption, while smaller animals are often transported whole for storage and consumption (e.g., Bettinger 1991; Binford 1981; Gifford-Gonzalez 1989; Lyman 1991, 1992a). The prey choice model predicts that as the availability of large-bodied taxa decline, predation on smaller-bodied and lower-ranked taxa increases (Stephens and Krebs 1986). This change in foraging efficiency and return rate must have affected decisions regarding

Human Impacts on Seals, Sea Lions, and Sea Otters: Integrating Archaeology and Ecology in the Northeast Pacific, edited by Todd J. Braje and Torben C. Rick.

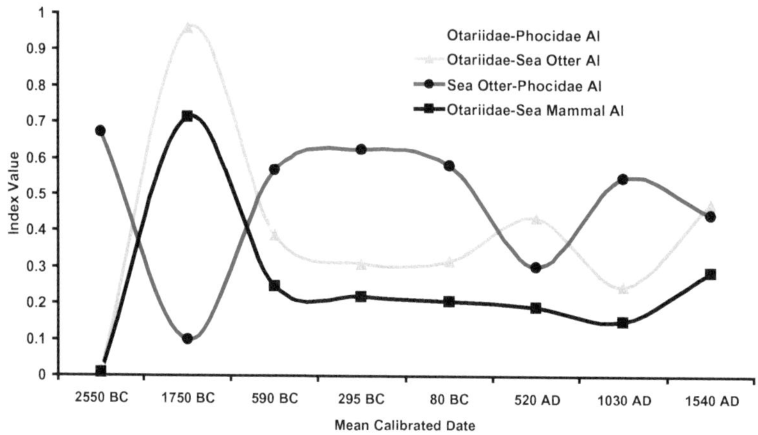

FIGURE 6.1. Otariid abundance compared to sea otter abundance (after Betts et al. this volume: Figure 5.4).

processing times of both large- and small-bodied prey. As described by Nagaoka (2006; see also Kopperl 2003; Lyman 1987, 1991; Rapson 1990), a decline in foraging efficiency and a shift toward hunting smaller-bodied prey may be associated with an increase in carcass processing time, as predators attempt to maximize energy returns through intensive processing. When large-bodied prey are abundant, and forging efficiency is high, processing times may be truncated, and the intensity of processing of animal carcasses should be lower (see discussion below).

In this chapter, we analyze element and cut mark frequencies to assess the response of Aleut processing and transport strategies to variations in marine mammal abundance. We propose that during times of declining encounter rates for large-bodied marine mammals, the inhabitants of Sanak Island exploited marine mammal carcasses more intensively, an activity that should be detectable in element distributions, cut mark location, and cut mark intensity. Frequencies of marine mammal remains in the last four temporal contexts in the Sanak Island sequence appear to be linked to climatic and population shifts (Betts et al. this volume). For this chapter we have chosen to focus our analyses on the latter half of the post-Neoglacial, where we might better speculate on the relationship between climatic changes and their impact on consumption strategies. Ultimately, our research illustrates the importance of investigating patterns of marine mammal transport and butchery as part of historical ecological analyses.

ALEUT PROCUREMENT AND PROCESSING STRATEGIES

The Aleut occupation of Sanak, characterized by evidence for the hunting of marine mammals, deep-sea fishing, intensive exploitation of intertidal zones, and semi-sedentary to sedentary villages, exhibits traits consistent with a "developed maritime culture" (Ames and Maschner 1999; Fitzhugh 2003; Price and Brown 1985; Workman and McCartney 1998). Yesner (1980; see also Maschner 1998, 1999) has previously described the propensity for Aleut costal settlements to be optimally located to benefit from multiple resource patches, similar to a "logistical," or collecting, strategy as defined by Binford (1980). On Sanak Island, this logistical site placement strategy appears to have taken three primary resource patches into account: (1) the location of marine mammal rookeries and haul-outs, (2) the location of rich intertidal resources, and (3) the location of groundfish stocks. Unlike the western Alaska Peninsula (e.g., Maschner 1998), salmon streams play a more limited role in determining site location on Sanak Island than elsewhere in the region.

Although all resources in this "triumvirate" were critical, marine mammals were perhaps fundamental to Aleut society. Not only did they supply a significant portion of the diet, they provided raw material (skin, bone, and ivory) for construction of Aleut technologies, and their hunting and processing provided a fulcrum around which status and gender-based relations of production where aligned. The hunting of marine mammals is a labor-intensive activity that requires careful scheduling and advanced planning by hunters, as well as complex technologies such as tools with points, barbs, hooks, and toggles (Johnson 2005; McCartney and Veltre 1999, Oswalt 1976; Yesner 1980). While a great deal of effort (measured as calories) is invested in an open-water, or offshore, marine mammal hunt, returns are usually high, as marine mammals are particularly fatty and provide an abundant source of calories and raw materials (e.g., Hildebrandt and Jones 1992; Johnson 2005; Yesner 1980). However, efficiently accessing these caloric and other returns requires a highly structured consumption (i.e., processing) strategy. Because significant energy is expended in the butchery of large-bodied marine mammals (e.g., the Steller sea lion), we assume that butchery techniques will change in relation to the availability of large-bodied (highly ranked) prey (e.g., Bettinger 1991; Binford 1978, 1981; Johnson 2005; Lyman 1991, 1992a, 2003,). For example, when large-bodied prey are abundant, less energy should be expended in processing individual carcasses (see discussion below), as processors have the luxury of avoiding difficult to process body sections. Concomitantly, when prey are less abundant, more-intensive processing will take place to maximize caloric return (see discussion below; see also Kopperl 2003; Nagaoka 2006).

The relationships between processing strategies and the abundance of prey should have cascading effects to other low-ranked animals in the diet breadth. As Betts et al. (this volume) describe, otariids, phocids, and sea otters were consistent members of the prehistoric diet breadth on Sanak, but their importance in terms of their return rate (caloric return based on searching, pursuit, and processing costs) was significantly different. Otariids were likely the highest-ranked noncetacean marine mammal, followed sequentially by phocid seals and sea otters (see Lyman 2003 for an in depth discussion of rank that considers sexual dimorphism; see also Johnson 2005). Of the three taxa, sea otters had a unique position in the diet breadth. In terms of return rate, their low-calorie meat and low-fat anatomy (see discussion in Banfield 1974:344–345) mean that they were likely significantly lower-ranked than even the smallest phocids.

Ethnographic accounts indicate that sea otters were exploited only during times of desperation, and were otherwise avoided because of their low-fat content and the poor taste of their meat (Laughlin 1980; Veniaminov 1984). Additionally, early ethnographic accounts (Johnson 2005; Liapunova 1996; Veniaminov 1984) reveal that sea otters were hunted using specific and much less efficient hunting methods. The Aleut believed that the sea otter was a transformed human being, and they would not use harpoons to hunt them; rather, the sea otter was tormented to the point of exhaustion and then clubbed when it could no longer dive (Black 1998; Ransom 1946). The combined effects of their anatomy and the added pursuit costs involved in their hunting meant that sea otters where likely low-ranked in comparison to other marine mammals in the Aleut diet, a fact that makes them an interesting case study both for changes in foraging strategies (see Betts et al. this volume) and shifts in consumption strategies relating to foraging shifts.

Betts et al. (this volume, Figure 5.1) constructed several abundance indices (AIs) to measure changes in the frequency of otariids and sea otters over a 4500-year sequence on Sanak Island and correlated these changes with climate and population levels. We utilize this analysis as a basic framework to assess changes in Aleut consumption patterns as reflected in element frequencies and cut mark patterns. Observing Betts et al.'s (this volume) sea otter–phocid AI, a decline in otariid populations results in an apparent increase in sea otter frequencies, illustrating substantial variability in procurement patterns over time. This inverse relationship in abundance provides an intriguing opportunity to explore the relationship between processing strategies, abundance, and resource rank in prehistoric Aleut subsistence strategies.

BUTCHERY PRACTICES AND THE ABUNDANCE OF HIGH-RANKED TAXA

Lyman (1987, 1992a) defines butchery as the human reduction and modification of an animal carcass into consumable and/or usable parts. When an individual animal is harvested, it immediately becomes a commodity, and the analysis of butchery marks, and to a lesser extent element frequencies, reveal how people utilized animals as commodities and how these commodities were turned into consumer units (portions of animals to be ingested, traded, stored, etc.). To better understand the relationship between procurement and consumption, we examined the butchery activity on skeletal remains in relation to shifts in otariid and sea otter abundances (Figure 6.1). By determining which animal carcasses were subjected to the most intense butchering, we hope to explore how butchery and consumption patterns responded to shifts in the frequency of a top-ranked resource in the middens.

Binford (1981; see also Lyman 1991) has observed three different processing activities that leave marks on bones. These marks derive from the different stages of processing an animal and normally result in a patterned distribution of marks that follow in sequential order: skinning, dismemberment, and filleting. As bones are subjected to more extensive processing, the more likely higher frequencies of cut mark activities are made obvious on the bones (Binford 1981; Fairnell and Barrett 2007; Lyman 1991, 1992a). We observed similar patterns on the marine mammal remains from Sanak; specimens that have undergone all stages of butchery (skinning to filleting) appear to show greater evidence of cut mark activity (measured as the number of independent "cut marks" on a bone) than those that have gone through only one or two stages.

Skinning is the most difficult activity to detect in cut mark analyses because there are only a few locations where the handling of the skin brings the butcher's tool in contact with the bone. The fattier the animal, the more difficult it will be to see signs of skinning, or any butchery activity for that matter, as the fat acts as a buffer between the tool and the bone (e.g., Binford 1981; Gifford-Gonzalez 1989; Lyman 1992a, 2005). Skinning is also less abrasive than disarticulation or filleting, and as a result, skinning marks are usually made at the "point of entry," as well as at locations that require cutting skins from limbs, where they will encircle the shafts of the bone. These points are most often located on the lower legs, the head, metacarpals, metatarsals, and phalanges (Binford 1981; Lyman 1991; Fairnell and Barrett 2007).

Dismemberment or disarticulation occurs during the primary stage of butchery and generally produces the most distinctive cut marks (see display of disarticulation activity in Figure 6.2). The goal of disarticulation is to butcher the animal into cuts of meat that can be more easily transported from the kill site back to the residential site, and as a result, cut marks are located at points of articulation and are cruder in form and generally more aggressive than skinning (Binford 1981).

Filleting is the most common method of preparing animal parts for consumption and is considered to be a secondary form of butchery. It usually takes place prior to transport, storage, or consumption (Binford 1981), and involves the removal of fillets of meat from "meaty" appendicular elements such as the limbs (as opposed to the axial skeleton). Filleting meat can produce two types of cut marks, long longitudinally oriented cuts (along the long axis of bones) and short oblique cuts, made to the ven-

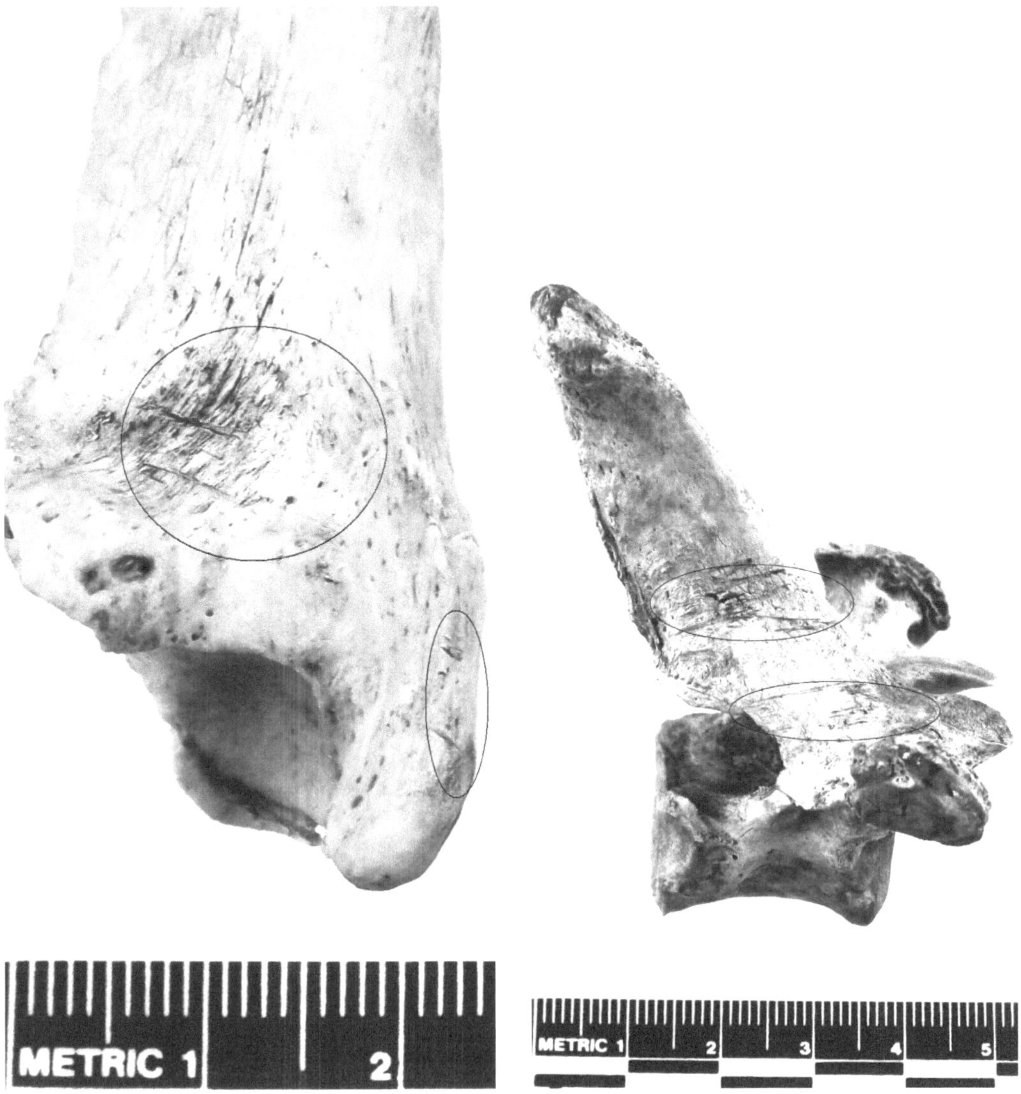

FIGURE 6.2. Cut mark activity observed on sea otter bones. The photograph on the left shows signs of disarticulation observed on a sea otter tibia from site XFP-056 Lower. The photograph on the right shows evidence of disarticulation and filleting activity on a sea otter thoracic vertebra from site XFP-056 Lower.

tral side of the bone to free it from the mass of meat and/or sever muscle insertions (Binford 1981; Lyman 1991).

How might the frequencies of these three types of cut marks be related to the abundance of top-ranked prey? Following Nagaoka (2006), we assume that when return rates decline, and the procurement of lower-ranked taxa increases, this shift should also be reflected in the intensity of exploitation of their carcasses (see detailed discussion below). That is, during times of resource stress, foragers will more intensively exploit animal carcasses to maximize energy return. For higher-ranked taxa, this should be associated with increased frequencies of filleting marks, especially on lower-utility elements. This should be especially prevalent on lower-ranked resources such as sea otters, which were presumably only taken for skins and rarely eaten (e.g. Laughlin 1980; Veniaminov 1984). Ultimately, we can link these potential processes to shifts in climate and ocean productivity by comparing our consumption patterns to shifts in the climatic model developed by Betts et al. (this volume).

METHODOLOGY

Betts et al. (this volume) provide details on the faunal database utilized in this paper. Here we provide a brief overview to provide the context for our analysis. Our samples are derived from eight discrete shell midden contexts that span a period from ca. 4500 to 400 cal BP (Betts et al. this volume, Figure 5.1), with a relatively even chronological distribution. Deposits were screened through 8 mm mesh, and analysis of the vertebrate remains indicates low levels of predepositional burning, and post-depositional fragmentation, carnivore gnawing, and weathering. Total sample sizes for all vertebrate remains range between 426 and 8339 NISP (Number of Identified Specimens, or count). Our analyses of cut marks were carried out using a low-power Luxo KFM 10× magnifier lamp.

Our samples are universally derived from cold season occupations (see Betts et al. this volume), and all are derived from sites located within a 12 to 22 km radius of known rookeries. As such, transport logistics related to season of occupation and/or distance to resource patches should have been fairly consistent between all sites (see below). Because the faunal samples were all recovered from midden contexts, we note that certain stages in the processing of carcasses may not be recoverable from these datasets. For example, if sea otters were being consumed, sea otter bones may have been discarded in other areas of the site. Some cut marks related to processing may not be present in the middens, especially if those elements ended up in floor deposits or in the fire. Therefore, cut marks relating to the final stage of consumption may be misrepresented (i.e., some, but perhaps not all, bones relating to this stage were not entering the midden). Regardless, we assume that general Aleut site-use patterns remained consistent through time, and therefore that the contexts being compared were formed by a similar set of procurement, transport, and processing behaviors.

While analysis of the tools associated with the faunal assemblages has yet to be completed, our preliminary analysis indicates that a similar suite of tool types were used to butcher all marine mammal remains from all periods. In particular, the artifacts indicate that both slate and chert cutting implements are found in association with the faunal samples on Sanak. Our analysis of the remains suggests that butchery strategies (i.e., the sequencing and location of cuts) remained remarkably consistent through time.

RESULTS

Otariids, phocids, and sea otters are the three most abundant mammalian taxa found in our samples. Figure 6.3 shows our observations of the percent cut NISP for skinning, disarticulation, and filleting for all of three taxa combined by time period. What is interesting about this graph is that, while suggesting variability, on a rank order it is remarkably consistent. As such, it documents an overall Aleut butchery strategy for marine mammals. We know that skinning should leave very low frequencies of cut marks on bone, because of the limited opportunity for cutting implements to come into contact with the skeleton (e.g., Binford 1981; Gifford-Gonzalez 1989; Lyman 1992a, 2005). If we assume that most marine mammals needed to be disarticulated before consumption, then we would expect that disarticulation marks would be universally high, as they are in Figure 6.3. We might expect filleting marks, which represent the next stage in processing, to also be low because not all butchered units may have the opportunity to be further processed, either because they spoiled or were discarded before processing could occur, or because they did not require further processing. If this is the case, then filleting, the end stage in the butchery process, should provide a relative indication of overall processing intensity.

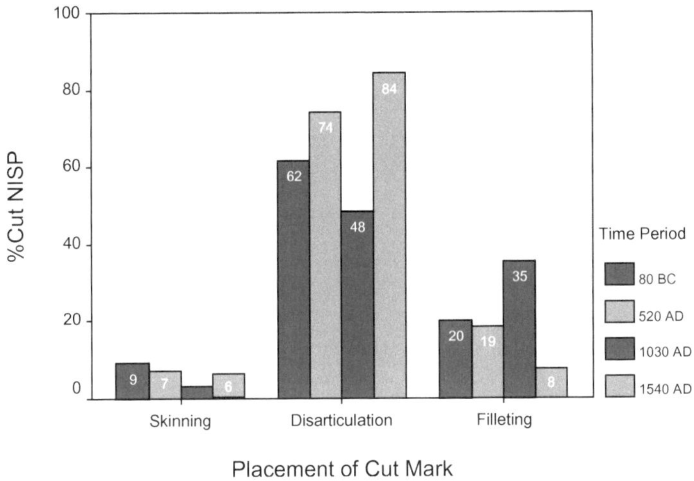

FIGURE 6.3. Bar graph displaying the %cut NISP for otariids, phocids, and sea otters by activity (skinning, disarticulation, and filleting) for the four time periods focused on for these analyses. Dark bars indicate warmer climatic periods and light bars represent cooler climatic periods.

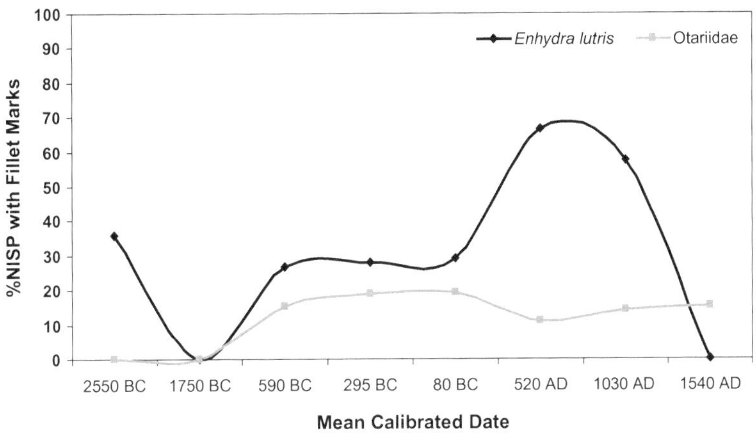

FIGURE 6.4. %NISP of all cut marked specimens that showed evidence of filleting activity for sea otter and otariids.

Across all taxa, Figure 6.3 suggests that during colder, more productive regimes, we see a relative increase in the frequency of disarticulation marks and a relative decline in filleting marks. Concomitantly, during warmer, less productive regimes, there is a decline in frequency of disarticulation and an increase in the frequency of filleting. If filleting is a good indication of butchery intensity, as we propose, then while disarticulation of carcasses increases during productive regimes when large-bodied mammals are abundant, the intensity of processing of their carcasses does not increase. In fact, Figure 6.3 suggests that increased processing is associated with warmer, less productive, regimes when the abundance of large-bodied marine mammals decline (Figure 6.1).

We believe these relationships may be especially prevalent in sea otters, the lowest-ranked taxon in the diet-breadth. Figure 6.4 shows our observations of filleting activity on sea otters throughout the 4500-year sequence. When compared to the otariid–sea otter abundance index (Figure 6.1), there is a direct association between increased sea otter abundance and the proportion of filleting marks on their bones. The only case of discordance occurs at AD 520, for reasons that are still unclear to us. Overall, however, this correspondence suggests that when otariids are more abundant in the middens, the filleting activity of sea otters declines and sea otters also become less frequent. In contrast, during periods when otariids are less abundant in middens, sea otter filleting activity increases. This increase in both the abundance of low-ranked prey (sea otters) and the processing of their carcasses is exactly what we might expect if declining otariid encounter rates reduced overall foraging efficiency and foragers responded by reorganizing their processing of carcasses to maximize energy return. Furthermore, after 80 BC, there appears to be an inverse relationship between filleting marks on otariids and filleting marks on sea otters. When filleting marks increase on sea otters, they decline on otariids, and when they decline drastically on sea otters, they increase slightly on otariids. We explore these relationships below.

OTARIID ELEMENT UTILITY

To investigate changes in otariid processing and economic transport decisions on Sanak, we used a utility index developed by Savelle et al. (1996) for otariid seals. The index, called a *meat utility index* (MUI), is based on the amount of edible meat associated with skeletal elements, as calculated from two California sea lions (family Otariidae), which we assume will be generally applicable to other otariid taxa (Steller sea lion and northern fur seal) found in the Sanak middens. We primarily utilize Savelle et al.'s (1996) *modified meat utility index* (MMUI), an index derived from the MUI but which accounts for "riders." Binford (1978) describes "riders" as low-ranked utility parts found in places where you would expect to find high-ranked body parts. Essentially, these lower-ranked parts "ride" along with higher-ranked utility elements (such as limbs) because the animal is butchered into parts or segments that include lower-ranked elements (e.g., a fibula riding along with a tibia; Lyman 1992b: Lyman et al. 1992; Metcalfe and Jones 1988). The %MMUI is calculated when the average of these flesh weights are normed on a scale of 1 to 100 (Savelle et al. 1996).

In Figure 6.5, values plotted display the %MAU frequency of otariids against %MMUI (see Savelle et al. 1996). The MAU (minimal animal unit) is calculated by dividing the MNE, or minimum number of elements (discounting sides and age), by the number of times that particular element is present in a complete skeleton. The %MAU is a normed value (Binford 1981; Savelle et al. 1996). While statistically there is no significant correlation between %MAU frequencies of otariids against %MMUI in any of our four time periods, the overall trend suggests a negative relationship across all time periods, except for the final context (XFP-110), dated to ca. AD 1540. This flattening of the utility relationship is intriguing, considering that the pattern appears to be caused by the relative increase in medium utility (skull, mandible, and scapula) elements in the middens.

Interpreting change through time from these graphs is cumbersome because of the nature of the graphing procedure. In the hope of assessing overall difference in trends in the transport of high- and low-utility elements, we amalgamated our element frequencies based on the following: (1) *low-ranked elements* with %MMUI values between 4.8% and 22.2% (the humerus, tibula and fibula, femur, radius and ulna, flippers, pelvis, and lumbar vertebra), (2) *medium ranked elements* with %MMUI values between 54.6% and 64.3% (the head, scapula, and sternum), and (3) *high-ranked elements* with %MMUI values between 94.8% and 100% (the cervical vertebra, thoracic vertebra, and ribs). We expect to see more desirable parts (those higher in utility) selected during periods of high otariid abundance and lower-ranked parts (less desirable) selected to compensate for food shortages during warmer, less productive regimes, when otariid encounter rates are low. For a complete distribution of elements into utility rank, see Table 6.1. We used an average of Savelle et al.'s (1996) %MMUI for the front and rear flippers because in some cases identification between the two was not possible. Table 6.2 displays the % elements which fall into these three categories over time on Sanak Island.

We conducted a chi-square test to see if there was any difference in the standard proportions of high, medium, and low utility elements in an otariid skeleton and those which we found in middens. A chi-square test was not statistically significant, $x^2(8)=11.71$, $p<.05$, indicating that in all time periods otariid elements were generally being brought back to the site in the same rank order in which they appear in an otariid skeleton. However, while the chi-square suggests this overall consistency, the frequencies of high-, medium-, and low-ranked elements sometimes differed between periods. For example, between AD 1030 and 1540, the transition from the Medieval Warm Period to the Little Ice Age, we see an increase in high-utility elements by over 12% and an increase in medium utility by 10%. This corresponds well with the increase in otariids during this period and suggests that when otariid frequencies increased, transport of higher-utility elements to the site increased. This is not apparent in the transition from early warm periods at 80 BC to the colder period represented by AD 520, when otariid abundances appear to increase, but lower-utility elements increase by over 20% in the middens. We are uncertain what mechanisms are driving this difference, but one explanation may be the sudden increase in human population ca. AD 520 (see Betts et al. this volume). If the frequency of lower-utility elements increased during this period, it may be a result of more intensive processing associated with the population increase. This is a rather parsimonious answer, because we know populations increased slightly at 80 BC as well, although this increase was not as large as the one that occurred in AD 520.

A COMPARISON OF SEA OTTER TO OTARIID %MAUS

Because otariids are larger and contain abundant edible tissue relative to sea otters, they require much more intensive processing and transport. Comparing element frequencies between these two very different animals enables us to develop a better understanding of Aleut butchery and transport patterns in relation to changing climatic conditions and fluctuations in foraging efficiency. Figure 6.6 displays a comparison of element frequencies for otariids and sea otters across all four time periods. Our assumption was that sea otters would have a fairly even element distribution because they would largely be transported back to sites whole due to their size. However, a very uneven distribution of sea otter skeletal parts is evident in some time periods. During the earlier contexts, ca. 80 BC and AD 520, there is good representation of all elements, relatively consistent with whole-animal transport. However, at AD 1030 there is an increase in forelimb elements relative to others; this pattern is intensified in the 1540 AD context, where there is an emphasis on limbs. This pattern might be consistent with an increase in density mediated attrition (e.g. Lyman 1992b; Wall 1983), although this is a difficult inference to reconcile against the generally excellent preservation of all classes of elements and the lack of carnivore gnawing associated with all the contexts. In short, however, we see a clear indication between changes in climate and changes in element representation for sea otters.

Scrutinizing the otariid portions of the graphs reveals a very interesting pattern. While the first two periods, 80 BC and AD 520, are generally consistent (although note the reverse in representations of humeri and radii between the two periods), the following two periods represent a radical departure both from previous periods and from each other. At AD 1030 there is a clear emphasis on low-utility but high-bone-density elements, while in the following period there is an emphasis on medium-utility elements. The correlation with climate here is intriguing; during warm, less productive regimes, when otariid populations are low, there is an increase in low-utility elements. During cold regimes, when otariid populations are abundant, there is an increase in higher-utility (medium-utility) elements.

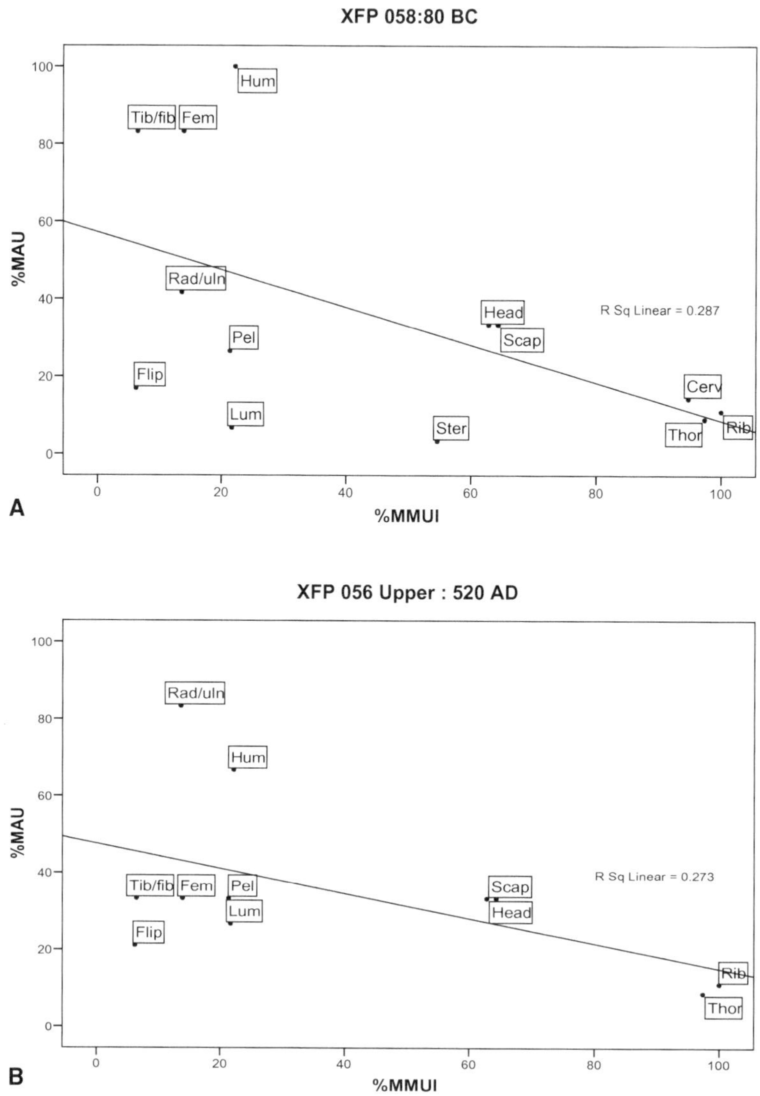

FIGURE 6.5. Scatterplots illustrating the %MAU frequencies of otariidae against the %MMUI for the following sites: (A) XFP 058, dated to 80 BC (a warm period); (B) XFP 056 Upper, dated to 520 AD (a cold period); (C) XFP 056 Lower, dated to 1030 AD (a warm period); and (D) XFP 110 dated to 1540 AD (a cold period).

DISCUSSION

Lupo and O'Connell (2002) theorize that the number of cut marks should be directly related to carcass size; the larger the animal, the more cuts are required to cut through the higher proportion of weight of attached tissues. However, contrary to this idea, Domínguez-Rodrigo (2003) demonstrates that absolute flesh weight has no quantitative correlation with the number of cuts required for processing an animal carcass. Rather, the opposite is apparent; smaller carcasses seem to produce higher frequencies of cut marks. Yet he argues this is not because larger mammals often have thicker, fattier tissues, making it more difficult for the tool to cut through to bone, but instead that for large animals it is simply not necessary (or efficient) to remove all the flesh from bones.

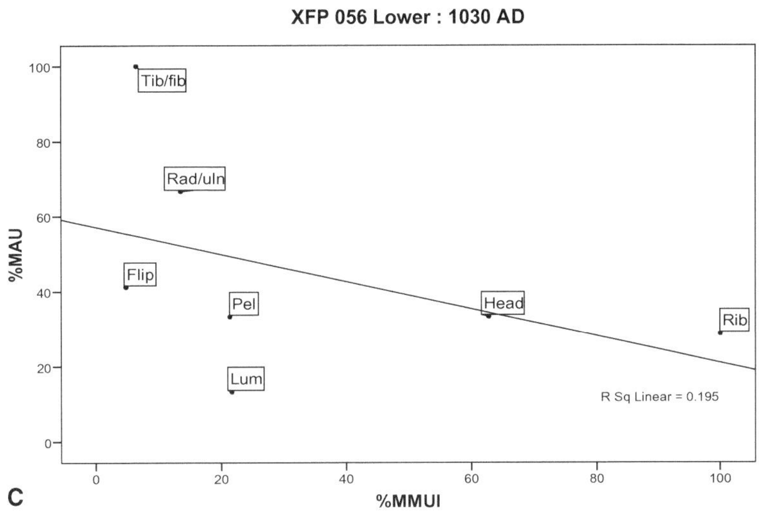

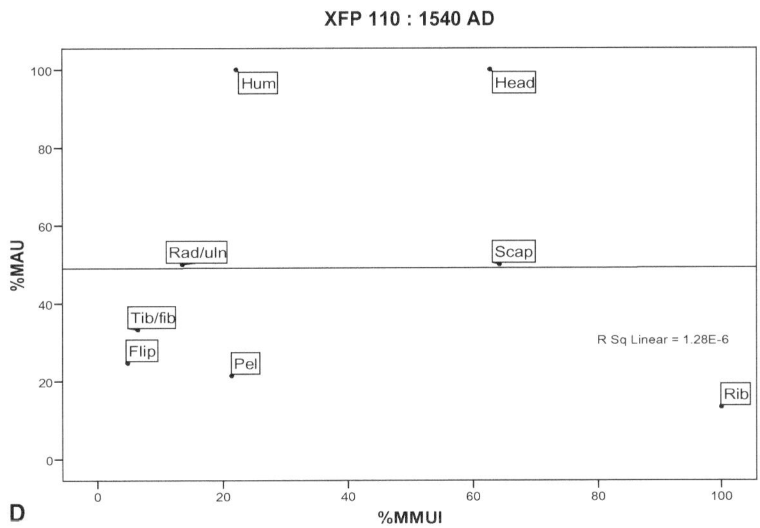

FIGURE 6.5. (*Continued*).

Domínguez-Rodrigo's (2003) explanation is interesting, but it does not explain the inverse relationship that we see in filleting activity between otariids (highly ranked taxon) and sea otters (lower-ranked taxon) in Figure 6.4. If Domínguez-Rodrigo's (2003) explanation applied here, consistently low filleting activity of otariids should be evident. However, this is not the case, and rather, we see variability—shifts that we believe pertain to the shifts we see in the otariid abundance index in Figure 6.1. Furthermore, when otariids are abundant during cooler periods, there is less filleting activity of sea otters (considered to be a desperate meal), and vice versa. Sea otter filleting activity increases when demand for otariids is high and their abundance is low.

Binford (1978) has described three different butchering and transport strategies based on economic utility and the relative frequency that animal parts were removed from kill/butchering sites by the Nunamiut: bulk, gourmet, and unbiased. A bulk strategy can be described as an effort to maximize the quantity of animal parts returned to a residential site, whereas the gourmet strategy attempts to

TABLE 6.1

Derivation of %MMUI from Flesh Weights with Utility Ranking

ANATOMICAL PORTION	AVERAGE FLESH WEIGHT (G)	%MUI	ELEMENT	MODIFIED FLESH WEIGHT (G)	%MMUI	UTILITY RANK
Head	7800	30.7	Head and cervical	15,942	62.8	Medium
Cervical vertebra	24,083	94.8	None	24,083	94.8	High
Thoracic vertebra	5599	22.0	Cervical and ribs	24,743	97.4	High
Lumbar vertebra	3499	13.8	Thoracic and pelvis	5512	21.7	Low
Pelvis, sacrum, caudal	5424	21.4	None	5424	21.4	Low
Rib	25,402	100.0	None	25,402	100.0	High
Sternum	2325	9.2	Sternum and rib	13,864	54.6	Medium
Scapula	7275	28.6	Scapula and rib	16,339	64.3	Medium
Humerus	4004	15.8	Humerus and scapula	5640	22.2	Low
Radius/ulna	2926	11.5	Radius/ulna and humerus	3465	13.6	Low
Front flipper	871	3.4	Front flipper and radius/ulna	1899	7.5	Low
Femur	995	3.9	Pelvis and tibia/fibula	3520	13.9	Low
Tibia/fibula	1615	6.4	None	1615	6.4	Low
Rear flipper	804	3.2	Rear flipper and tibia/fibula	1210	4.8	Low
*Average both flippers	838	3.3	None			

SOURCE: Savelle et al. (1996:706–707)

TABLE 6.2

Cross-tabulation of occupational period and utility rank

OCCUPATIONAL PERIOD	COOL OR WARM PERIOD	UTILITY RANK (% WITHIN OCCUPATIONAL PERIOD)			TOTAL
		HIGH	MEDIUM	LOW	
80 BC	Warm	25.7	7.9	66.3	100.0
AD 520	Cool	16.7	3.3	80.0	100.0
AD 1030	Warm	13.8	8.6	77.6	100.0
AD 1540	Cool	25.9	18.5	55.6	100.0
Observed	—	23.1	23.1	53.8	100.0
Total	—	21.4	9.6	69.0	100.0

NOTE: The "observed" period represents our standard and its total percentage of utility ranking parts calculated from Savelle et al.'s (1996) %MMUI calculations.

simply maximize the quality (the quality of meat). The unbiased strategy is characterized by the removal of animal parts in proportion to their economic value (Metcalfe and Jones 1988). Ideally, one would expect to see a bulk strategy during a resource depression when demand is high and a gourmet strategy when demand is low and resources are in abundance. Our analysis in Figure 6.5 suggests no apparent patterns for otariids. While the distribution at 80 BC might be described as a bulk strategy during a warm period, the same general strategy is apparent in the succeeding cold period (ca. AD 520). In fact, all the periods suggest an even or slightly negative relationship between utility and frequency (Figure 6.5), which is not significantly correlated. Figure 6.6 does expose some variability in later periods, which is interesting, but the pattern is far from robust.

Smaller animals, such as sea otters, do not present major transport problems (Lyman 1991), and as such, we expect the distribution of skeletal parts to represent a bulk strategy as whole carcasses are transported back to sites. This strategy seems consistent with the first two contexts in our sequence, at ca. 80 BC and 520 AD. However, in the later two contexts the distribution of sea otter elements suggests an emphasis on low-utility elements. Lyman (1992b) and others attribute these inverse utility curves (higher frequencies of low-utility parts) to bone density mediated attrition. Yet, still confusing is that with the Sanak fauna we have such excellent bone preservation, not only with mammal but also in the less dense classes of avian and fish. In fact, bone preservation in the Sanak middens is so great that even the most fragile fish elements are both abundant and typically complete (including such delicate cranial elements as the operculum, entopterygoid, basipterygium, and fin rays and spines). In addition to excellent bone preservation, the majority of long bones are whole, thereby limiting the effects that any bone density destruction may have had on the faunal assemblages. One possible explanation might be that scavengers may also contribute to this under-representation of low-utility animal parts (Erlandson and Moss 2001). We discovered and excavated an ancient fox den on Sanak that included a high proportion of sea otter remains. This is an interesting discovery because it suggests impacts on middens from scavengers (e.g., that density mediated attrition occurred), and that sea otters may have been left

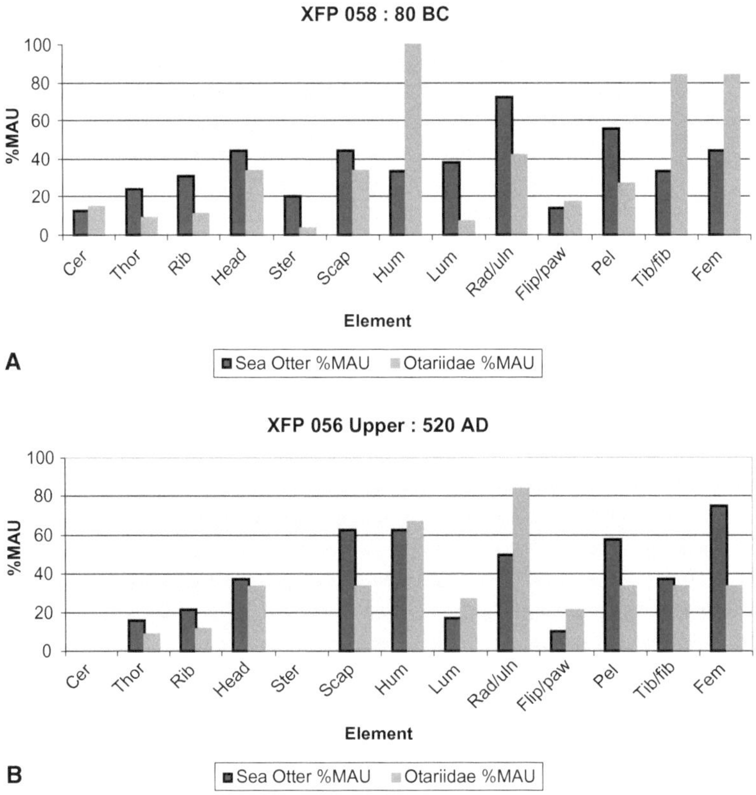

FIGURE 6.6. Charts comparing the %MAU frequencies of sea otters to otariids for the following sites: (A) XFP 058, dated to 80 BC (a warm period); (B) XFP 056 Upper, dated to 520 AD (a cold period); (C) XFP 056 Lower, dated to 1030 AD (a warm period); and (D) XFP 110 dated to 1540 AD (a cold period).

in middens with significant amounts of edible tissue attached (no otariid remains were found in the fox dens).

VARIABILITY IN THE SANAK ISLAND SEQUENCE

Betts et al. (this volume) believe that the variability in otariid abundance in the Neoglacial, between 2550 BC and 80 BC, reflects an array of predator and prey relationships. The frequencies of otariids in the four contexts representing the post-Neoglacial sequence, which represents the focus of this paper, appear to correlate well with changes in oceanic productivity and climate. Betts et al. (this volume) interpret the ca. 80 BC context as representing the tail end of a period of relative stability in otariid abundance (and human predation rates). This is reflected in our cut mark analysis, where both otariid and sea otter filleting activity is stable during this period (relative to earlier periods). In addition to cut mark activity, our analysis of otariid utility in Figure 6.5 shows that we have a good representation of low- to high-utility elements.

During the cooler period from ca. AD 300 to 900, there is an increase in otariid abundance, as reflected in the otariid–Phocidae index (see Figure 6.1). We also observed that the overall % fillet NISP marks on otariid remains decrease slightly, but increase substantially for sea otters. We expect to find higher-utility elements present in greater quantities in middens as demand should be low, but there is no significant change in otariid utility. The incongruence between sea otter abundance and filleting activity is confusing, as our model

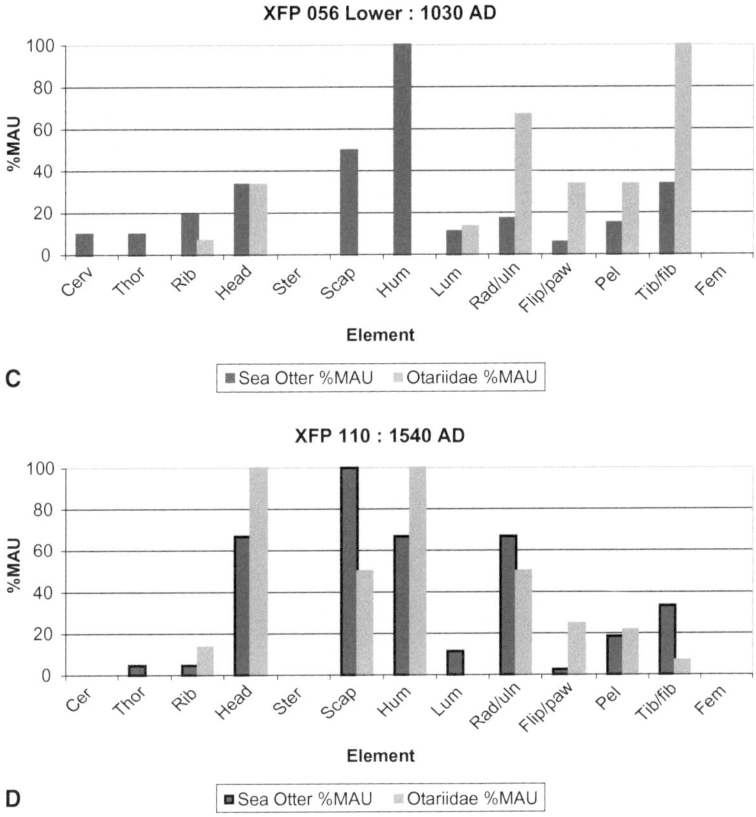

FIGURE 6.6. (*continued*).

predicts that when large-bodied mammals are abundant the intensity of processing on lower-ranked taxa should decline. The fact that only one of our otariid abundance indices suggests an increase in otariids during this period may imply that something more complex is occurring here. Betts et al. (this volume) have demonstrated a rise in human population during this time, and population pressure may provide a possible explanation for this increased processing of lower-ranked resources during this period.

During the Medieval Warm Period (ca. AD 900 to 1250), there is a decline in otariid abundance. We also observed a slight increase in filleting activity on otariid remains and high frequencies of filleting activity on sea otter remains. There is no significant change in otariid utility as there continues to be fairly even distribution of low to high otariid utility elements, although low-ranked elements tend to dominate.

The last site in our sequence was occupied during the Little Ice Age (ca. AD 1250 to 1850). During this period, there is an increase in otariid frequency associated with cooler temperatures, correlated with an increase in oceanic and marine productivity (Betts et al. this volume). During this time, otariid filleting activity remains very low and is entirely absent altogether on sea otter remains (see Figure 6.4). This is exactly what we would expect when otariid populations are abundant and there is little need to maximize energy return. It is interesting that there is a slight increase in medium-ranked utility elements (head, neck, and scapula) during this period, suggesting a small shift toward the preferential transport of higher-utility elements.

CONCLUSIONS

Our analyses demonstrate that there is a greater amount of filleting activity during less productive climatic periods and more evidence for disarticulation during cold climatic periods, evidence which suggests butchery patterns changed as a result of resource stress. Our research indicates that during periods of declining foraging efficiency, which seem to occur during warmer climatic periods, the inhabitants of Sanak more intensively exploited marine mammal carcasses, as seen in cut mark location and cut mark intensity. However, overall, there does not appear to be a significant relationship between element utility and time period. For example, our analysis of otariid element utility instead implies that a roughly "bulk" strategy was employed, consistent with returning the entire animal to the residential site. Given the clear fluctuations in foraging efficiency that occurred over the sequence, and given the transport costs involved with large-bodied marine mammals such as otariids, this transport decision may represent some form of communal sharing and a desire for nothing to be wasted regardless of resource pressure.

Perhaps most intriguing about this research is what it reveals about the relationship between Aleuts and sea otters. A long and detailed ethnohistoric record, and much ethnographic data, play down or completely deny that sea otters played any role in Aleut subsistence, yet our analysis of butchering patterns shows this to be unfounded (see also Corbett et al. 2008; Simenstad et al. 1978). Rather, we suspect the numbers of people in the region had been so decimated by disease and violence when the ethnographic data were recorded and the productivity of the Little Ice Age was supporting abundant higher-ranked marine mammals that the relationship between sea otters and humans was profoundly altered. Recently, many management techniques have been established in the North Pacific to help stabilize and rejuvenate marine mammal populations. However, our knowledge of historical conditions and records is inadequate for understanding long-term ecosystem changes, and as such, assessment of paleorecords is essential to characterize ecosystem variability and function (Corbett et al. 2008). This study indicates that human consumption patterns, as revealed through the study of animal remains, may also provide primary evidence of changes in ancient animal populations. Such research links the history of marine mammal consumption to changes in oceanic productivity, climate, and sea mammal behavior. These data are beneficial for assessing the long-term and future relationships between indigenous harvesting/consumption of otariids and climate change/modern management policies.

This study contributes directly to foraging theory, evolutionary ecology, and studies of the long-term health of the North Pacific. The relationship between the intensity of butchering sea otters and the encounter rates of other species, specifically otariids, and by default measures of population numbers of these other species, creates another proxy for understanding long-term ecosystem dynamics. As such, we might consider this to be another measure of the health and historical ecology of the North Pacific; that is, if we find evidence of intensive sea otter butchery, then those data are, by proxy, an independent measure of regional productivity in the vicinity of sites where no other paleoecological proxy measures are available. Furthermore, this study highlights, the potential cascading effects of removing a high ranked species (such as the Steller sea lion) from an ecosystem and the downward harvesting spiral that may follow as dominant predators move down the ranks of primary prey (e.g. Estes et al. 1998).

ACKNOWLEDGMENTS

This research was supported by a grant from the National Science Foundation (ARC 0508101) and by Idaho State University and the Canadian Museum of Civilization. We gratefully acknowledge the support and access provided by the Sanak Corporation and the Pauloff Harbor Tribe. We would especially like to thank the many undergraduate and graduate students who helped with the laboratory analyses used in this paper, and Todd Braje and Torben Rick for inviting us to participate in a wonderful session at the SAA. All opinions, findings, and conclusions or recommendations expressed in this material are those of the authors and do not necessarily reflect the views of the National Science Foundation.

REFERENCES CITED

Ames, K., and H. D. G. Maschner
1999 *Peoples of the Northwest Coast: Their Prehistory and Archaeology*. Thames and Hudson, London.

Banfield, A.
1974 *Mammals of Canada*. University of Toronto Press, Toronto, Canada.

Bettinger, R. L.
1991 *Hunter-Gatherers: Archaeological and Evolutionary Theory (Interdisciplinary Contributions to Archaeology)*. Plenum Press, New York.

Binford, L. R.
1978 *Nunamiut Ethnoarchaeology*. Academic Press, New York.
1980 Willow Smoke and Dogs' Tails: Hunter-Gatherer Settlement Systems and Archaeological Site Formation. *American Antiquity* 45(1): 4–20.
1981 *Bones: Ancient Men and Modern Myths*. Studies in Archaeology. Academic Press, New York.

Black, L. T.
1998 Animal World of the Aleuts. *Arctic Anthropology* 35(1): 126–135.

Corbett, D. G., D. Clausey, M. Clementz, P. L. Koch, A. Doroff, C. Lefèvre, and D. West
2008 Aleut Hunters, Sea Otters, and Sea Cows: Three Thousand Years of Interactions in the Western Aleutian Islands, Alaska. In *Human Impacts on Ancient Marine Ecosystems: A Global Perspective*, edited by T. C. Rick and J. M. Erlandson, pp. 43–75. University of California Press, Berkeley.

Domínguez-Rodrigo, M.
2003 On Cut Marks and Statistical Inferences: Methodological Comments on Lupo & O'Connell (2002). *Journal of Archaeological Science* 30: 381–386.

Estes, J. A., M. T. Tinker, T. M. Williams, and D. F. Doak
1998 Killer Whale Predation on Sea Otters Linking Oceanic and Nearshore Ecosystems. *Science* 282(5388): 473.

Erlandson, J. M., and Moss, M. L
2001 Shellfish Eaters, Carrion Feeders, and Aquatic Adaptations. *American Antiquity* 66(3): 413–432.

Fairnell, E. H., and J. H. Barrett
2007 Fur-Bearing Species and Scottish Islands. *Journal of Archaeological Science* 34:463–484.

Fitzhugh, J. B.
2003 *The Evolution of Complex Hunter-Gatherers: Archaeological Evidence from the North Pacific*. Kluwer-Plenum, New York.

Gifford-Gonzalez, D.
1989 Ethnographic Analogues for Interpreting Modified Bones: Some Cases from East Africa. In *Bone Modification*, edited by R. Bonnichsen and M. H. Sorg, pp. 179–246. Center for the Study of the First Americans, University of Marine, Orono.

Hildebrandt, W., and T. L. Jones
1992 Evolution of Marine Mammal Hunting: A View from the California and Oregon Coasts. *Journal of Anthropological Archaeology* 11: 360–401.

Johnson, L. L.
2005 *Aleut Sea-Mammal Hunting: Ethnohistorical and Archaeological Evidence*. In *The Exploitation and Cultural Importance of Marine Mammals*, edited by G. G. Monks, pp. 39–61. Proceedings of the 9th ICAZ Conference, Durham 2002, Oxbow Books, Oakville, CT.

Kopperl, R. E.
2003 *Cultural Complexity and Resource Intensification on Kodiak Island, Alaska*. Unpublished Ph.D. dissertation, Department of Anthropology, University of Washington.

Laughlin W. S.
1980 *Survivors of the Bering Land Bridge*. Holt, Rinehart and Winston, New York.

Liapunova, R. G.
1996 *Essays on the Ethnography of the Aleuts: At the End of the Eighteenth and the First Half of the Nineteenth Century*. The Rasumson Library Historical Translation Series IX. University of Alaska Press, Fairbanks, AK.

Lupo, K. D., and J. F. O'Connell
2002 Cut and Tooth Mark Distributions on Large Animal Bones: Ethnoarchaeological Data from Hadza and Their Implications for Current Ideas about Early Human Carnivory. *Journal of Archaeological Science* 29:85–109.

Lyman, R. L.
1987 Archaeofaunas and Butchery Studies: A Taphonomic Perspective. In *Advances in Archaeological Method and Theory*, vol. 10, edited by M. B. Schiffer, pp. 249–337. Academic Press, San Diego.
1991 *Prehistory of the Oregon Coast: The Effects of Excavation Strategies and Assemblage Size on Archaeological Inquiry*. Academic Press, San Diego.
1992a Prehistoric Seal and Sea-Lion Butchering on the Southern Northern Coast. *American Antiquity* 57(2): 246–261.

1992b Anatomical Considerations of Utility Curves in Zooarchaeology. *Journal of Archaeological Science* 19:7–22.

2003 Pinniped Behavior, Foraging Theory, and the Depression of Metapopulations and Nondepression of a Local Population on the Southern Northwest Coast of North America. *Journal of Anthropological Archaeology* 22:376–388.

2005 Analyzing Cut Marks: Lessons from Artiodactyl Remains in the Northwestern United States. *Journal of Archaeological Science* 32: 1722–1732.

Lyman, R. L., J. M. Savelle, and P. Whitridge

1992 Derivation and Application of a Meat Utility Index for Phocid Seals. *Journal of Archaeological Science* 19:531–555.

Maschner, H. D. G.

1998 Salmon Run Volatility, Subsistence, and the Development of North Pacific Societies. In *Proceedings of the 12th International Abashiri Symposium: Salmon Fishery in the North and its Change through Time*, edited The Hokkaido Museum of Northern Peoples, pp. 11–28. Abashiri, Hokkaido, Japan.

1999 Sedentism, Settlement and Village Organization on the Lower Alaska Peninsula: A Preliminary Assessment. In *Settlement Pattern Studies in the Americas: Fifty Years Since Viru*, edited by B. Billman and G. Feinman, pp. 56–76. Smithsonian Institution Press, Washington D.C:

Maschner, H. D. G., M. Betts, K. Reedy-Maschner, and A. Trites

2008 A 4500-Year Time Series of Pacific Cod (*Gadus macrocephalus*): Archaeology, Regime Shifts, and Sustainable Fisheries. *Fishery Bulletin* 106(4): 386–394.

Maschner, H. D. G., B. Finney, J. Jordan, K. L. Reedy-Maschner, A. Tews, N. Misarti, and G. Knudsen

n.d. Did the North Pacific Ecosystem Collapse in AD 1250? In *The Northern World AD 900–1400*, edited by H. Maschner, O. Mason, and R. McGhee, editors. The University of Utah Press, Salt Lake City.

McCartney, A. P., and D. W. Veltre

1999 Aleutian Island Prehistory: Living in Insular Extremes. *World Archaeology* 30(3): 503–515.

Metcalfe, D., and K. T. Jones

1988 A Reconstruction of Animal Body-Part Utility Indices. *American Antiquity* 53(3): 486–504.

Nagaoka, L.

2006 Prehistoric Seal Carcass Exploitation at the Shag Mouth Site, New Zealand. *Journal of Archaeological Science* 33:1474–1482.

Oswalt, W.

1976 *An Anthropological Analysis of Food-Getting Technology*. Willey, New York.

Price, T. D., and J. A. Brown, (editors)

1985 *Prehistoric Hunter-Gatherers: The Emergence of Cultural Complexity*. Studies in Archaeology, Academic Press, Orlando.

Rapson, D. J.

1990 *Pattern and Process in Intra-Site Spatial Analysis: Site Structural and Faunal Research at the Bugas-Holding Site*. Unpublished Ph.D. dissertation, Department of Anthropology, University of New Mexico, Albuquerque.

Ransom, J. E.

1946 Aleut Natural-Food Economy. *American Anthropologist, New Series* 48(4):607–623.

Savelle, J. M., T. M. Friesen, and R. L. Lyman

1996 Derivation and Application of an Otariid Meat Utility Index. *Journal of Archaeological Science* 23:705–712.

Simenstad, C. A., J. A. Estes, and K. W. Kenyon

1978 Aleuts, Sea Otters, and Alternate Stable-State Communities. *Science* 200(4340): 403–411.

Stephens, D., and J. Krebs

1986 *Foraging Theory*. Princeton University Press, Princeton.

Tews, A. M.

2005 *Prehistoric Eastern Aleut Subsistence: 5000 Years of Faunal Variations*. Unpublished master's thesis, Department of Anthropology, Idaho State University.

Veniaminov, I.

1984 *Notes on the Islands of the Unalaska District* [1840]. L. T. Black and R. H. Goeghegan, trans., R. A. Pierce, ed. Alaska History, No. 27. The Limestone Press, Kingston.

Wall, W. P.

1983 The Correlation between High Limb-Bone Density and Aquatic Habits in Recent Mammals. *Journal of Paleontology* 57:197–207.

Workman, W. B., and A. P. McCartney

1998 Coast to Coast: Prehistoric Maritime Cultures in the North Pacific. *Arctic Anthropology* 35(1): 361–370.

Yesner, D. R.

1980 Maritime Hunter-Gatherers: Ecology and Prehistory. *Current Anthropology* 21(6): 727–750.

7

Toward a Historical Ecology of Pinniped and Sea Otter Hunting Traditions on the Coast of Southern British Columbia

Iain McKechnie and Rebecca J. Wigen

MARINE MAMMALS (pinnipeds, cetaceans, and sea otters [*Enhydra lutris*]) have been important to First Nations people in coastal British Columbia for millennia, but their archaeological distribution is poorly known. While archaeological evidence of marine mammal hunting is known for numerous locations over the past 10,000 years of human occupation on the British Columbia Coast (e.g., Carlson 2003; Cannon 1991; Fedje et al. 2005; Matson 1976), it is remarkable that few studies have examined archaeological evidence of mammalian hunting traditions on broad regional and/or temporal scales. Considering the importance of these animals to the modern marine ecosystem, understanding the long-term human use and past distribution of marine mammals can add considerable perspective to contemporary knowledge of these ecologically important species and this highly valued marine ecosystem (cf. Jackson et al. 2001; Lotze and Worm 2009; Pitcher 2005).

The occurrence of marine mammal bones in archaeological contexts reflects the direct use and long-term occupation of this region by coastal First Nations people. Such occurrences indicate that humans have been participants in this ecosystem for at least the past 10,000 years and as such, likely directly and indirectly affected the distribution, growth, behavior, and relative densities of marine mammals and their prey (cf., Crockford et al. 2002; Etnier 2002a; Gifford-Gonzalez et al. 2005; Lyman 2003a). In this chapter, we compile archaeological data on the distribution of pinnipeds and sea otters from archaeological assemblages along the coast of southern British Columbia. We evaluate the spatial and temporal extent of human hunting and explore the possible influence humans may have had on this aspect of the marine environment and, conversely, discuss the potential significance that hunting these animals had to First Nations cultures in the region. Drawing on archaeological and ethnographic information, we ask three questions regarding the use of marine mammals by First Nations people over the past 8000 years:

- What marine mammals did aboriginal people in southern British Columbia most commonly utilize?
- How similar or how different are species occurrences and proportions relative to today?
- Is there evidence of specialized or regional hunting traditions and if so, what might have been the potential impacts of these activities on the ancient marine ecosystem?

CONTEXT

Recent assessments of Northwest Coast archaeology have emphasized the need to adopt a regional approach to characterizing patterns of aboriginal resource use (Butler and Campbell 2004; Cannon 2001). In discussing the role of salmon on the central Northwest Coast, Cannon has noted:

> More important than improved recovery and analysis of fish remains from individual site locations is the pressing need to expand research strategies beyond the individual sites to encompass a variety of site locations within specific regions. This is the only way to gain a better appreciation of the extent of regional, seasonal, and longer-term temporal variability in the focus and intensity of local fisheries. (2001:185)

Although Cannon's suggestion is directed toward the archaeological assessment of fisheries, it is broadly applicable to archaeological research on the Northwest Coast. To date, only a few regional studies of indigenous resource use have been conducted (Butler and Campbell 2004; Crockford et al. 2002; Hanson 1991; Hanson and Kusmer 2001; Hebda and Frederick 1990; Hobson and Driver 1989; Moss 2008; Orchard and Clark 2005). These efforts have provided a host of insights but have been hindered by the small number of published zooarchaeological studies, a methodological concern on sampling adequacy and taxonomic identification, and a focus on characterizing the "full range" of animal consumption, including birds, fish, mammals, and shellfish (e.g., Croes and Hackenberger 1988; Driver 1993; Hanson 1991; Mitchell 1990). Additionally, many of these studies have focused on demonstrating quantitative differences in tabular form making it difficult to assess the potential continuity and spatial associations between adjacent sites and regions.

In this chapter, we adopt an explicitly spatial approach to examining the archaeological expression of aboriginal hunting practices in southern British Columbia, focusing specifically on pinnipeds and sea otters. We explore compositional patterns using zooarchaeological assemblage data from 75 temporally distinct assemblages from 58 sites spanning the past 8000 years. We group assemblages into four broad temporal categories variously corresponding to broadly defined archaeological cultural historical "periods" (300–1200; 1200–2400; 2400–5000; 5000–8000 calibrated years BP) following Lepofsky et al. (2005, 2007) and Ames and Maschner (1999). These data provide context for considering the regional extent and potential influence that First Nations hunting practices had on the marine and terrestrial environments of southern British Columbia.

THE MODERN MARINE ECOSYSTEM

Our study area includes the western and eastern coasts of southern Vancouver Island, the Gulf Islands, and the British Columbia mainland and adjacent areas on the Olympic Peninsula in Washington State (approximately 40,000 km^2; Figure 7.1). This region is temperate with occasional winter snowfall at sea level, high annual rainfall on western Vancouver Island (260–340 cm), and much lower rainfall on southeastern Vancouver Island and the Gulf Islands (70–110 cm; Environment Canada 2009). The terrestrial environment is heavily forested with the densest forest cover on western Vancouver Island. The least densely forested areas are on southeastern Vancouver Island and the Gulf

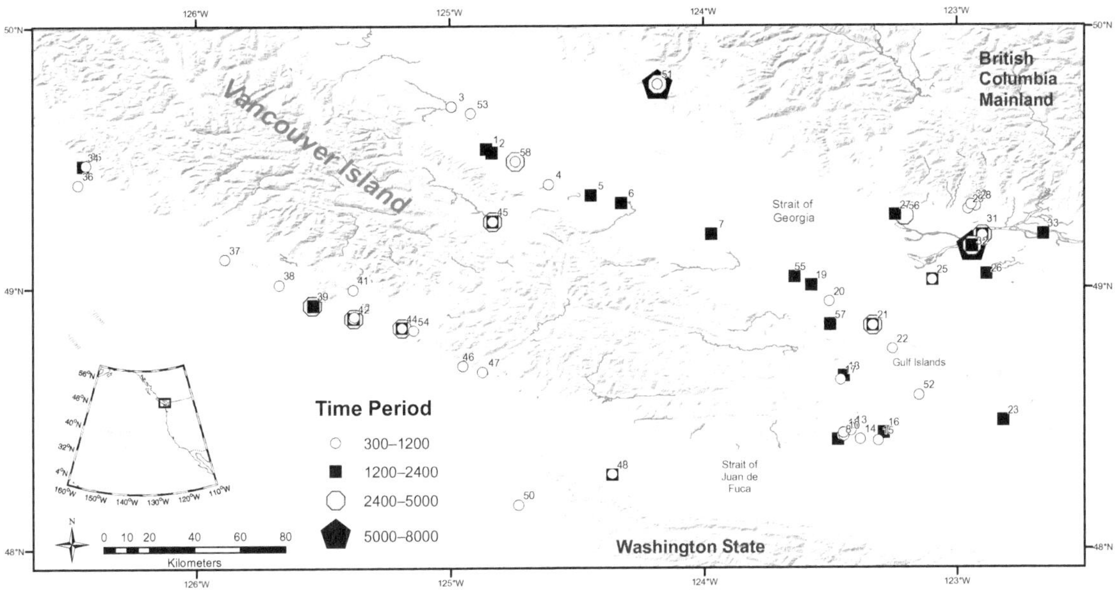

FIGURE 7.1. Map of the study area showing the distribution of archaeological assemblages used in the analysis and their chronological age range.

Islands, which contain patchy oak savanna woodlands and prairies with evidence of anthropogenic fire regimes extending back at least 2000 years (Brown and Hebda 2002).

The marine waters of the Strait of Georgia, the Strait of Juan de Fuca, and Puget Sound (recently renamed the "Salish Sea" by a transboundary commission) are heavily influenced by the seasonal input of freshwater from the Fraser River and other major rivers (Thomson 1981). Marine primary productivity is further influenced by nutrient-rich deep-water drawn into the Strait of Juan de Fuca from the north flowing California counter-current. This hyper-saline and oxygenated water is vertically mixed by the strong tidal currents in the Gulf Islands at which point this mixed surface water then flows back out the strait and northwards along the outer coast of Vancouver Island (Thomson et al. 2007). The greatest annual primary productivity in the study area occurs off the coast of southwestern Vancouver Island (Ware and Thomson 2005), but strongly increases during summer when wind-driven upwelling and prolonged sunlight along the continental shelf break facilitate massive blooms of plankton that attract a host of marine predators (Allen et al. 2001).

The contemporary distribution of pinnipeds on the southern British Columbia coast is represented by six species occupying a range of habitats and exhibiting seasonal fluctuations in abundance (Table 7.1). Harbor seals (*Phoca vitulina*) are the most numerous and ubiquitous species that are present year round (Department of Fisheries and Oceans 2010). Northern fur seals (*Callorhinus ursinus*) are slightly more abundant but do not come ashore and are found almost exclusively on the continental shelf off western Vancouver Island, primarily between December and May (Olesiuk 2009a:57). Steller sea lions (*Eumetopias jubatus*), while resident year round, do not have rookeries in the study area and are only seasonally present in the Salish Sea during winter and spring (Bigg 1988; Olesiuk 2009b). California sea lions (*Zalophus californianus*) are present in fewer numbers, and elephant seals (*Mirounga angustirostris*) migrate to the southern BC coast seasonally (Table 7.1). Elephant seals have recently established a year-round haul-out on Race Rocks on the southern tip of Vancouver Island where at least two elephant seals pups have been born in 2009 and 2010 (Racerocks.com 2010).

The contemporary British Columbia sea otter population consists of 4182 animals (Nichol et al.

TABLE 7.1

Ecological Characteristics for Pinnipeds and Sea Otters in British Columbia

	NORTHERN FUR SEAL	HARBOR SEAL	STELLER SEA LION	CALIFORNIA SEA LION	ELEPHANT SEAL	SEA OTTER
Species Name	*Callorhinus ursinus*	*Phoca vitulina richardsi*	*Eumetopias jubatus*	*Zalophus californianus*	*Mirounga angustirostris*	*Enhydra lutris*
Mean Female Wt. (kg)	35–45	58	200–300	94	650	44
Mean Male Wt. (kg)	100–200	72.5	400–800	375	1800	33
Modern Period of Pup Dependency	7–8 months	1 month	12–24 months	12–24 months	1 month	5–8 months
Modern Birthing Season, Location	July–November, Bearing Sea and California	May–June, Throughout BC	May–June, Central Northern BC	May–July, California and Oregon	December–March, California and Mexico	Year-round with peak in Spring
Modern Seasonal Occurrence in BC	March–June, with peak in May	Resident in BC year-round	Resident in BC year-round	Year-round with peak October–May	Recently present year-round	Resident year-round
Prevalence in Archaeological Sites in Southern BC	Ubiquitous off western Vancouver Island	Ubiquitous	Common	Rare	Rare but increasing	Present on Western VI, rare/absent on Eastern VI
Estimated Contemporary Population in BC	123,000	105,000	18,400–19,700	2,000–3,000	Fewer than 100	4712
Management Status in Canada (2009) (COSEWIC)	Threatened	Not at Risk	Special Concern	Not at Risk	Not at Risk	Special Concern
IUCN Red List Status (2009)	Vulnerable	Least Concern	Endangered	Least Concern	Least Concern	Endangered
References	Banfield 1974; DFO 2007; Gentry 1998	Baird 2001: 668; DFO 2010:1; Jefferies et al. 2003	Bigg 1988; DFO 2003; Olesiuk 2009b	Bigg 1985:17; Hancock 1970; NOAA Fisheries Service 2009	Racerocks.com 2010; Stewart 1999: 217-218	Estes 1999:182; Gregr et al. 2008; Nichol et al. 2009

2009), which have increased from a reintroduced population of 89 animals transported from the Aleutian Islands to northern Vancouver Island between 1969 and 1972 (Bigg and MacAskie 1978). This population has been increasing annually at a rate of 15% but has yet to entirely recolonize its former habitat (Nichol et al. 2009). A recent population reconstruction (Gregr et al. 2008), incorporating habitat characteristics and contact-era (ca. AD 1774–1825) fur trade records, suggests that the British Columbia coast may have once supported a population of between 34,000 and 74,000 otters. Interestingly, this estimate excludes the Salish Sea due to the absence of sea otters in published archaeological assemblages (i.e., Hanson and Kusmer 2001).

Until 1970, the Canada Department of Fisheries and Oceans supported "population control" programs, which offered incentives to reduce pinniped populations (particularly Harbor seals and Steller sea lions; Jeffries et al. 2003; Olesiuk 2009b). These controls ended when the marine mammal protections were implemented in 1970 (Olesiuk 2009a). Since this time, harbor seal populations have increased dramatically from 10,000 animals to 105,000 animals (DFO 2010). Harbor seals in particular were viewed as competitors with human fisheries, especially salmon, despite the fact that the bulk of their prey is composed of nonsalmonids (Olesiuk et al. 1990).

HISTORIC CHANGES TO THE MARINE ECOSYSTEM

Marine mammals, particularly the trade of their highly valued pelts, were central throughout the process of early historic contact between First Nations and Europeans on the northwest coast of North America (Fisher 1977). Sea otters, northern fur seals, and then whales sequentially became the focus of international commercial industries, which rapidly diminished these culturally and ecologically significant animals over several decades. The corresponding impact of the sudden removal of these ecologically important animals is poorly known (due in part to the paucity of archaeological data and lack of long-term ecological studies), and this issue is a contemporary research concern for ecologists and managers throughout the northern Pacific (e.g., Springer et al. 2003; Misarti et al. 2009).

The early "contact period" on the Northwest Coast (ca. AD 1774–1812) focused on the trade of sea otter pelts obtained from the western coast of Vancouver Island as well as the northern and central British Columbia coast (Busch and Gough 1997; Dick 2006). Sea otter pelts were recognized as wealth items and chiefly regalia within aboriginal communities prior to contact with European trading vessels. Aboriginal exchange of pelts for highly valued European trade goods was therefore immediately incorporated into the indigenous economy and caused a large expansion of this particular hunting activity (Dick 2006; Lutz 2008). Over a few decades, the burgeoning trade led to an increasing scarcity of sea otters and increasing conflict between European and American traders and First Nation communities. European trading ships ceased to seek sea otter pelts on the southern British Columbia coast after 1812 following a series of dramatic incidents involving the destruction of European ships and crew as well as reprisal attacks on aboriginal villages (e.g., McMillan 1999:188).

After a period of several decades of infrequent interaction, the European colonial presence on the coast increased substantially during the mid 19th century. From the 1880's through the early 1900's, commercial demand for seal pelts in the European market fueled a commercial sealing industry, which targeted northern fur seals off the British Columbia coast and Washington State (C. Crockford 1996). Dozens of schooners operating out of Victoria and Seattle employed hunters from First Nations communities on western Vancouver Island and the Olympic Peninsula to travel offshore where hunters would launch traditional dugout canoes and use harpoons to approach and capture fur seals "sleeping" in the surface waters (Swan 1883). In 1880 alone, over 12,000 fur seals were harvested off the coast of Vancouver Island (Swan 1887:397). Growing international pressure concerning the declining stocks of fur seals at their breeding grounds in the Bearing Sea in the late 19th century (Elliot 1886) eventually resulted in one of the first international treaties protecting marine wildlife in 1911, and ended the commercial harvest of fur seals in Canadian waters (Gay 1987).

Also during this time, a nascent whaling industry developed on the British Columbia coast during the mid-19th century but expanded significantly after 1905, regionally diminishing whale populations, particularly sperm whales (*Physetermacrocephalus*), fin whales (*Balaenopteraphysalus*), and humpbacks (*Megapteranovaeangliae*; Gregr et al. 2000). Nonmigratory populations of humpback whales residing in coastal inlets were among the first whales targeted (Merilees 1985; Webb 1988). Whaling continued on the British Columbia coast until it was halted in 1967.

HISTORIC CHANGES AMONG COASTAL FIRST NATIONS

The dramatic changes described above occurred in the context of tumultuous and tragic social, economic, and demographic changes within aboriginal communities over the last two centuries. Particularly devastating were introduced diseases (particularly smallpox and measles), which swept through the coastal population in the 1790s and later in the 1850s and 1860 s (Boyd 1999). These epidemics severely reduced the aboriginal population and correspondingly destabilized the highly structured social and political networks and dramatically reduced the scale and extent of the aboriginal subsistence economy. A large influx of European and Asian colonists in the mid-to-late 19th century introduced a host of new opportunities and challenges for sustaining and generating wealth in aboriginal communities. Many aboriginal people began to participate in the developing wage economy in forestry, fisheries, and agriculture on a seasonal basis (Lutz 2008). Among the highest wages were those offered to aboriginal hunters employed in the fur seal industry who earned seasonal incomes well above European counterparts (Lutz 2008). This industry persisted until the 1911 international fur seal treaty between America, Japan, Russia, and Canada mandated a halt to this once central economic activity.

Also during the late 19th century, anthropological fieldworkers began to document aboriginal peoples of the coast (e.g., Boas 1887). Numerous researchers compiled detailed descriptions and oral historical accounts of aboriginal lifeways including marine mammal hunting practices and rituals associated with hunting (e.g., Sapir and Swadesh 1939; Waterman 1920). Whaling was a particularly honored tradition among Nuu-chah-nulth peoples on the west coast of Vancouver Island and Makah peoples on the Olympic Peninsula (McMillan 1999), but sealing was also a respected skill in coastal aboriginal communities on both the east and west coasts of Vancouver Island (Drucker 1951; Elemendorf 1960; Suttles 1987a). Hunters were typically chiefs or men of high status who underwent elaborate ritual and physical preparation (Arima 1988; Drucker 1951:168; Elmendorf 1960:86; Suttles 1987a; Waterman 1920). Hunter's wives were obligated to observe ritual practices prior to and during the hunt and contributed labor to butchering and cooking as well as preparing hunting technology (e.g., preparing sinew, seal-skin floats, cedar bark rope, etc.). Hunting took place from dugout canoes using harpoon technology with seal-skin floats or from land using clubs. A host of historic and ethnographic accounts are available for further investigating hunting practices and techniques, some of which are described further in the sections below.

METHODS

To investigate the regional expression of marine mammal hunting practices, we compiled available zooarchaeological data from academic papers, graduate student theses, and provincially reviewed archaeological permit reports on file at the British Columbia Archaeology Branch in Victoria and federally reviewed reports on file with Cultural Resource Management Services Division of Parks Canada, Calgary, Alberta (Table 7.2). All sites are within 500 m of the coastline or along the banks of lower Fraser River (a major tidally influenced river draining into the Salish Sea). Sites that are more than 500 m from the ocean ("inland" sites) are not included in the present analysis, because they have not been investigated to the same degree as shoreline shell-midden sites and generally contain poorly preserved faunal assemblages.

All assemblages included in the analysis are from precontact archaeological deposits (ca. > AD 1774) and date to the past 8000 years. However, the majority of the assemblages date to the last 2400 years. Most sites in the study sample (>90%) are from deposits which have been radiocarbon dated. Assemblages without radiocarbon ages are either from the upper layers of ethnographically identified village locations (assumed to date within the past 1200 years) or are associated with temporally diagnostic artifact assemblages. To enable comparability between regions, we combined assemblages into one of four broad temporal categories (300–1200; 1200–2400; 2400–5000; 5000–8000 calibrated years BP; Table 7.2).

Zooarchaeological assemblages included in the analysis were identified using comparative osteological collections based at large universities in the region (e.g., Simon Fraser University, University of British Columbia, University of Victoria, and Washington State University) and/or at the Royal BC Museum in Victoria (RBCM). Individual bone specimens were identified to the lowest taxonomic level possible on the basis of their morphological similarly to comparative specimens. A majority of analyses have been conducted by or with the consultation of a few individuals who have a long history of work in the region (e.g., Susan Crockford, Jon Driver, Gay Frederick [formerly Calvert & Boehm, Gregory Monks, Rebecca Wigen). Researchers at the University of Victoria and the RBCM have assembled a comprehensive comparative collection of vertebrates in the North Pacific including marine and terrestrial mammals from multiple individuals of different age and sex classes. Other universities have less-comprehensive collections but include common mammals like deer and harbor seal. Identification and quantification criteria have been reviewed and standardized for consistency. For identifications that appear questionable, as in the case of northern fur seals identified from Burrard Inlet in North Vancouver (i.e., Galdikas-Brindamour 1972), revised assessments of earlier identifications are utilized (cf. Trost 2005).

Faunal remains from the majority of sites included in the analysis were recovered from controlled excavations using standard ¼" (6.35 mm) mesh. However, in one instance of salvage data recovery, mammal bones were collected by hand from excavated sediments (Site 46, Table 7.2). Mammal remains recovered from fine-mesh bulk samples were not included in the analysis due to the small sample sizes and their lack of comparability with fauna from larger volumes of excavated sediment. In most cases, mammal bones were not recovered or were present in very low numbers in these fine-mesh bulk samples.

Faunal identification data are quantified by number of identified specimens (NISP), which includes individual skeletal elements or fragments that can be confidently attributed to taxonomic categories of family-, genus-, or species-level designation (e.g., cervidae, deer, or mule deer). Highly fragmented remains that cannot be confidently assigned to family, genus, or species level are not included in the analysis (e.g., unidentified sea mammal). Quantification of relative abundance is evaluated by comparing the percentage of identified specimens (NISP) from individual taxonomic categories to all other positively identified specimens (e.g., %NISP of pinnipeds).

Species- or genus-specific taxonomic identifications are shown in Table 7.2, but for the

TABLE 7.2

Site Names, Locations, References, and Number of Identified Specimens (NISP) Used in the Analysis.
All Mammal Totals Exclude Whales and Domestic Dogs.

NUM.	REGION	SITE #	PERIOD	SITE NAME	REFERENCE	HARBOR S.	STELLER SL	N. FUR SEAL	C. SEA LION	ELEPHANT SEAL	SEA OTTER	TERRESTRIAL MAMMALS	ALL MAMMALS
1	ecVI	DjSf-13	1200–2400	Buckley Bay	Wigen 1980	4	10	0	0	0	0	221	235
2	ecVI	DjSf-14	1200–2400	Tsable River	Wigen 1980	2	4	0	0	0	0	194	203
3	ecVI	DkSg-2	300–1200	Sandwick midden	Capes, 1964:90	5	0	0	0	0	0	66	75
4	ecVI	DiSc-1	300–1200	Little Qualicum River	Bernick and Wigen 1990	18	1	0	0	0	0	243	263
5	ecVI	DiSc-26	1200–2400	Qualicum Beach Golf Course	Willows et al. 2008:75	0	0	0	0	0	0	10	11
6	ecVI	DhSb-3	1200–2400	Dogwood St., Parksville	Wilson et al. 2006:58	0	0	0	0	0	0	12	12
7	ecVI	DhRx-16	1200–2400	Departure Bay	Wilson and Crockford 1994:118	6	0	0	0	0	0	733	743
8	scVI	DcRu-74	1200–2400	Esquimalt Lagoon Wet site	McKechnie, 2004:51, table 19	3	0	0	0	0	0	63	67
9	scVI	DcRu-2	300–1200	Esquimalt Lagoon	Hanson 1991, table 6 citing Stevenson 1978	9	3	0	0	0	0	254	266

10	scVI	DcRu-2	300–1200	Esquimalt/ Ft Rodd Hill	Crockford 1997b	17	0	0	0	0	0	337	358
11	scVI	DcRu-78	300–1200	Ft Rodd Hill	Mitchell 1981	0	5	0	0	0	0	88	96
12	scVI	DcRu-136	300–1200	Esquimalt Harbour	Wigen 2002	5	0	0	0	0	0	53	58
13	scVI	DcRu-4	300–1200	Kosapsom	Stewart and Wigen 2003	0	0	0	0	0	4	19	23
14	scVI	DcRu-75	300–1200	Dallas Road	Wilson et al. 2003:81	14	0	0	0	0	0	126	143
15	scVI	DcRt-16	300–1200	King George Terrace	Wilson et al. 2004a:107	3	4	0	0	0	0	111	119
16	scVI	DcRu-71	1200–2400	Eagles Nest	Eldridge 2000:40	11	2	0	0	0	0	54	77
17	scVI	DdRu-18	300–1200	Eventide Rd.	Weathers et al. 2007:29	2	5	0	0	0	0	37	44
18	scVI	DdRu-5	1200–2400	Patricia Bay	Kanipe et al. 2006:264	9	10	0	0	0	0	145	165
19	Gulf Is	DgRu-3	1200–2400	Dionisio Point	Ewonus, 2006:39	3	0	0	0	0	0	88	96

(continued)

TABLE 7.2 *(continued)*

NUM.	REGION	SITE #	PERIOD	SITE NAME	REFERENCE	HARBOR S.	STELLER SL	N. FUR SEAL	C. SEA LION	ELEPHANT SEAL	SEA OTTER	TERRESTRIAL MAMMALS	ALL MAMMALS
20	Gulf Is	DfRv-106	300–1200	Galiano Island	Mason et al. 1995:21	0	1	0	0	0	0	20	21
21.1	Gulf Is	DfRu-8	300–1200	Helen Point	Boucher 1976: assemblages 12, 13, & 14	12	1	0	0	0	0	180	194
21.2	Gulf Is	DfRu-8	1200–2400	Helen Point	Boucher 1976: assemblages 5 & 6	18	5	0	0	0	0	93	119
21.3	Gulf Is	DfRu-8	2400–5000	Helen Point	Boucher 1976: assemblages 1, 2, 3, & 4	187	54	0	0	0	1	360	664
22	Gulf Is	DhRt-1	300–1200	Pender Canal	Hanson, 1995:34	62	0	0	0	0	0	406	525
23	Gulf Is	45SJ169	1200–2400	Decatur Island-169	Lyman, 2003a	245	0	0	0	0	1	264	298
24	Gulf Is	45SJ165	1200–2400	Decatur Island-165	Lyman, 2003	10	0	0	0	0	0	185	202
25.1	Mnlnd	DgRr-2	300–1200	Tsawwassen	Butler and Campbell 2004:353 citing Kusmer 1994	4	0	0	0	0	0	16	20

25.2	Mnlnd	DgRr-2	1200–2400	Tsawwas-sen	Butler and Campbell 2004:353 citing Kusmer 1994	1	0	0	0	0	0	5	6
26	Mnlnd	ɔgRr-1	1200–2400	Crescent Beach	Hanson 1991, table 6 citing Ham 1982	2	1	0	0	0	0	805	820
27	Mnlnd	DhRr 18	1200–2400	Point Grey	Coupland 1991:89	0	0	0	0	0	0	16	16
28	Mnlnd	DhRr-6	300–1200	Belcarra (*Tum-tu-may-whueton*)	Trost 2005:116 citing Charlton 1977:226 and Galdikas-Brindamour 1972	10	0	0	0	0	0	227	239
29	Mnlnd	DhRr-8	300–1200	Cates Park (*Whey-Ah-Wichen*)	Trost 2005:116 citing Williams 1974	14	0	0	0	0	0	227	251
30	Mnlnd	DhRr-18	300–1200	Cove Cliff (*Say-Umiton*)	Trost 2005:112	7	0	0	0	0	0	123	133
31.1	Mnlnd	DgRr-2	300–1200	St. Mungo - late	Boehm 1973, table VIII	4	0	0	0	0	0	15	19
31.2	Mnlnd	DgRr-2	1200–2400	St. Mungo - mid	Boehm 1973, table VIII	3	0	0	0	0	0	6	9

(continued)

TABLE 7.2 *(continued)*

NUM.	REGION	SITE #	PERIOD	SITE NAME	REFERENCE	HARBOR S.	STELLER SL	N. FUR SEAL	C. SEA LION	ELEPHANT SEAL	SEA OTTER	TERRESTRIAL MAMMALS	ALL MAMMALS
31.3	Mnlnd	DgRr-2	2400–5000	St. Mungo - early	Boehm 1973, table VIII	23	0	0	0	0	0	212	235
32.1	Mnlnd	DgRr-6	1200–2400	Glenrose Cannery-late	Imamoto 1976, table 2.1	6	0	0	0	0	0	70	76
32.2	Mnlnd	DgRr-6	2400–5000	Glenrose Cannery-mid	Imamoto 1976, table 2.1	23	0	0	0	0	0	193	216
32.3	Mnlnd	DgRr-6	5000–8000	Glenrose Cannery-early	Imamoto 1976, table 2.1	3	0	0	0	0	0	32	35
33	Mnld	DhRp-17	1200–2400	Port Hammond	Rousseau et al. 2003:101	3	0	0	0	0	0	264	267
34	wcVI	DiSo-9	1200–2400	Hesquiat DiSo9Combined	Calvert 1980, tables 43 and 47	38	1	100	2	0	68	88	328
35	wcVI	DiSo-16	300–1200	Hesquiat rockshelter	Calvert 1980, table 39	0	0	0	0	0	0	66	126
36	wcVI	DiSo-1	300–1200	Hesquiat Village combined	Calvert, 1980, tables 51, 55, 59, 63, and 67	86	41	228	29	0	90	155	844

37	wcVI	DgSl-67	300–1200	Chester-man Beach	Wilson 1994, table 5	0	0	25	12	0	0	6	52
38	wcVI	DgSk-1	300–1200	Long Beach	Wigen 2003, table 2	0	1	0	0	0	0	0	5
39	wcVI	DfSj-100	240–5000	Little Beach	Weathers et al. 2008; Wigen 2008, table 5	0	1	19	0	0	0	3	25
40	wcVI	DfSj-57	1200–2400	Spring Cove, Ucluelet	Spady and Wigen 2008, table 15	1	0	49	0	0	0	5	57
41	wcVI	DfSi–5	300–1200	Ma'acoah	Monks 2006:224	32	2	84	14	1	6	71	224
42.1	wcVI	DfSi-16 + 17	300–1200	Tsishaa Late	McKechnie 2007a, tables 17 and 18, including Frederick and Crockford 2005	28	13	214	1	0	3	29	344
42.2	wcVI	DfSi-16 + 17	1200–2400	Tsishaa Mid	McKechnie 2007a, tables 17 and 18, including Frederick and Crockford 2005	14	0	49	0	1	1	25	109

(continued)

TABLE 7.2 *(continued)*

NUM.	REGION	SITE #	PERIOD	SITE NAME	REFERENCE	HARBOR S.	STELLER SL	N. FUR SEAL	C. SEA LION	ELEPHANT SEAL	SEA OTTER	TERRESTRIAL MAMMALS	ALL MAMMALS
42.3	wcVI	DfSi 16 + 17	2400–5000	Tsishaa Early	McKechnie 2007a, tables 17 and 18, including Frederick and Crockford 2005	2	7	35	0	0	2	131	408
43	wcVI	DfSi-26	300–1200	Clarke Island	McKechnie 2007a, tables 17 and 18	6	7	56	0	0	0	9	80
44.1	wcVI	DfSh-7	300–1200	Huu7ii late	Frederick et al. 2006, tables 6, 8, and 10	21	24	46	0	0	12	49	298
44.2	wcVI	DfSh-7	1200–2400	Huu7ii mid	Frederick et al. 2006, tables 6, 8, and 10	5	1	1	0	0	0	3	17
44.3	wcVI	DfSh 7	2400–5000	Huu7ii Early	Frederick et al. 2006, tables 6, 8, and 10	9	16	39	0	0	3	65	169
45.1	wcVI	DhSe-2	300–1200	Shoemaker Bay Late	Calvert and Crockford 1982:186	75	0	11	0	0	0	781	889

45.2	wcVI	DhSe-2	1200–2400	Shoemaker Bay Mid	Calvert and Crockford 1982:186	13	2	0	0	0	0	182	201
45.3	wcVI	DhSe 2	2400–5000	Shoemaker Bay Early	Calvert and Crockford 1982:186	7	0	1	0	0	0	91	101
46	wcVI	DeSf-6	300–1200	Klanawa	McKechnie, 2007b:7	7	0	8	0	0	0	59	75
47	wcVI	DeSf-2	300–1200	Tsuxwk-waada	McKechnie 2005, table 1	3	0	185	0	0	0	3	191
48	WA	45CA213	1200–2400	Hoko R. Wet/Dry	Croes 1995: 71	7	1	3	0	0	0	10	22
49	WA	45CA21	300–1200	Hoko R. Rockshelter	Wigen 2005, table 4.4	114	40	2425	0	1	40	423	3318
50	WA	45CA24	300–1200	Ozette	Huelsbeck 1994:28 with additions from Etnier 2007:200	377	10	47296	1	24	501	708	51007
51.3	MnInd	DkSb-30	5000–8000	Saltery Bay	Pegg et al. 2007, table 20	17	0	0	0	0	0	108	129
51.2	MnInd	DkSb-30	2400–5000	Saltery Bay	Pegg et al. 2007, table 20	6	2	0	0	0	0	76	163

(continued)

TABLE 7.2 *(continued)*

NUM.	REGION	SITE #	PERIOD	SITE NAME	REFERENCE	HARBOR S.	STELLER SL	N. FUR SEAL	C. SEA LION	ELEPHANT SEAL	SEA OTTER	TERRESTRIAL MAMMALS	ALL MAMMALS
51.1	MnInd	DkSb-30	300–1200	Saltery Bay	Pegg et al. 2007:table 20	3	0	0	0	0	0	14	27
52	Gulf Is	45-SJ-24	300–1200	English Camp	Pegg 1999:69	20	0	0	0	0	1	310	372
53	ecVI	DkSf-4	300–1200	Comox Harbour Site	Simonsen 1990:35, table 11	2	7	0	0	0	2	262	292
54	wcVI	DfSg-2	300–1200	Aguilar House	Engish 2006:28	2	4	19	0	0	2	6	36
55	Gulf Is	DgRv-2	1200–2400	Shingle Point	Matson et al. 1999:61–71	4	0	0	0	0	0	107	116
56	Mnlnd	DhRt-6	2400–5000	Locarno Beach	Brolly and Muir 1993:52	1	0	0	0	0	0	67	68

57	Gulf Is	DfRu-3	1200–2400	Harbour House, Saltspring	Brolly et al. 1993:72, table 8	0	1	0	0	0	0	127	129
58.1	ecVI	DiSe-10	300–1200	Denman - Southern Rckshltr	Eldridge 1987:5-57, tables 5–9	0	0	0	0	0	0	7	7
58.2	ecVI	DiSe-10	2400–5000	Denman - Northern Rckshltr	Eldridge 1987:5-59, tables 5–10	2	0	0	0	0	0	57	59
58.3	ecVI	DiSe-10	2400–5000	Denman - Blufftop	Eldridge 1987:5-67, table 5–16	1	0	0	0	0	0	26	27

NOTE: ecVI-East Coast Vancouver Island, scVI-South Coast Vancouver Island, Gulf Is-Gulf Islands, wcVI-West Coast Vancouver Island, Mnlnd-Mainland of British Columbia, WA-Washington State.

TABLE 7.3

Sex Distribution of Steller Sea Lion Remains from Selected Sites

REGION	SITE	TOTAL NISP	MALE	FEMALE	REFERENCE
ecVI	DiSe-7, Deep Bay	1	1		Wilson et al. 2004a
scVI	DcRt-16, King George Terrace	4	4		Wilson et al. 2004b:94
scVI	DdRu-5, Pat Bay	10	10		Kanipe et al. 2007:269
scVI	DcRu-136, Esquimalt Lagoon	2	2		Wigen 2002
Gulf Is	DfRv-106, Galiano Island	1	1		Mason et al. 1995:22
wcVI	DiSo-9-II, Hesquiat	1		1	Calvert 1980:155
wcVI	DiSo-1-I, Hesquiat	13	2		Calvert 1980:156
wcVI	DiSo-1-II, Hesquiat	1	1		Calvert 1980:157
wcVI	DiSo-1-III, Hesquiat	12	3		Calvert 1980:158
wcVI	DiSo-1-IV, Hesquiat	14	2	1	Calvert 1980:159
wcVI	DjSf-100, Little Beach	1		1	Wigen 2008:11
wcVI	DfSi-26, Clarke Island	7	4		McKechnie 2007a:30
wcVI	DfSi-16, Ts'ishaa	20	7		Unpublished data in Frederick and Crockford 2005
wcVI	DfSh-7, Huu7ii	9	7	2	Unpublished data in Frederick et al. 2006
	Total	96	44 (90%)	5 (10%)	

purpose of this analysis, specimens have been lumped into larger categories of "marine" and "terrestrial" mammals. The marine category includes all pinnipeds (seals and sea lions), delphinids (porpoises, dolphins), and sea otters. River otters (*Lontra canadensis*) are also included in the marine mammal category, as they inhabit coastal waters throughout the study area (Banfield 1974). Whales have been excluded due to the highly fragmentary nature of these large elements, which make identification and quantification problematic (Huelsbeck 1988; Monks 2001). Canids and domestic dogs (the only domesticated animal in the region) have been excluded from this analysis since it is not clear they are used for dietary consumption (Crockford 1997a).

Age and sex classification (based on epiphyseal fusion, eruption and wear of teeth) is not consistently recorded for most of the assemblages in this study. However, sex information is available for a small selection of the examined sites for adult-sized Steller sea lions (summarized in Table 7.3). Ideally, future analyses will examine incremental skeletal structures, such as growth lines in nondeciduous teeth as well as measuring the dimensions of diagnostic elements, which has been published for a handful of sites (e.g., Crockford et al. 2002; Etnier 2002a).

The compiled assemblage data have been entered into to a geographic information system (GIS) and plotted over a series of geospatial layers depicting the major river drainages, the

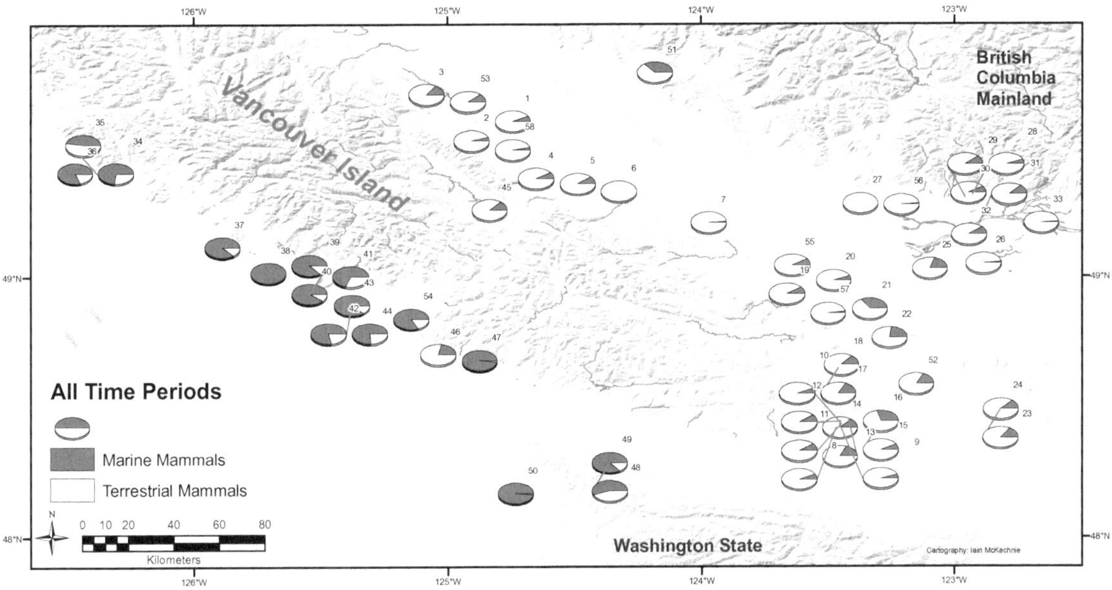

FIGURE 7.2. Relative composition of marine and terrestrial mammals (%NISP) for all sites in the study area ($n=58$). Chronologically distinct periods have been combined in this figure.

terrestrial topography, and the 200 m and 500 m bathymetric contour lines (Figure 7.2). Mammalian fauna from individual sites are represented as pie charts showing the relative abundance of mammals for different taxonomic categories. Our analysis examines the differences in the relative composition between marine and terrestrial mammals for all sites and time periods and then examines species-specific patterns for marine mammals.

RESULTS

Seventy-five temporally distinct assemblages from 58 archaeological sites are included in this analysis, representing a combined total of 67,500 mammalian skeletal specimens (Table 7.2). Fifty-one assemblages are from the Salish Sea region, 21 are from the west coast of Vancouver Island, and three are from the Olympic Peninsula in Washington State. Site distribution is relatively more concentrated in areas impacted by modern development and sparse in other areas, particularly north of the city of Vancouver and on southwestern Vancouver Island (Figure 7.1). Slightly more than half of the sites (53%, $N=31$) were excavated as part of academically driven research projects, while the remainder were cultural resource management projects mandated under heritage legislation in the province of British Columbia. The most recent temporal period (ca. 300–1200 cal BP) contains the greatest number of assemblages ($N=36$), and this progressively decreases with assemblage age (1200–2400 cal BP, $N=26$; 2400–5000, $N=11$; 5000–8000, $N=2$). This trend reflects a sampling bias toward younger sites associated with modern sea levels and a lack of research devoted to locating and sampling earlier Holocene archaeological assemblages (cf. Fedje et al. 2004). Nine sites span multiple temporal periods whereas the remaining sites fall into one of the four broadly defined temporal periods (Table 7.2). Assemblage size (NISP) ranges widely from a low of 5 mammalian elements (Site 38) to a high of 51,937 (Site 50), but over 70% of the assemblages have more than 50 identified specimens. The largest assemblage from Ozette (Site 50) was the product of a decades-long collaborative project between the Makah tribe and a team of archaeologists from Washington State University (Huelsbeck 1994; Samuels and Daugherty 1991). Sites with small assemblages (<50 NISP)

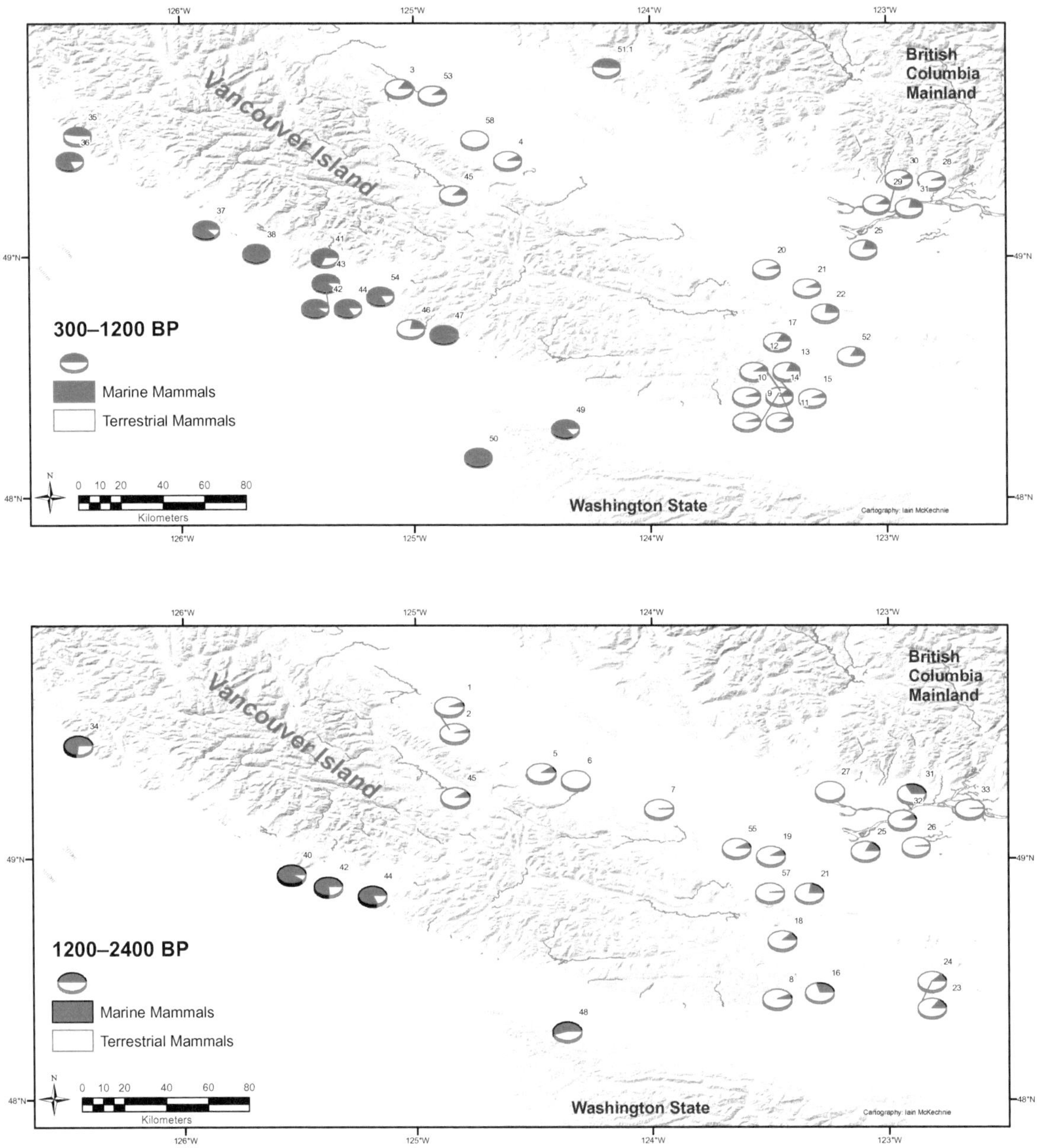

FIGURE 7.3. Relative composition of marine and terrestrial mammals (%NISP) for all sites in the study area according to temporal period.

are included in the analysis as they represent additional locations with confidently identified taxonomic information.

SPATIAL AND TEMPORAL PATTERNING

Figure 7.2 illustrates the relative abundance of marine and terrestrial mammals among all examined assemblages spanning the past 8000 years and shows a sharply defined regional difference in the use of marine mammals and terrestrial mammals between the west and east coasts of Vancouver Island. Sites show remarkable consistency, with terrestrial mammals comprising the majority of all mammalian assemblages. Sites on western Vancouver Island, in contrast, are composed primarily of assem-

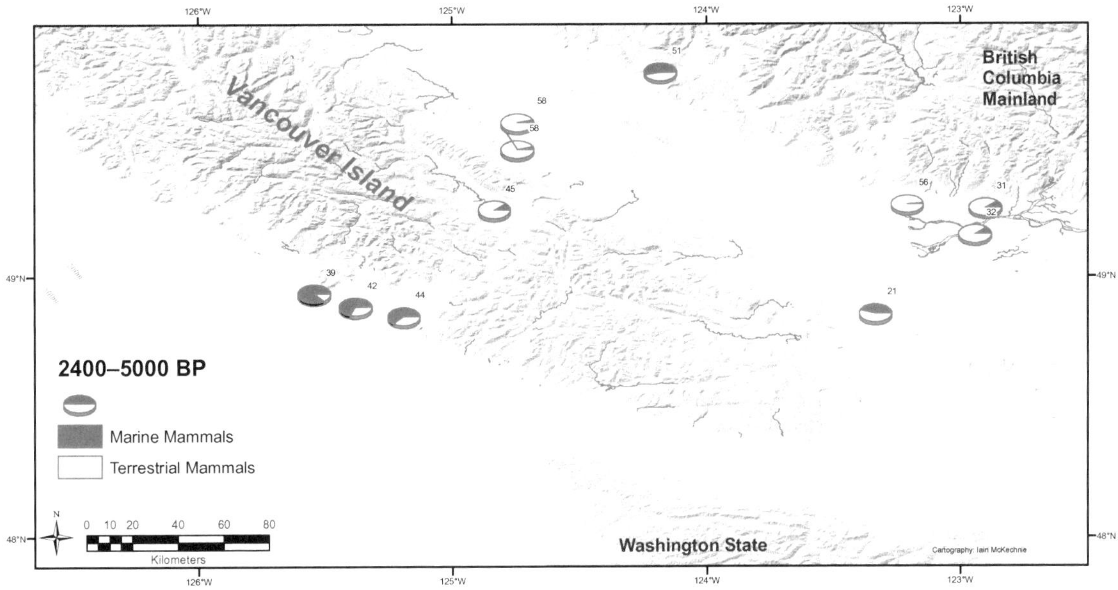

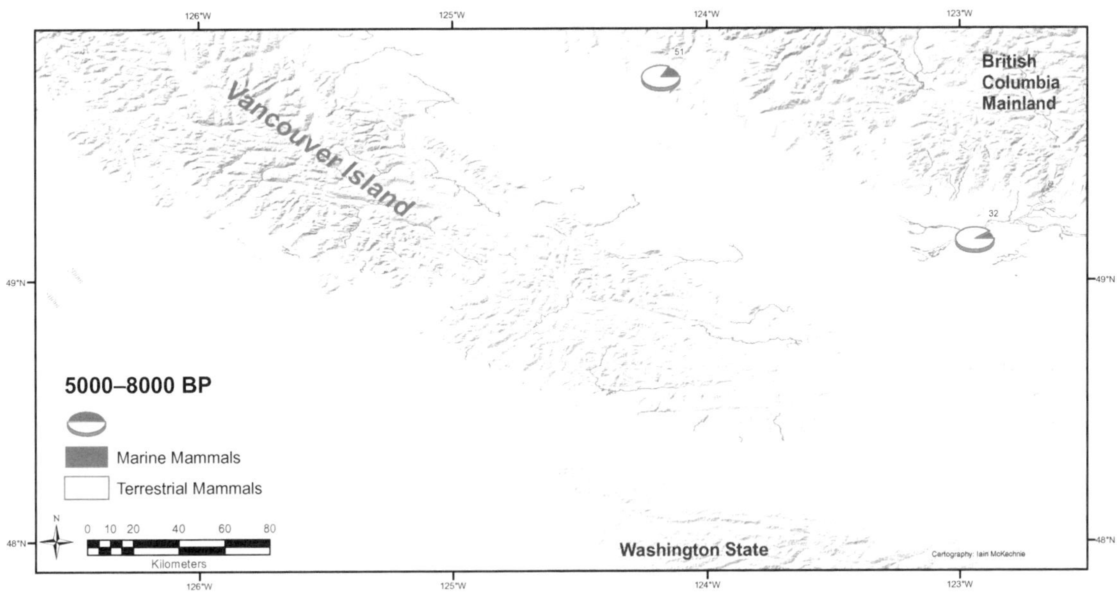

FIGURE 7.3. *(continued)*

blages with more than 75% marine mammals (Figure 7.2). Notable exceptions to this pattern occur at two sites on western Vancouver Island, which contain a higher percentage of terrestrial than marine mammals. One of these sites is a rockshelter in a sheltered area of Hesquiat Harbour (Site 35), and the other is located on a highly exposed linear coastline near a river that is navigable for 10 km upstream but where the surf makes it hazardous to launch canoes in most conditions (Site 46). In the Salish Sea, sites that show the greatest percentage of marine mammals are in the Gulf Islands and the Saanich and Tsawwassen peninsulas (Sites 15–18, 21–24). Sites situated on the east coast of Vancouver Island show a consistently low percentage of marine mammals relative to elsewhere in the Salish Sea. Shoemaker Bay (Site 45), a site at the head of an inlet, which deeply incises the western coast of Vancouver Island, has an assemblage dominated by terrestrial mammals that appears more similar to eastern Vancouver Island assemblages. Saltery Bay (Site 51), a multicomponent site in the northern portion of the study area, shows a notably high percentage of marine mammals relative to other sites in the Salish Sea.

Figure 7.3 depicts the chronological sequence of the relative composition of marine and terrestrial mammals over the four distinct time periods. Terrestrial mammals represent a majority of the two earliest assemblages in the Salish Sea, and this pattern follows in subsequent time periods with the notable exception of Saltery Bay (Site 51). Similarly, sites on western Vancouver Island have a majority of marine mammals and this is consistently represented over the three temporal periods with available data (Figure 7.3). These persistent compositional patterns in regionally distributed mammal assemblages indicate remarkable continuity over long periods of time. However, temporal trends within particular sites are also apparent, although more subtle. Some individual sites have suggestive chronological changes in the relative composition of marine versus terrestrial mammals. For instance, Helen Point (Site 21) shows a relative decrease in marine mammal composition over three successive time periods, which speculatively correlates with greater terrestrial mammal composition on southeastern Vancouver Island (Sites 8–18). In contrast, sites on southwestern Vancouver Island and Washington State appear to show a progressive increase in marine mammal percentages over the three time periods (sites 34–50). While these trends are suggestive of regionalized subsistence practices, it is not within the scope of this paper to evaluate each of these possible local patterns in detail.

Figure 7.4 shows the same temporal sequence of analyses shown in Figure 7.3 but depicts only the relative composition of the three most common pinniped species (harbor seal, Steller sea lion, northern fur seal) and sea otter. This analysis purposefully excludes terrestrial and cetacean (delphinid) specimens, which correspondingly reduces sample size and alters the representation of relative percentage (i.e., %NISP refers to just these species). Nevertheless, this analysis indicates that assemblages on western Vancouver Island are predominantly composed of northern fur seal, followed by Steller sea lion and harbor seal. Northern fur seals are strikingly absent from all sites in the Salish Sea and harbor seals dominate the marine mammal assemblages in the Strait of Georgia, followed distantly by Steller sea lions. Sea otters are present at eight sites on western Vancouver Island and Washington and comprise substantial proportions of four of these assemblages (Sites 34, 36, 44, and 45). Sea otters are present at five sites in the Salish Sea dating to between 2400 and 300 yr BP (Sites 13, 23, 52, and 53), which contrasts with the findings of Hanson and Kusmer (2001), who observed a near absence of sea otters in this region. Where sea otters are present, they represent small proportion of the assemblages except for one small marine mammal assemblage (NISP = 4) that is composed entirely of sea otter (Site 13). Steller sea lions also are present at sites on southern and western Vancouver Island but are found in greater densities on western Vancouver Island and in lower percentages along the Salish Sea (Figure 7.4).

California sea lions are identified in only six sites in the study area, all on the exposed Pacific coast (Table 7.2). Elephant seals are similarly present in small numbers at four sites in the western portion of the study area (Sites 41, 42, 48, and 49). Guadalupe fur seals, whose skeletal morphology closely resembles the northern fur seal (Etnier 2002b), are only identified at one site (NISP/34) which also has the largest identified assemblage (Ozette, Site 50).

AGE AND SEX DATA

Due to the striking disparity between the archaeological and modern distributions of northern fur seal, age and sex data for this once commercially significant species have been discussed elsewhere in great detail (Burton et al. 2001; Calvert 1980; Crockford et al. 2002; Etnier 2002a, 2007; Gifford-Gonzalez et al. 2005; Gustafson 1968; Moss et al. 2006; Newsome et al. 2007). These studies have identified adult males and females as well as immature pups less than 6 months of age in a variety of sites in the study area. Combining archaeological data from Ts'ishaa (Site 42) with historic accounts from schooner captains, aboriginal hunters, and merchants involved in the sealing industry of the 1880s (Swan 1883, 1887), Crockford et al. (2002:170) argue for the presence of a nonmigratory breeding population of fur seals capable of giving birth in kelp beds or kelp rafts at sea. Subsequent studies have used stable isotopes to posit the existence a mid-latitude breeding distribution in the Pacific Northwest (Newsome et al. 2007). However, definitive evidence of a terrestrially based northern fur seal rookery remains to be identified.

Age and sex data from few species other than fur seal have been compiled. For sites in the study sample, a small portion of Steller sea lion skeletal remains from a limited number of sites has been confidently sexed. Table 7.3 shows the breakdown of sexed elements for sites where the information was available from controlled excavations. Although the sample is small (NISP=19), 95% of the sexed elements are from adult males, with only two elements out of 40 either unknown or female. This pattern is similar to data from the Oregon coast, which indicates that Steller sea lions from archaeological contexts are primarily adult males (Lyman 2003b). The incidence of females and/or juveniles of either sex are very low, with only one definitively female element found on southeast Vancouver Island (Table 7.3).

Additional Steller sea lion sex and age information comes from fauna collected from archaeological contexts during monitoring of construction projects (Table 7.3). Monitoring samples are collected without screening and have only general provenience. However, males dominate four large assemblages from the southern Salish Sea region. For example, at Patricia Bay (Site 18), at least 74 male Steller sea lion elements have been identified, in contrast to only one female and three juveniles of indeterminate sex (Wigen 2007:24). Explanation for this pattern of male dominance could relate to seasonal availability of Steller sea lions and the differential behavior of the females and males at rookeries and haul-out sites and/or preference on the part of the human hunters.

Steller sea lions breed at communal rookeries in northern British Columbia during the summer but spend the winter at haul-out sites throughout British Columbia (Olesiuk 2009b). Since the first survey in British Columbia in 1913 (Newcombe and Newcombe 1914), the most southerly sea lion breeding rookery is located on the north end of Vancouver Island. Nonbreeding animals of both sexes use a limited number of established year-round haul-outs found mainly on the west coast of Vancouver Island. In the winter, individuals of both sexes use a wider range of haul-outs, including the Strait of Georgia and southern Vancouver Island. This pattern suggests both males and females should be available to hunters in the winter, so it appears seasonal distribution does not explain the hunting pattern. Recent studies of the behavioral responses to human disturbance of hauled-out Steller sea lions note that females and pups are the first to leave, followed by the subadult males and, finally, the bulls (COSEWIC 2003; Szaniszlo 2005). The dominance of males in the archaeological sites may be a result of this behavioral pattern, with the males being easier for the hunters to target (Lyman 2003b). Additionally, haul-outs of "bachelor males" are commonly present near rookeries and may have also been targeted by hunters. The extremely large size of males and these behavioral characteristics suggest that males were hunted while onshore rather than offshore, as documented for fur seals.

Harbor seals and sea otters are more difficult to sex, as sexual dimorphism is less extreme. In most cases, no information has been recorded about the sex of individuals. Age for harbor seals has been recorded in a few site

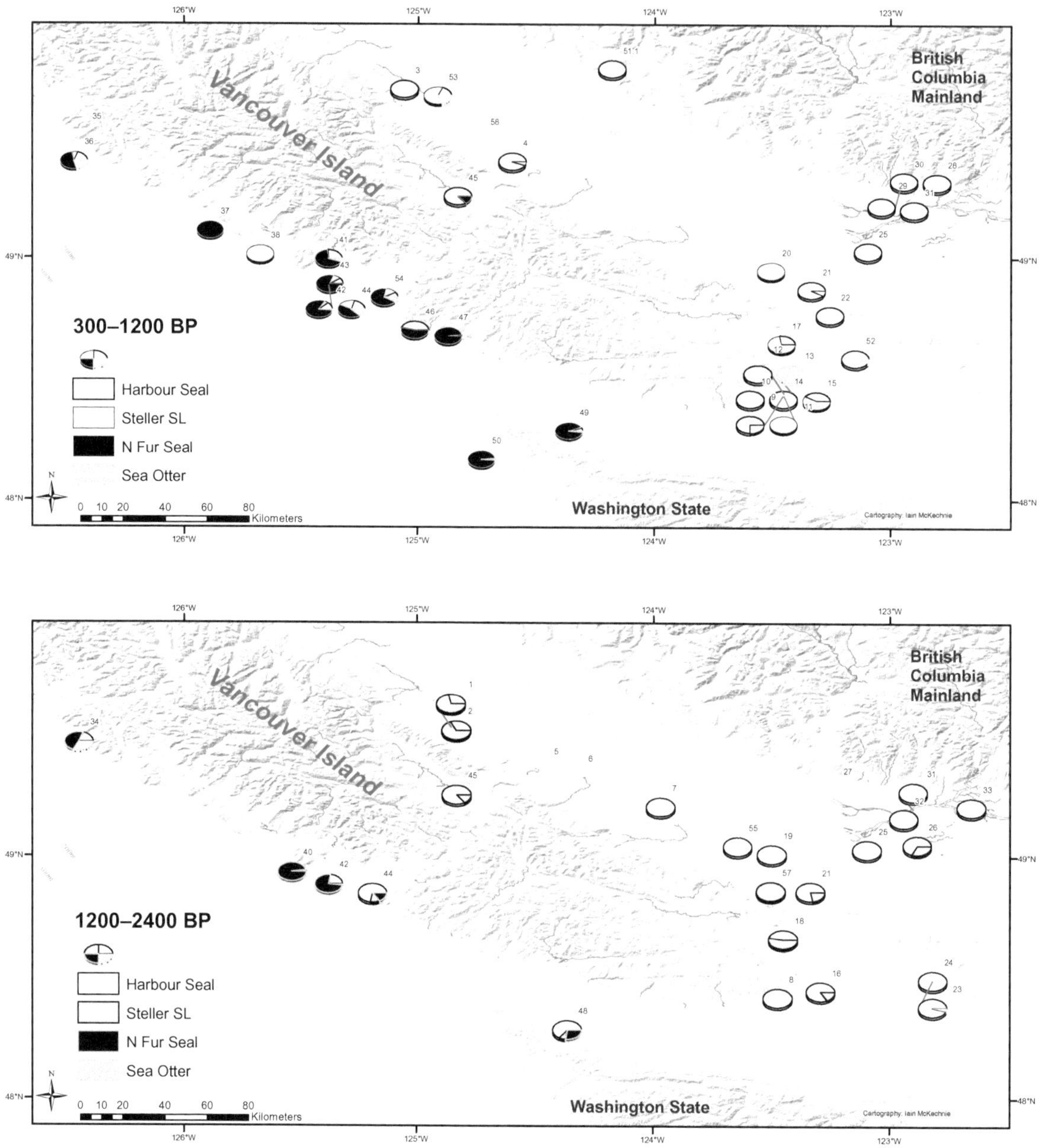

FIGURE 7.4. Relative abundance of marine versus terrestrial mammals for sites dating to within the period 1200 to 2500 BP. Size of the pie charts is scaled to assemblage size.

reports (e.g., Frederick and Crockford 2005; Frederick et al. 2006; Wigen 2007), indicating that harbor seals of all ages were regularly hunted in all areas. No age or sex information is available for the few identified California sea lion or elephant seal elements.

Steller sea lions are present in sites along the southwestern portion of the study area, suggesting they may have been moderately abundant in the ancient environment. Elephant seals, California sea lions, and Guadalupe fur seals rarely occur in the assemblages, suggesting they were not commonly present at this latitude. However, taxonomic identifications of male California sea lions may be confused with female Steller sea lions (personal observations),

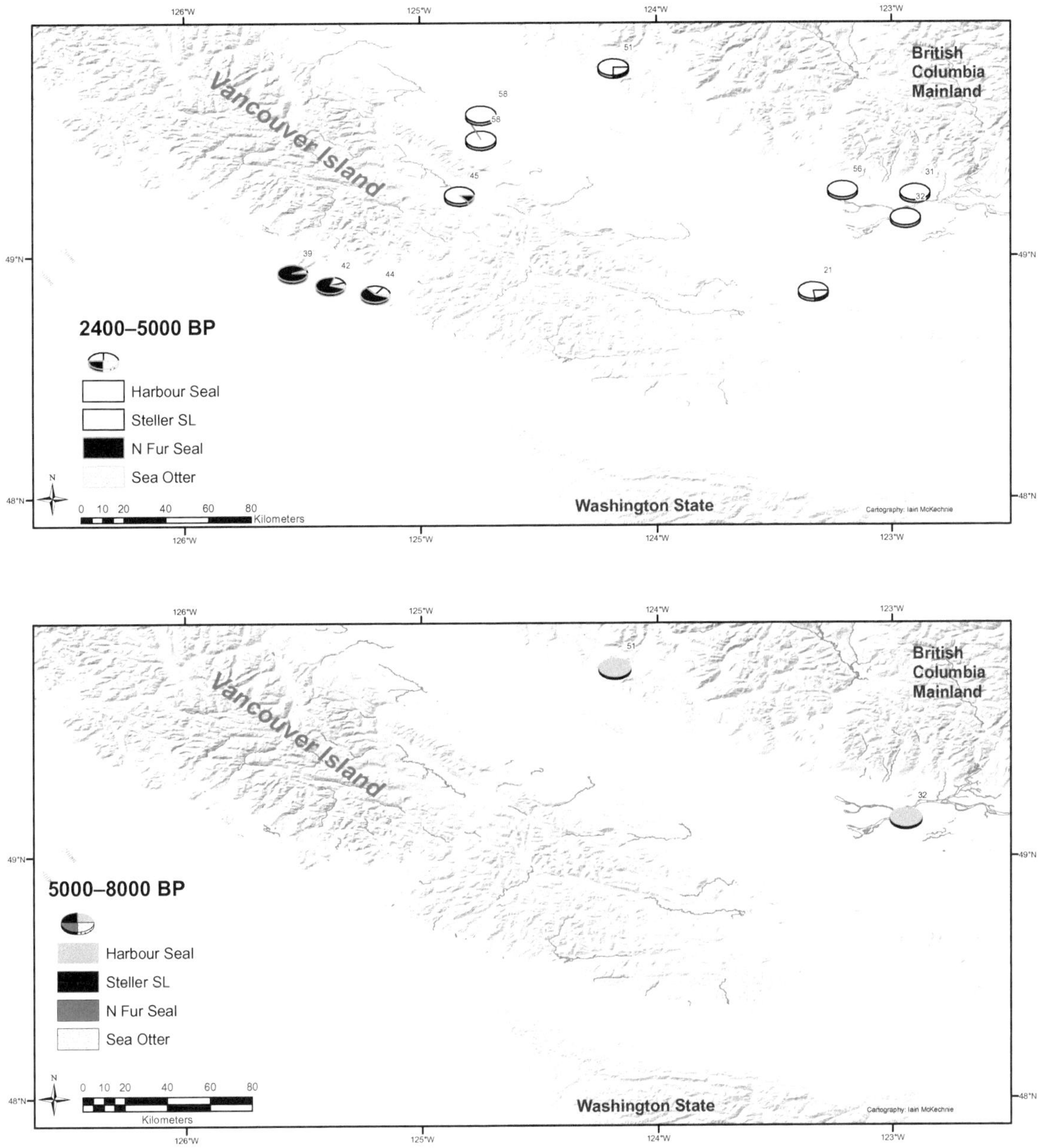

FIGURE 7.4. *(continued)*

and, similarly, Guadalupe fur seals appear similar to northern fur seals (Etnier 2002b).

DISCUSSION

This analysis has identified distinct regional patterns in marine and terrestrial mammal assemblages from archaeological sites on western Vancouver Island and the Salish Sea. These differences provide a basis for inferring that aboriginal hunting practices dramatically differed in focus between these two regions. In particular, aboriginal peoples in the Salish Sea region did not hunt marine mammals to the same degree as was practiced by aboriginal peoples along the Pacific coast of Vancouver Island and Washington State. Rather, terrestrial mammals appear to be relatively more important in the Salish Sea region and relatively less so on the exposed Pacific coast.

These geographic patterns appear robust at least over the past 5000 years and have implications for a number of perspectives in anthropology and ecology. Of particular anthropological relevance is the apparent coherence and continuity of aboriginal marine and terrestrial hunting patterns within each of the two areas. This difference is most strongly apparent in the relative composition of marine versus terrestrial mammals (Figure 7.2 and Figure 7.3) and among the individual marine species (Figure 7.4). For specific localities within these two regions, variation in marine mammal use also appears to be spatially patterned, such as the slightly higher abundance of marine mammals in assemblages along southeastern Vancouver Island and in the Gulf Islands (Figure 7.2). These archaeological patterns are an outcome of the direct participation of generations of aboriginal people in this marine ecosystem. Such patterns reflect a continuity of skillful cultural knowledge and practical engagement with these marine environments (cf. Ingold 2000).

These regional differences are consistent with a host of ethnographic accounts and observations (e.g., Boas 1887; McMillan 1999; Suttles 1987b), which identify the aboriginal peoples who occupy these two regions as belonging to two culturally and linguistically distinct groups (Wakashan-speaking peoples [Nuu-chah-nulth, Ditidaht, Makah] and Salish-speaking peoples). During the postcontact era (ca. AD 1774–1860) and continuing among First Nations today, these cultural differences include distinct patterns of residence, conventions of artistic representation, ritual practice, and a host of other regionally unique forms of cultural expression (ibid.). Our analysis provides insight into an aspect of these rich cultural histories, suggesting that cultural distinctions between regions and cultural similarities within regions have great antiquity in regards to hunting practice. Indeed, since our data relate to an important subsistence activity (hunting), such regional expressions can be hypothesized to have contributed structure to cultural history and community identity. This is illustrated in a 1922 quote discussing the biography of Tom Sayach'apis (ca. 1835–1927), a respected cultural historian of the Tseshaht, a Nuu-chah-nulth Nation on southwestern Vancouver Island:

> Tom ate very little meat of land mammals in his early days. Indeed, like most of the [west] Coast people, he had a prejudice against deer meat and it was not until as a middle-aged man, he had come into contact with some of the deer hunting tribes of the interior of the island, that he learned to prize it, though to this day, venison has not for him the toothsome appeal of a chunk of whale meat. (Sapir 1922:304)

From an ecological perspective, such strong geographic patterning as seen through the lens of human hunting activity is relevant for considering the long-term ecological role of marine mammal species in the study area. Our data represent a unique and otherwise unattainable record of marine mammal abundance and distribution over the Holocene. This information helps contextualize contemporary marine mammal ecology by greatly extending temporal and geographic knowledge of these important species. In this regard, it is notable that the three most abundant pinnipeds in the archaeological assemblages (northern fur seals, harbor seals, and Steller sea lions) are also the three most numerically abundant in the contemporary environment (Table 7.1). The similarity in rank-ordered abundance between our archaeological data and contemporary pinniped population estimates indicates these three species have exhibited remarkable demographic resilience to human harvesting.

The stable long-term trends observed in the archaeological data stand in contrast to the dynamic historical events documented over the past 200 years that resulted in the rapid extirpation of the sea otter (ca. AD 1774–1811) and culling of pinniped populations (ca. AD 1868–1970). In this context, it seems important to note that the sudden cessation of marine mammal hunting over the past 40 years appears to be unprecedented in a record spanning at least the past 5000 years. The closest possible parallels to this sudden reduction in hunting intensity and corresponding population increase likely occurred after European diseases dramatically reduced aboriginal populations in the early contact era (ca. AD 1774–1811) and again in the mid-19th century, just prior to the beginning of the commercial fur seal industry (Boyd 1999).

While it is striking to note the ongoing demographic consequences of historic-era commercial hunting, equally important are possible behavioral and ecological changes that occurred among targeted taxa over the Holocene. For instance, our compiled archaeological results indicate that aboriginal peoples regularly hunted northern fur seals for a minimum of 5000 years on the exposed Pacific coasts of Vancouver Island and northwestern Washington. These observations correspond with the locations of northern fur seal hunting mentioned in historic-era harvest logbooks (Olesiuk 2009a:42) indicating that the continental shelf off southwestern Vancouver Island has long been a major focus of fur seal foraging activity. However, many of these same historic-era harvest records (ca. AD 1891–1911) also indicate a strongly seasonal presence of northern fur seals at this latitude (January–May; Olesiuk 2009a:54), which contrasts with zooarchaeological data including osteometric evidence for localized breeding (Crockford et al. 2002; Etnier 2002a; Gustafson 1968), and geochemical evidence for nonmigratory midlatitude foraging and prolonged pup weaning dating from the mid-Holocene (Newsome et al. 2007). Tantalizingly suggestive historic accounts compiled from interviews with commercial fur sealers conducted a decade prior to the historic log-book-based records mentioned above, describe nonmigratory behavior involving giving birth at sea in kelp beds and observing fur seals in the Strait of Georgia during late summer when the migratory populations are in the Bering Sea (Crockford et al. 2002 citing Swan 1883). These data suggest historic-era commercial harvests may have eliminated or altered a behaviorally unique subpopulation of fur seals. Archaeological data yield important insight into the previously more diverse ecology of this currently strictly pelagic pinniped.

Although the absolute number of animals harvested cannot be quantified with our zooarchaeological data alone, our analysis has several implications concerning potential human impacts on pinniped and sea otter populations. First, our data indicate a regional focus on specific pinniped species, which implies that aboriginal peoples in these regions targeted specific taxa with regularity (e.g., annually or seasonally) as suggested by ethnographic accounts (e.g., Arima 1988; Suttles 1987a). Second, our data represent only a small percentage of the approximately 4000 currently documented coastal shell-midden sites present in the study area (British Columbia Archaeology Branch n.d.). Thus, the collective impact of aboriginal hunting may be considerably greater than previously recognized if these regional patterns apply to these several thousand unexamined sites. Finally, the temporal and spatial consistency of hunted marine mammals in these two regions indicates a degree of stability and continuity to aboriginal resource harvesting practices, which demonstrates the capacity for sustained harvesting in antiquity (cf. Etnier 2007).

Repetitive and consistent human participation in an ecosystem provides structure to ecological interactions (Balee 2006; Hobbs and Fowler 2008; Liu et al. 2007). However, determining and/or identifying "human impacts" in such an archaeological context is challenging as these ecosystems may represent the endpoint of centuries or millennia of human hunting. Future analyses that can incorporate more fine-grained osteometric information on age and sex distribution will likely yield greater insight into these ecological impacts.

Considering the ecological dimensions of human subsistence activity in coastal environments, Fedje et al. (2004) offer a distinction that seems to encompass the regional differences we observe in our analysis of mammalian hunting practices; the concept of "maritime" and "coastal adaptations." These authors define a "maritime adaptation" as one that is:

> heavily reliant upon marine and coastal resources for the majority of subsistence needs. In contrast, a "coastal adaptation" involves the fluent use of coastal and marine resources for at least some, and probably a significant portion, of the subsistence needs of a group of people. Coastal adaptations should therefore include the possibility of considerable use of terrestrial resources and inland areas. (Fedje et al. 2004:112)

This synthetic statement helps characterize the regional differences observed in our analysis, suggesting that aboriginal peoples in the Salish Sea adopted and practiced terrestrial mammal hunting much more extensively than aboriginal peoples on western Vancouver Island. This observation is consistent with ethnographic and archaeological information from Coast Salish peoples in the Salish Sea region, where widespread aboriginal burning practices aimed to enhance and expand habitat for economically important plants such as camas (*Camassia quamash*) and berries as well as to entice foraging ungulates into recently burned areas with young regrowth (Brown and Hebda 2002; Suttles 1987b; Weiser and Lepofsky 2009). Conversely, on western Vancouver Island, the logistics of hunting pinnipeds in a pelagic environment likely facilitated other forms of offshore subsistence activities (e.g., deep-water fishing and birding) as well as voyaging and trading (Arima 1988; McMillan 1999).

However, while the concept of "adaptation" (defined as "the suiting of one thing to another") may help characterize human hunting activity expressed over thousands of years, a term that is more appropriately scaled to human lifetimes is "tradition" (defined as "the act of handing down"). Considering long-term hunting practices as "traditions" accommodates the intergenerational transference of complex hunting knowledge in a way that anthropological uses of "adaptation" often underspecify or ignore altogether (cf. Moss 2008).

The regionally consistent patterns of terrestrial and marine mammal use were additionally constrained or enabled by environmental differences between the two regions: for instance, the open prairie oak savanna woodlands of the Salish Sea versus the very moist dense coastal rainforest on western Vancouver Island (Suttles 1987b), or the higher primary productivity on the exposed Pacific coast versus the Salish Sea (Ware and Thompson 2005). However, while it is essential to consider the environmental factors that may contribute to how aboriginal peoples may have focused on marine or terrestrial mammals, these factors alone cannot adequately explain how or why generations of people choose to participate in an ecosystem in a particular manner. Rather, explanations that posit an environmental imperative for human action risk imposing an unnecessarily deterministic assumption. Thus, although the relative contribution of marine and terrestrial hunting was undoubtedly constrained to some degree by the presence or absence of particular animals, human hunting may also have facilitated and/or enhanced abundance of certain species. For instance, the ethnographically observed harvesting of starchy, carbohydrate-rich camas bulbs in the Coast Salish region may have nutritionally facilitated higher levels of ungulate hunting (which are rich in protein but are relatively low in fat). Likewise, the widespread association of camas harvesting with aboriginal burning likely increased ungulate forage (and presumably abundance) while also expanding patchy savanna habitat conducive to terrestrial hunting (Weiser and Lepofsky 2009). On the exposed Pacific Coast, however, the observed focus on pinniped hunting (foods that are rich in fat and oil) may have nutritionally compensated for a relative lack of dietary carbohydrates and enabled the elaboration of other forms of pelagic hunting practice while also expanding the capacity for trade and regional interaction. Rather than a simple consequence of "available resources," these hunting practices reflect socially and logistically elaborate traditions that likely have considerable ecological significance to contemporary ecosystem.

In contrast to an environmentally determined model of human subsistence, aboriginal peoples of the Northwest Coast are increasingly recognized as highly active in "managing" subsistence resources through explicit cultural definitions of harvesting practice and etiquette including selective harvesting, seasonal restrictions on use and/or consumption, and proprietorship over resources that was contingent on sustained productivity (Berkes and Turner 2006; Hunn et al. 2003; Weiser and Lepofsky 2009; Trosper 2009). The data presented here suggest that generations of people actively chose to pursue an elaborate, highly structured, and specialized activity—primarily pelagic marine mammal hunting or primarily terrestrial mammal hunting in different regions. These hunting traditions embody the skill, training, and cooperation, that develop over years and decades among groups of individual hunters and which are passed on to subsequent generations in a community of continuous social practice (cf. Bourdieu 1977). The consistency and continuity with which certain species were utilized in the archaeological assemblages over 5000 years suggests that similar cultural practices informed aboriginal peoples' hunting behavior in the archaeologically documented past.

CONCLUSIONS

We posed three questions at the beginning of this chapter:

- What marine mammals did aboriginal people in southern British Columbia most commonly utilize?
- How similar or how different are species occurrences and proportions relative to today?
- Is there evidence of specialized or regional hunting traditions and if so, what might have been the potential impacts of these activities on the ancient marine ecosystem?

In answer to the first question, northern fur seals, harbor seals, and Steller sea lions respectively appear to be the most heavily utilized pinnipeds on the southern BC coast over the last 8000 years. Sea otters appear to be utilized by peoples on western Vancouver Island and the Olympic Peninsula, and are present only in low numbers at several sites in the Salish Sea. The least utilized pinnipeds appear to be California sea lions, elephant seals, and Guadalupe fur seals, which are rare but are found in sites along the southwestern margin of the study area. These are relative differences however, and, considering these data represent small samples from much larger sites, even small relative percentages may in fact represent large numbers of harvested animals.

In regards to the second question, our data expand knowledge of the range and relative composition of marine mammal species that are currently threatened or endangered (Table 7.1). Examining the archaeologically known distribution of these species provides context for understanding their recovery since historical overexploitation and the extent to which aboriginal peoples may have influenced the marine ecosystem before the historic-era fur trade. Considering the fact that the 58 examined sites represent less than 1% of the approximately 4000 currently documented coastal shell-midden sites present in the study area (BC Archaeology Branch n.d.), the perspective gained here hints at the potential cumulative influence of large indigenous human populations in the region. The extent and significance of this influence is unknown at this time, but did not appear to have resulted in a dramatical change their hunting patterns. Such basic insights help refine our understanding of the influence of human activity on the environment and, conversely, the influence of animals on human activity.

And finally, this study indicates that aboriginal people on the outer coast of Vancouver Island and Washington State developed a specialized hunting tradition targeting northern fur seals. This conclusion is further supported by several ethnographic and historic sources (e.g.,

Arima 1988; Sapir and Swadesh 1955; Swan 1887; Waterman 1920) and by artifactual data from precontact archaeological sites in the study area (cf. McMillan 1999; Samuels and Daugherty 1991). Ethnographically, such hunting activity involved paddling considerable distances offshore, sometimes beyond sight of land, to stealthily approach "sleeping" fur seals to within range for launching a harpoon. Such specialized journeys entailed considerable physical risks and logistical preparation. Ethnographic accounts emphasize how participants ritually engaged in elaborate physical and spiritual training and how hunters utilized specifically designed canoes, sharpened paddles, unique harpoons, yew-wood throwing sticks, seal skin floats to tire harpooned animals, as well as long lengths of cedar-bark and elk-hide rope (Arima 1988; Sapir and Swadesh 1955; Waterman 1920).

In contrast to western Vancouver Island and northwestern Washington, aboriginal peoples inhabiting the Salish Sea appear to have developed marine mammal hunting traditions that targeted harbor seals and to a lesser extent Steller sea lions. As discussed previously, harbor seals are nonmigratory, nearshore foragers who regularly haul out on land and are the most common seal present on the coast today (Table 7.1). Ethnographic accounts describe hunting hauled-out harbor seals from boats, by hand with clubs, or by stealthily placing nets in the water below haul-out areas, particularly outside sea caves (Suttles 1987a; Waterman 1920). Steller sea lions, on the other hand, occur seasonally in the Salish Sea during winter and spring when herring aggregate prior to spawning (COSEWIC 2003). Ethnographically, sea lions were hunted by numerous Coast Salish peoples, although a few groups were recognized as particularly skilled specialists (Suttles 1987a). Ethnographic descriptions of hunting both Steller sea lions and harbor seals were associated with elaborate forms of physical and spiritual preparation, which occurred prior to and during the hunt and were observed by both hunters and their wives (Suttles 1987a; Elmendorf 1960).

While hunting practices documented in archaeological contexts represent only a small portion of the rich ancestral pasts of coastal First Nation communities, zooarchaeological data represent a readily available source of information that broadens our contemporary understanding of ancient coastal lifeways and cultural environments. Such ecologically significant marine data is essentially nonexistent outside of archaeological contexts, and future research will benefit from additional analyses of new sites, regions, and time periods as well as more detailed osteometric observations. The compilation of zooarchaeological information explored in this paper suggests coherent patterns indicating the persistence of hunting traditions practiced continuously over 5000 years. These patterns help unravel the complex interwoven cultural and ecological histories on the Northwest Coast.

ACKNOWLEDGMENTS

We thank editors Todd Braje and Torben Rick for the opportunity to submit this paper, for their editorial comments, and for their patience in waiting for its arrival. We thank the dozens of individuals who have contributed to the zooarchaeological record on the BC Coast over the past 50 years. We additionally thank Mike Blake and the UBC Laboratory of Archaeology for providing us with mapping software and guidance, as well as Susan Crockford, Quentin Mackie, and Nicole Smith for encouraging helpful comments (and relieve them of any responsibility for our final product). Thanks also to Parks Canada, Pacific Identifications, IR Wilson Consultants, and the Provincial Archaeological Report Library of BC for bibliographic information. McKechnie acknowledges support from the Charles and Alice Borden fellowship for Archaeology at the University of British Columbia and the Social Sciences and Humanities Research Council of Canada.

REFERENCES CITED

Allen, S. E., C. Vindeirinho, R. E. Thomson, M. G. G. Foreman, and D. L. Mackas
2001 Physical and Biological Processes over a Submarine Canyon during an Upwelling Event. *Canadian Journal of Fisheries and Aquatic Sciences* 58:671–684.

Ames, K. M., and H. D. G. Maschner
1999 *Peoples of the Northwest Coast: Their Archaeology and Prehistory.* Thames and Hudson, London.
Arima, E. Y.
1983 *The West Coast People: The Nootka of Vancouver Island and Cape Flattery.* Special Publication 6. British Columbia Provincial Museum, Victoria, B.C.
1988 Notes on Nootkan Sea Mammal Hunting. *Arctic Anthropology* 25(1):16–27.
Baird, R. W.
2001 Status of Harbour Seals in Canada. *Canadian Field Naturalist* 115(4):663–675.
Balee, W.
2006 The Research Program of Historical Ecology *Annual Review of Anthropology* 35:5.1–5.24.
Banfield, A. W. F.
1974 *The Mammals of Canada.* National Museum of Natural Science, University of Toronto Press, Toronto.
Berkes, F., and N. Turner
2006 Knowledge, Learning and the Evolution of Conservation Practice for Social-Ecological System Resilience. *Human Ecology* 34(4):479–494.
Bernick, K., and R. J. Wigen
1990 Seasonality of the Little Qualicum River Wet Site. *Northwest Anthropological Research Notes* 24(2):153–159.
Bigg, M. A.
1985 Status of Steller Sea Lion (*Eumetopias jubatus*) and California Sea Lion (*Zalophus californianus*) in British Columbia. Canadian Special Publication of Fisheries and Aquatic Sciences 77. Pacific Biological Station, Department of Fisheries and Oceans, Nanaimo.
1988 Status of the Steller Sea Lion, *Eumetopias jubatus,* in Canada. *Canadian Field Naturalist* 102(2):315–336.
Bigg, M. A., and I. B. MacAskie
1978 Sea Otters Reestablished in British Columbia. *Journal of Mammalogy* 59(4):874–876.
Boas, F.
1887 The Coast Tribes of British Columbia. *Science* 9(216):288–289.
Boehm, S. G.
1973 *Cultural and Non-Cultural Variation in the Artifact and Faunal Samples from the St. Mungo Cannery Site, DgRr-2.* M.A. thesis, University of British Columbia.
Boucher, N. D.
1976 *Prehistoric Subsistence at the Helen Point Site.* M.A. thesis, Simon Fraser University.
Bourdieu, P.
1977 *Outline of a Theory of Practice.* Cambridge University Press, Cambridge.
Boyd, R. T.
1999 *The Coming of the Spirit of Pestilence: Introduced Infectious Diseases and Population Decline Among Northwest Coast Indians, 1774–1874.* University of Washington Press, Seattle.
British Columbia Archaeology Branch
n.d. Remote Access to Archaeological Data Application. Accessed March 15, 2010. http://www.tsa.gov.bc.ca/archaeology/.
Brolly, Richard, and Robert J. Muir
1993 Archaeological Investigations at the Locarno Beach Site (DhRJ-6), Vancouver, B.C. Arcas Consulting Archaeologists report submitted to the Archaeology Branch, Ministry of Small Business, Tourism and Culture.
Brolly, R., R. J. Muir, and R. Schulting
1993 *Archaeological Investigations at the Harbour House Site (DjRu 3), Saltspring Island, B.C., Permit 1993-83.* ARCAS Associates Report Prepared for the Ministry of Transportation and Highways and submitted to the Archaeology Branch of the Ministry of Small Business, Tourism and Culture.
Brown, K. J., and R. J. Hebda
2002 Ancient Fires on Southern Vancouver Island, British Columbia, Canada: A Change in Causal Mechanisms at about 2000 YBP. *Environmental Archaeology* 7:1–13.
Burton, R. K., J. J. Snodgrass, D. Gifford-Gonzalez, T. Guilderson, T. Brown, and P. L. Koch
2001 Holocene Changes in the Ecology of Northern Fur Seals: Insights from Stable Isotopes and Archaeofauna. *Oecologia* 128:107–115.
Busch, B. C., and B. M. Gough
1997 *Fur Traders from New England: The Boston Men in the North Pacific, 1787–1800, The Narratives of William Dane Phelps, William Sturgis and James Gilchrist Swan.* Arthur H. Clarke Company, Spokane.
Butler, V. L., and S. K. Campbell
2004 Resource Intensification and Resource Depression in the Pacific Northwest of North America: A Zooarchaeological Review. *Journal of World Prehistory* 18(4):327–405.
Calvert, S. G.
1980 *A Cultural Analysis of Faunal Remains from Three Archaeological Sites in Hesquiat Harbor, B.C.* Ph.D. dissertation, University of British Columbia.
Cannon, A.
2001 Was Salmon Important in Northwest Coast Prehistory? In *People and Wildlife in Northern North America: Essays in Honor of R. Dale Guthrie,*

edited by S. C. Gerlach and M. S. Murray, pp. 178–187. International Series, vol. S944. British Archaeological Reports, Oxford, UK.

Capes, K. H.

1964 *Contributions to the Prehistory of Vancouver Island* 15. Idaho State University, Occasional Papers, Pocatello, ID.

Carlson, C. C.

2003 The Bear Cove Fauna and the Subsistence History of Northwest Coast Maritime Culture. In *Archaeology of Coastal British Columbia: Essays in Honour of Professor Philip M. Hobler*, edited by R. L. Carlson, pp. 65–86. Archaeology Press, Simon Fraser University, Burnaby, B.C.

Charlton, A. S.

1977 *The Archaeology of the Belcarra Park Site: A Contribution to Strait of Georgia Prehistory*. M.A. thesis, Simon Fraser University.

Committee on the Status of Endangered Wildlife in Canada (COSEWIC)

2003 *COSEWIC Assessment and Update Status Report on the Steller Sea Lion, Eumetopias jubatus, in Canada*. Committee on the Status of Endangered Wildlife, Ottawa.

Coupland, G.

1991 The Point Grey Site: A Marpole Spring Village Component. *Canadian Journal of Archaeology* 15(1):73–96.

Crockford, C. E.

1996 *Nuu-chah-nulth Labour Relations in the Pelagic Sealing Industry, 1868–1911*.M.A. thesis, University of Victoria.

Crockford, S. J.

1997a *Osteometry of Makah and Coast Salish Dogs*. Archaeology Press, Simon Fraser University, Burnaby.

1997b *Faunal Analysis Report, DcR-u 2 Esquimalt Lagoon, Fort Rodd Hill National Historic Park Victoria*. Pacific Identifications report submitted to Parks Canada Western and Northern Service Centre, Calgary.

Crockford, S. J., S. G. Frederick, and R. J. Wigen

2002 The Cape Flattery Fur Seal: An Extinct Species of *Callorhinus* in the Eastern North Pacific? *Canadian Journal of Archaeology* 26(2): 152–174.

Croes, D. R.

1995 *The Hoko River Archaeological Site Complex: The Wet/Dry Site (45CA213), 3000–1700 B.P.* Washington State University Press, Pullman.

Croes, D. R., and S. Hackenberger

1988 Hoko River Archaeological Complex: Modeling Prehistoric Northwest Coast Economic Evolution. In *Prehistoric Economies of the Pacific Northwest Coast*, edited by B. Issac, pp. 19–85. Research in Economic Anthropology, vol. 3. JAI Press, Greenwich CT.

Department of Fisheries and Oceans

2003 *Population Assessment for Steller Sea lions (*Eumetopias jubatus*)*. Science Advisory Report 2003/37. Canadian Science Advisory Secretariat, Ottawa.

2007 *Population Assessment for Northern Fur Seals (*Callorhinus ursinus*)*. Science Advisory Report 2007/52. Canadian Science Advisory Secretariat, Ottawa.

2010 *Population Assessment Pacific Harbour Seal (*Phoca vitulina richardsi*)*. Science Advisory Report 2009/011. Canadian Science Advisory Secretariat, Ottawa.

Dick, L.

2006 *The Maritime Fur Trade in Southern Haida Gwaii (Queen Charlotte Islands), ca. 1787–1920*. Report on file, Western and Northern Service Centre, Parks Canada, Vancouver, British Columbia, Canada.

Driver, J. C.

1993 Zooarchaeology in British Columbia. *BC Studies* 99:77–104.

Drucker, P.

1951 *The Northern and Central Nootkan Tribes*. Bureau of American Ethnology Bulletin 144. Smithsonian Institution, Washington D.C.

Eldridge, M.

1987 Mitigative Excavation at DiSe-10, Denman Island, B.C. Rockshelters and Blufftop Hunting Magic. Report on file at the Archaeology and Registry Services Branch.

2000 *Eagles Nest Archaeological Data Recovery DcRt-71*. Report prepared by Millennia Research Ltd for Ken Ujimoto, International Integration Inc. (agent for M. Wakao) and Songhees First Nation. Report on file at the Archaeology and Registry Services Branch.

Elliott, H. W.

1886 *Our Arctic Province: Alaska and the Seal Islands*. 1897 ed. Library of Contemporary Exploration and Adventure. Charles Scribner's Sons, New York.

Elmendorf, W. W.

1960 *Twana Narratives: Native Historical Accounts of a Coast Salish Culture*. University of Washington Press, Seattle.

Engisch, C.

2006 *Report for the Post-Construction Archaeological Impact Assessment of the Aguilar House, Bamfield BC Heritage Inspection Permit 2006-402*. Baseline Archaeological Services Report prepared for Mark Leriger and submitted to the

Huu-ay-aht First Nation and the BC Archaeology Branch.
Environment Canada
2009 Canadian Climate Normals or Averages 1971–2000. In *National Climate Data Archive, http://www.climate.weatheroffice.ec.gc.ca/climate_normals/index_e.html.*Government of Canada, Ottawa.
Estes, J. A.
1999 Sea Otter, *Enhydralutris.* In *Smithsonian Book of North American Mammals,* edited by D. E. Wilson and S. Ruff, pp. 180–183. UBC Press and Smithsonian Institution Press, Vancouver and Washington, D.C.
Etnier, M. A.
2002a *The Effects of Human Hunting on Northern Fur Seal (Callorhinus ursinus) Migration and Breeding Distributions in the Late Holocene.* Ph.D. dissertation, University of Washington.
2002b Occurrences of Guadalupe Fur Seals (*Arctocephalus townsendi*) on the Washington Coast over the Past 500 Years. *Marine Mammal Science* 18(2):551–557.
2007 Defining and Identifying Sustainable Harvests of Resources: Archaeological Examples of Pinniped Harvests in the Eastern North Pacific. *Journal for Nature Conservation* 15(3): 96–207.
Ewonus, P. A.
2006 *The Social Economy of a Northwest Coast Plank House in Perspective.* M.A. thesis, McMaster University.
Fedje, D. W., Q. Mackie, E. J. Dixon, and T. H. Heaton
2004 Late Wisconsin Environments and Archaeological Visibility on the Northern Northwest Coast. In *Entering America: Northeast Asia and Beringia before the Last Glacial Maximum,* edited by D. B. Madsen, pp. 97–138. University of Utah Press, Salt Lake City.
Fedje, D. W., A. P. Mackie, R. J. Wigen, Q. Mackie, and C. Lake
2005 KilgiiGwaay: An Early Maritime Site in the South of Haida Gwaii. In *Haida Gwaii: Human History and Environment from the Time of Loon to the Time of the Iron People,* edited by D. W. Fedje and R. W. Mathewes, vol. 187–203. UBC Press, Vancouver.
Fisher, R.
1977 *Contact and Conflict: Indian-European Relations in British Columbia, 1774–1890.* UBC Press, Vancouver.
Frederick, G., and S. J. Crockford
2005 Analysis of the Vertebrate Fauna from Ts'ishaa village, DfSi-16, Benson Island. In *Ts'ishaa: Archaeology and Ethnography of a Nuu-chah-nulth Origin Site in Barkley Sound,* edited by A. D. McMillan and D. E. St. Claire, pp. 173–205. Archaeology Press, Simon Fraser University, Burnaby.
Frederick, G., I. McKechnie, and R. J. Wigen
2006 *Faunal Report for the Huu7ii Archaeology Project: Results from the 2004 Huu7ii Village Excavations.* Report submitted to the Huu-ay-aht First Nation (Anacla, BC.), Denis St. Claire, and Alan McMillan.
Galdikas-Brindamour, B.
1972 Faunal Material from Eight Archaeological Sites: A Preliminary Report. In *Salvage '71: Reports on Salvage Archaeology Undertaken in British Columbia in 1971,* edited by R. L. Carlson, pp. 199–209. Department of Archaeology, Simon Fraser University, Burnaby.
Gay, J. T.
1987 *American Fur Seal Diplomacy: The Alaskan Fur Seal Controversy.* Peter Lang, New York.
Gentry, R. L.
1998 *Behavior and Ecology of the Northern Fur Seal.* Princeton University Press, Princeton, NJ.
Gifford-Gonzalez, D. P., S. D. Newsome, P. L. Koch, T. P. Guilderson, J. J. Snodgrass, and R. K. Burton
2005 Archaeofaunal Insights on Pinniped-Human Interactions in the Northeastern Pacific. In *The Exploitation and Cultural Importance of Sea Mammals,* edited by G. Monks, pp. 19–38. Oxbow Books, Oxford.
Gregr, E. J., L. Nichol, J. K. B. Ford, G. Ellis, and A. W. Trites
2000 Migration and Population Structure of Northeastern Pacific Whales off Coastal British Columbia: An Analysis of Commercial Whaling Records from 1908–1967. *Marine Mammal Science* 16(4):699–727.
Gregr, E. J., L. M. Nichol, J. C. Watson, J. K. B. Ford, and G. M. Ellis
2008 Estimating Carrying Capacity for Sea Otters in British Columbia. *The Journal of Wildlife Management* 72(2):382–388.
Gustafson, C. E.
1968 Prehistoric Use of Fur Seals: Evidence from the Olympic Coast of Washington. *Science* 161(3836):49–51.
Ham, L. C.
1982 *Seasonality, Shell Midden Layers, and Coast Salish Subsistence Activities at the Crescent Beach Site, DgRr 1.* Ph.D. dissertation, University of British Columbia.
Hancock, D.
1970 California Sea Lion as a Regular Winter Visitant off the British Columbia Coast. *Journal of Mammalogy* 51(3):614.

Hanson, D. K.
1991 *Late Prehistoric Subsistence in the Strait of Georgia Region of the Northwest Coast.* Ph.D. dissertation, Simon Fraser University.
1995 Subsistence during the Late Prehistoric Occupation of Pender Canal, British Columbia (DeRt-1). *Canadian Journal of Archaeology* 19(1):29–48.

Hanson, D. K., and K. D. Kusmer
2001 Sea Otter Scarcity and the Prehistoric Environment of the Strait of Georgia, British Columbia. In *People and Wildlife in Northern North America: Essays in Honor of R. Dale Guthrie,* edited by S. C. Gerlach and M. S. Murray, pp. 58–66. BAR Press, Oxford.

Hebda, R. J., and S. G. Frederick
1990 History of Marine Resources of the Northeast Pacific since the Last Glaciation. *Transactions of the Royal Society of Canada* 1(1):319–342.

Hobbs, L., and C. W. Fowler
2008 Putting Humans in Ecology: Consistency in Science and Management. *Ambio* 37(2):119–124.

Hobson, K. A., and J. C. Driver
1989 Archaeological Evidence for Use of the Strait of Georgia by Marine Birds. In *The Ecology and Status of Marine and Shoreline Birds in the Strait of Georgia, British Columbia,* edited by K. Vermeer and R. W. Butler, pp. 168–173. Canadian Wildlife Service, Ottawa.

Huelsbeck, D. R.
1988 Whaling in the Pre-contact Economy of the Central Northwest Coast. *Arctic Anthropology* 25(1):1–15.
1994 Mammals and Fish in the Subsistence Economy of Ozette. In *Ozette Archaeological Project Research Reports Volume 2, Fauna,* edited by S. R. Samuels, pp. 17–91.Reports of Investigations, vol. 66. Department of Anthropology, Washington State University, Pullman.

Hunn, E. S., D. R. Johnson, P. N. Russell, and T. F. Thornton
2003 Huna Tlingit Traditional Environmental Knowledge, Conservation, and the Management of a "Wilderness" Park. *Current Anthropology* 44(Suppl.):S79–S103.

Imamoto, S.
1976 An Analysis of the Glenrose Faunal Remains. In *The Glenrose Cannery Site,* edited by R. G. Matson, pp. 21–41. Mercury Series, vol. 52. Archaeological Survey of Canada, National Museum of Man, Ottawa.

Ingold, T.
2000 The Perception of the Environment: Essays on Livelihood, Dwelling and Skill. Routledge, London.

Jackson, J. B. C., M. X. Kirby, W. H. Berger, K. A. Bjorndal, L. W. Botsford, B. J. Bourque, R. H. Bradbury, R. Cooke, J. Erlandson, J. A. Estes, T. P. Hughes, S. Kidwell, C. B. Lange, H. S. Lenihan, J. M. Pandolfi, C. H. Peterson, R. S. Steneck, M. J. Tegner, and R. R. Warner
2001 Historical Overfishing and the Recent Collapse of Coastal Ecosystems. *Science* 293(5530): 629–638.

Jeffries, S., H. Huber, J. Calambokidis, and J. Laake
2003 Trends and Status of Harbor Seals in Washington State: 1978–1999. *Journal of Wildlife Management* 67(1):208–219.

Kanipe, H. E., R. J. Wigen, R. A. Spady, A. Hickok, I. R. Wilson, and J. Nord
2007 Archaeological Excavation and Monitoring Pat Bay Sewer Main Installation, DdRu-5 North Saanich, B.C. Heritage Conservation Act Site Investigation Permit 2006-038 and Site Alteration Permit 2006-107 Prepared for District of North Saanich.

Kusmer, K. D.
1994 Changes in Subsistence Strategies at the Tsawwassen Site, a Southwestern British Columbia Shell Midden. *Northwest Anthropological Research Notes* 28(2):189–210.

Lepofsky, D., T. Trost, and J. Morin
2007 Coast Salish Interaction: A View from the Inlets. *Canadian Journal of Archaeology* 31(2): 190–223.

Liu, J., T. Dietz, S. R. Carpenter, C. Folke, M. Alberti, C. L. Redman, S. H. Schneider, E. Ostrom, A. N. Pell, J. Lubchenco, W. W. Taylor, Z. Ouyang, P. Deadman, T. Kratz, and W. Provencher
2007 Coupled Human and Natural Systems. *Ambio: A Journal of the Human Environment* 36(8): 639–649.

Lotze, H. K., and B. Worm
2009 Historical Baselines for Large Marine Animals. *Trends in Ecology and Evolution* 24(5): 254–262.

Lutz, J. S.
2008 *Makúk: A New History of Aboriginal White Relations.* UBC Press, Vancouver.

Lyman, R. L.
2003a Appendix C. Zooarchaeology of Sites 45SJ169 and 45SJ165. In *Archaeological Investigations at Sites 45SJ169 and 45SJ165, Decatur Island, San Juan County, Washington,* edited by S. L. Walker. Reports in Archaeology and History, vol. 100–118. Eastern Washington University, Spokane.
2003b Pinniped Behavior, Foraging Theory, and the Depression of Metapopulations and Nondepression of a Local Population on the Southern

Northwest Coast of North America. *Journal of Anthropological Archaeology* 22(4):376–388.

Mason, A. R., I. R. Wilson, R. J. Wigen, and S. Keen
1995 *Heritage Resource Mitigation at DfRv 106, Lot 14, Section 67, Plan 23765, Cowichan District, Galiano Island, B.C. Permit 1994-155*. Report Submitted to the Ministry of Tourism, Culture and the Arts, Victoria.

Matson, R. G. (editor)
1976 *The Glenrose Cannery Site*. National Museums of Canada, Ottawa.

Matson, R. G., J. Green, and E. McLay
1999 *Houses and Households in the Gulf of Georgia: Archaeological Investigations of Shingle Point (DgRv-2), Valdes Island, British Columbia*. Report Submitted to the Archaeology Branch Victoria, B.C. Lyacksun First Nation, and the Social Science and Humanities Research Council of Canada, Ottawa.

McKechnie, I.
2004 Appendix 2: Faunal Analysis—Methodology and Results. In *Excavation and Monitoring at DcRu-74: Colwood Trunk Sewer Phase V, South Esquimalt Lagoon, HCA Site Alteration Permit 2002-398*, edited by D. Mathews and P. Dady, pp. 84–87. Prepared for the City of Colwood, Submitted to The Esquimalt First Nation, the Songhees First Nation, and the Archaeology and Registry Services Branch, Victoria, B.C.
2005 *Zooarchaeological Analysis of Fauna Recovered from Site 291T (DeSf-2), Tsuxwkwaada, Tsuquanah I.R. 2, Western Vancouver Island*. Report Submitted to Cultural Resource Services, Western Service Centre, Parks Canada, Calgary.
2007a *Vertebrate faunal analysis at Himayis (205T, DfSi-17) and the Clarke Island Defensive Site (212T, DfSi-26), Broken Group Islands, Pacific Rim National Park Reserve*. Report submitted to Parks Canada Cultural Resource Services, Western Canada Service Centre, Calgary.
2007b *Vertebrate Faunal Analysis from Tl'aadiiwa village (39T, DeSf-6), West Coast Trail Unit, Pacific Rim National Park Reserve*. Prepared by Millennia Research for Parks Canada Cultural Resource Services, Western and Northern Canada Service Centre, Calgary.

McMillan, A. D.
1999 *Since the Time of the Transformers: The Ancient Heritage of the Nuu-chah-nulth, Ditidaht, and Makah*. Pacific Rim Archaeology Series. UBC Press, Vancouver.

Merilees, B.
1985 Humpbacks in Our Strait. *Waters: the Journal of the Vancouver Aquarium* 8:7–24.

Misarti, N., B. Finney, H. D. G. Maschner, and M. J. Wooller
2009 Changes in Northeast Pacific Marine Ecosystems over the Last 4500 years: Evidence from Stable Isotope Analysis of Bone Collagen from Archeological Middens. *The Holocene* 19:1139–1151.

Mitchell, D. H.
1981 DcRu-78: A Prehistoric Occupation of Fort Rodd Hill National Park. *Syesis* 14:129–150.
1990 Coast Salish Subsistence Studies and a Methodological Barrier. *Northwest Anthropological Research Notes* 24(2):239–247.

Monks, G. G.
2001 Quit Blubbering: An Examination of Nuu'chah'nulth (Nootkan) Whale Butchery. *International Journal of Osteoarchaeology* 11: 136–149.
2006 The Fauna from Ma'acoah (DfSi–5), Vancouver Island, British Columbia: An Interpretive Summary. *Canadian Journal of Archaeology* 30: 215–244.

Moss, M. L.
1992 Relationships between Maritime Cultures of Southern Alaska: Rethinking Culture Area Boundaries. *Arctic Anthropology* 29:5–17.
2008 Outer Coast Maritime Adaptations in Southern Southeast Alaska: Tlingit or Haida? *Arctic Anthropology* 45(1):41–60.

Moss, M. L., D. Yang, S. D. Newsome, C. Speller, I. McKechnie, A. McMillan, R. J. Losey, and P. Koch
2006 Historical Ecology and Biogeography of North Pacific Pinnipeds: Isotopes and Ancient DNA from Three Archaeological Assemblages. *Journal of Island and Coastal Archaeology* 1(2):165–190.

Newcombe, C. F., and W. A. Newcombe
1914 *Sea-Lions on the Coast of British Columbia*. Report of the Commission of Fisheries for 1913, pp. 131–139, Province of British Columbia.

Newsome, S. D., M. A. Etnier, D. Gifford-Gonzalez, D. L. Phillips, M. v. Tuinen, E. A. Hadly, D. P. Costa, D. J. Kennett, T. P. Guilderson, and P. L. Koch
2007 The Shifting Baseline of Northern Fur Seal Ecology in the Northeast Pacific Ocean. *Proceedings of the National Academy of Sciences, USA* 104(23):9709–9714.

Nichol, L. M., M. Boogaards, and R. Abernethy
2009 Recent Trends in the Abundance and Distribution of Sea Otters (*Enhydralutris*) in British Columbia. DFO Canadian Scientific Advisory Secretariat Stock Status Report, vol. 2009/016.

NOAA Fisheries Service
2009 California Sea Lion (Zalophus californianus), http://www.nmfs.noaa.gov/pr/species/mammals/pinnipeds/californiasealion.htm. Office of

Protected Resources, National Oceanic and Atmospheric Administration, Silver Spring, MD.

Olesiuk, P. F.

2009a Preliminary Assessment of the Recovery Potential of Northern Fur Seals *(Callorhinus ursinus)* in British Columbia, Research Document 2007/076. Canadian Science Advisory Secretariat, Ottawa.

2009b Abundance of Steller Sea Lions (*Eumetopias jubatus*) in British Columbia. Research Document 2008/063. Canadian Science Advisory Secretariat, Ottawa.

Olesiuk, P. F., M. A. Bigg, G. M. Ellis, S. J. Crockford, and R. J. Wigen

1990 An Assessment of the Feeding Habits of Harbour Seals (*Phoca vitulina*) in the Strait of Georgia, British Columbia, Based on Scat Analysis. *Can. Tech. Rep. Fish. Aquat. Sci.* No. 1730.

Orchard, T. J., and T. Clark

2005 Multidimensional Scaling of Northwest Coast Faunal Assemblages: A Case Study from Southern Haida Gwaii, British Columbia. *Canadian Journal of Archaeology* 29(2):88–112.

Pegg, B.

1999 *The Taphonomic History of the Vertebrate Faunal Assemblage from British Camp, San Juan Islands, Washington.* M.A. thesis, Simon Fraser University.

Pegg, B., A. Mason, and G. Wada

2007 *Archaeological Mitgation of DkSb-30, Saltery Bay, B.C. Heritage Conservation Act Permit 2004-120.* Golder and Associates Report Submitted to TELUS Corporation (Burnaby, BC and the BC Archaeology Branch.

Pitcher, T. J.

2005 Back-to-the-Future: A Fresh Policy Initiative for Fisheries and a Restoration Ecology for Ocean Ecosystems. *Philosophical Transactions of the Royal Society B* 360:107–121.

Racerocks.com

2010 http://racerocks.ca/racerock/mirounga2/feb72010newpup/070210pup.htm. Accessed Feb 15th, 2010.

Rousseau, M. K., L. Seip, P. Ewonus, and S. Kaltenrieder

2003 Port Hammond Revisited. In *Archaeology of Coastal British Columbia: Essays in Honour of Professor Philip M. Hobler,* edited by R. L. Carlson, pp. 87–108. Archaeology Press, Simon Fraser Univeristy, Burnaby.

Samuels, S. R., and R. D. Daugherty

1991 Introduction to the Ozette Archaeological Project. In *Ozette Archaeological Project Research Reports, Vol. 1, House Structure and Floor Midden,* edited by S. R. Samuels, pp. 1–27. Reports of Investigations, vol. 63. Washington State University, Pullman.

Sapir, E.

1922 Saayach'apis, a Nootka Trader. In *American Indian Life, By Several of Its Students,* edited by E. C. Parsons, pp. 297–323. B.W. Huebsch, New York.

Sapir, E., and M. Swadesh

1939 *Nootka Texts: Tales and Ethnological Narratives with Grammatical Notes and Lexical Materials.* 1978 AMS Press, edition. Special Publications of the Linguistic Society of America. Linguistic Society of America, University of Pennsylvania, Philadelphia.

Sapir, Edward, and Morris Swadesh

1955 *Native Accounts of Nootka Ethnography.* 1978 ed. AMS Press, New York.

Simonsen, B. O.

1990 Archaeological Investigations at the Comox Harbour Site (DkSf-4): Permit 1990-25. Bastion Group Heritage Consultants report submitted to the Archaeology Branch Ministry of Municipal Affairs Recreation and Culture.

Spady, R., and R. J. Wigen

2008 Archaeological data recovery and monitoring DfSj-57 Storm Outfall Construction, 890 Peninsula Road, Spring Cove, Ucluelet, B.C. Permit 2008-088. IR Wilson Report submitted to the B.C. Archaeology Branch, Ministry of Tourism, Culture and the Arts, Victoria.

Springer, A. M., J. A. Estes, G. B. v. Vliet, T. M. Williams, D. F. Doak, E. M. Danner, K. A. Forney, and B. Pfisterg

2003 Sequential Megafaunal Collapse in the North Pacific Ocean: An Ongoing Legacy of Industrial Whaling? *Proceedings of the National Academy of Sciences, USA* 100(21): 12223–12228.

Stewart, B. S.

1999 Northern Elephant Seal. In *Smithsonian Book of North American Mammals,* edited by D. E. Wilson and S. Ruff, pp. 217–218. Smithsonian Institution Press, Washington, D.C.

Stewart, K. M., and R. J. Wigen

2003 Screen Size and the Need for Reinterpretation: A Case Study from the Northwest Coast. *Florida Museum of Natural History Bulletin* 44(1):27–34.

Suttles, W. P.

1987a Notes on Coast Salish Sea Mammal Hunting. In *Coast Salish Essays,* pp. 233–246. University of Washington Press, Seattle.

1987b Variation in Habitat and Culture on the Northwest Coast. In *Coast Salish Essays,* pp. 26–44. University of Washington Press, Seattle.

Swan, J. G.
1883 Report of Investigations at Neah Bay, Wash., Respecting the Habits of Fur Seals of that Vicinity, and to Arrange for Procuring Specimens of Skeletons of Cetacea. *Bulletin of the United States Fish Commission* 3:201–207.
1887 The Fur Seal Industry of Cape Flattery, Washington Territory. In *Fisheries and Fishery Industries of the United States*, edited by G. B. Goode, pp. 393–400, vol. 7 and 8. U.S. Government Printing Office, Washington, D.C.

Szaniszlo, W. R.
2005 *California Sea Lion (*Zalophus californianus*) and Steller Sea Lion (*Eumetopias jubatus*) Interactions with Vessels in Pacific Rim National Park Reserve: Implications for Marine Mammal Viewing Management.* MSc thesis, University of Victoria.

Thomson, R. E.
1981 *Oceanography of the British Columbia Coast.* Canadian Special Publication of Fisheries and Aquatic Sciences, vol. 56. National Research Council, Ottawa.

Thomson, R. E., S. F. Mihály, and E. A. Kulikov
2007 Estuarine Versus Transient Flow Regimes in Juan de Fuca Strait. *Journal of Geophysical Research-Oceans* 112:C09022.

Trosper, R. L.
2009 *Resilience, Reciprocity and Ecological Economics: Northwest Coast Sustainability.* Routledge, London and New York.

Trost, T.
2005 *Forgotten Waters: A Zooarchaeological Analysis of the Cove Cliff Site (DhRr-18), Indian Arm, British Columbia.* M.A. thesis, Simon Fraser University.

Ware, D. M., and R. E. Thomson
2005 Bottom-Up Ecosystem Trophic Dynamics Determine Fish Production in the Northeast Pacific. *Science* 308:1280–1284.

Waterman, T. T.
1920 *Whaling Equipment of the Makah Indians.* Publications in Political and Social Science Vol. 1. University of Washington, Seattle.

Weathers, B., M. Larsson, and J. Nord
2007 Permit 2006-444 Archaeological Excavation and Monitoring, DdRu-18 Eventide Road Outfall, North Saanich, B.C. Report by I. R. Wilson Consultants Ltd. Prepared for District of North Saanich.

Weathers, B., R. J. Wigen, and K. Kristensen
2008 Archaeological Investigation and Data Recovery. Peninsula Road Upgrade Little Beach Site (DfSj-100). Ucluelet, B.C. Permit 2007-332. Report Submitted to the Ministry of Tourism, Culture and the Arts, Victoria.

Webb, R. L.
1988 *On the Northwest: Commercial Whaling in the Pacific Northwest (1790–1967).* University of British Columbia, Vancouver.

Weiser, A. L., and D. Lepofsky
2009 Ancient Land Use and Management of Ebey's Prairie, Whidbey Island, Washington. *Journal of Ethnobiology* 29(2):184–212.

Wigen, R. J.
1980 *A Faunal Analysis of Two Middens on the East Coast of Vancouver Island.* M.A. thesis, University of Victoria.
2002 Faunal Report for DcRu-136, Canadian Forces Base, Esquimalt, Colwood Property. Pacific Identifications report submitted to Millennia Research Limited, Victoria BC.
2003 Faunal Report for DgSk-1 (311T) Long Beach Unit, Pacific Rim National Park Reserve. Report submitted to Parks Canada Cultural Resource Services, Northern and Western Service Centre, Calgary.
2005 Vertebrate Fauna. In *The Hoko River Archaeological Site Complex: The Rockshelter (45CA21), 1000–100 B.P., Olympic Peninsula, Washington,* edited by D. R. Croes, pp. 71–104. Washington State University Press, Pullman, WA.
2007 Pat Bay Project #1836 Faunal Report. Pacific Identifications report submitted to IR Wilson Consultants, July 4, 2007.
2008 Little Beach Faunal Report. Pacific Identifications report submitted to IR Wilson Consultants, March 5, 2008.

Williams, J.
1974 Faunal Remains from the Cates Park Site (DhRr–8), North Vancouver, British Columbia. In *Archaeological Investigations at the Cates Park Site (DhRr-8) Heritage Conservation Act Permit 1974-24 by Arthur S. Charlton,* pp. 21–52. Report on File at the Archaeology Branch, Victoria, B.C.

Willows, E., A. Hickok, and R. J. Wigen
2008 Archaeological Monitoring and Excavation DiSc-26, Memorial Golf Course Clubhouse, Qualicum Beach, B.C. Heritage Conservation Act Site Alteration Permit 2006-384. Report by IR Wilson Consultants submitted to the Archaeology Branch, Victoria.

Wilson, I. R.
1994 Archaeological Investigations at DgSl-67 Chesterman Beach, West Coast Vancouver Island, Permit 1993-148. Report prepared for Sur La Mer Development and submitted to the Archaeology and Outdoor Recreation Branch, Victoria.

Wilson, I. R. , and S. J. Crockford
1994 Public Archaeological Excavations at the Departure Bay Midden, DhRx-16; Permit 1992–9.

Report prepared for Nanaimo District Museum, Nanaimo and the BC Archaeology Branch, Victoria.
Wilson, I. R., S. Bond, and C. Lake
2003 Archaeological Excavation & Monitoring DcRu-75, 86-88 Dallas Road, Victoria, B.C. Permits 2003-187 and 2003-219. Report submitted to Culture Library, Ministry of Sustainable Resource Management, Victoria.
Wilson, I. R., R. J. Wigen, and M. Rogers
2004b Archaeological Column Sample Excavation DiSe-7, Deep Bay Parking Lot, Deep Bay, B.C., Permit 2004-55. IR Wilson Consultants report prepared for Department of Fisheries and Oceans, Pacific Region.
Wilson, I. R., S. Bond, C. Lake, and M. Rogers
2004b Archaeological Excavation and Monitoring DcRt-16, King George Terrace, Victoria, B.C. Permit 2003-232. IR Wilson Consultants report submitted to the BC Archaeology Branch, Victoria.
Wilson, I. R., R. J. Wigen, I. Streeter, A. Hickok, E. Willows, and J. Storey
2006 Archaeological studies, DeRt-4, Poets Cove at Bedwell Harbour, Pender Island, B.C. Permit 2002-388. IR Wilson Consultants report submitted to the BC Archaeology Branch, Victoria.

8

Native American Use of Seals, Sea Lions, and Sea Otters in Estuaries of Northern Oregon and Southern Washington

Madonna L. Moss and Robert J. Losey

OVER THE LAST 25 YEARS, archaeologists have compiled a substantial zooarchaeological record of pinnipeds on the Washington, Oregon, and California coasts (Burton et al. 2001, 2002; Colten 2002; Etnier 2002a, 2002b; Gifford-Gonzalez et al. 2005; Hildebrandt 1984a, 1984b; Hildebrandt and Jones 1992, 2002, 2004; Jones and Hildebrandt 1995; Lyman 1988, 1989, 1991a, 1991b, 1995, 2003; Porcasi et al. 2000; Walker et al. 1999). These and other studies (e.g., Cooper and Stewart 1983; Howorth 1993; McClenachan 2002; Pearson 1968; Pyle et al. 2001; Stewart et al. 1993) have established that during the 19th and 20th centuries, commercial hunting, target practice, etc., caused dramatic changes in the abundance and distribution of various pinniped species. Archaeological studies have also shown that late-20th-century patterns of pinniped species abundance and distribution do not necessarily reflect conditions in the ancient past, while recent biological studies demonstrate range expansion of some species subsequent to the Marine Mammal Protection Act of 1972 (e.g., Hodder et al. 1998; LeBoeuf 1996).

Beyond these points of consensus, archaeologists disagree over what the zooarchaeological record reveals about Native American hunting during precontact times. Did pinnipeds regularly use mainland breeding areas prior to human use? Did Native Americans club pinnipeds on their terrestrial breeding areas or use watercraft and specialized technologies to hunt at sea? Did Native Americans overexploit particular pinnipeds or sea otters? Did Native American hunting cause changes in pinniped or sea otter behavior and migration patterns? Did such hunting methods change over time? Did Native American hunting contribute to the long-term survival of pinnipeds and sea otters? What can these archaeological data tell us about the present day status of pinnipeds and otters in the region? Different investigators have compiled slightly different data sets from various subregions of Oregon and California to address these questions. They have categorized data in different analytical units and used both primary and secondary zooarchaeological data to support their cases for particular interpretations. Over the time period during which these studies have occurred, different middle-range theoretical models have been employed, hence the interpretations have shifted with respect to emerging trends in the discipline of archaeology.

In this chapter, we do not tackle all the issues for all pinniped species represented in Oregon and Washington archaeological sites. We focus our analysis on the zooarchaeological record of coastal southwest Washington and northwest Oregon, summarizing competing interpretations of key investigators who have interpreted the data. Then we assess these in light of contemporary abundance and distribution data of key pinnipeds on the southern Northwest Coast today. While we recognize that these data are themselves the result of historical processes (i.e., population growth and range expansion subsequent to severe depletion), taking these data seriously helps us more fully recognize the behavioral flexibility of pinnipeds along with real environmental constraints with regard to the location of suitable breeding areas. We then present the results of zooarchaeological analyses at four estuarine sites, relying on recent research by Losey (2002; n.d.), Colten (2002), Fancher (2001) and Bovy (2005). Although we will not address all questions raised in earlier debates, we hope this effort will move discussion forward in new ways.

THE CONTOURS OF PREVIOUS DEBATES

Lyon (1937), Gustafson (1968a, 1968b), and Walker and Craig (1979) were some of the first archaeologists working along the Pacific Coast to recognize discrepancies between the content of archaeological samples and contemporary patterns of marine mammal abundance. William Hildebrandt's interest in precontact pinniped hunting can be traced to his dissertation fieldwork in northern California (Hildebrandt 1981), where he found ample evidence of the taking of northern fur seals (*Callorhinus ursinus*) at Stone Lagoon in Humboldt County, suggesting that a haul-out for this species was located nearby as little as 1000 years ago (Hildebrandt 1984a).

In 1988, R. Lee Lyman discussed three key Oregon assemblages: Umpqua-Eden, Whale Cove, and Seal Rock. His data led him to propose that Umpqua-Eden was near a harbor seal (*Phoca vitulina*) rookery for 3000 years continuing into the 20th century. He inferred that Whale Cove was near a Steller sea lion (*Eumetopias jubatus*) haul-out and a small northern fur seal rookery between 3000 and 2500 years ago, but it was abandoned by fur seals prior to 1000 BP. He characterized Seal Rock as a "major prehistoric rookery" for Steller sea lions that was abandoned about 100 years ago. He also suggested that California sea lions (*Zalophus californianus*) and northern fur seals had bred and pupped at Seal Rock. In 1988, Lyman believed that precontact marine mammal hunting had not adversely affected local pinnipeds. He wrote, "The apparent continuity of Oregon coast use by all pinniped taxa and the sea otter between 3000 B.P. and approximately 100 B.P. lends indirect support to the hypothesis that prehistoric exploitation of these taxa did not adversely affect and may have enhanced individual survival probabilities by reducing inter- and intraspecific competition" (Lyman 1988:260). In 1989, using the same data, Lyman pushed his interpretations to suggest that mainland rookeries and haul-outs were the foci of ancient hunting and that oceangoing canoes were unnecessary to procure marine mammals. He thought it doubtful that Oregon Native Americans did much offshore hunting because offshore rocks were relatively few (Lyman 1989:86). At the end of this paper, Lyman (1989:92) suggested people could have been exploiting drift pinniped carcasses. Lyman (1991a:190) reiterated his position that complex technology, extensive cooperation, and complex social organization were not required for taking Oregon coast pinnipeds. Clubs, perhaps harpoons, and one or two hunters working together were all that was needed to exploit a sea lion rookery.

In 1992, Hildebrandt and Jones responded with a markedly different portrait of Native American resource use. They asserted that prehistoric hunting caused pinnipeds to move from mainland rookeries to remote offshore rocks. In turn, Native Americans developed watercraft and specialized hunting gear to exploit vulnerable offshore rookeries (Hildebrandt and Jones 1992:389). Because of high pursuit and transport costs, however, pinniped populations maintained themselves. "Technological co-evolution" allowed for resource intensification that developed along with cultural complexity. Hildebrandt and Jones proposed that species they termed migratory breeders (Steller sea lion, northern fur seal, California sea lion) decreased over time relative to those they termed resident breeders (harbor seal, sea otter [*Enhydra lutris*]). At Whale Cove, they saw this shift as evidence of animals driven off a mainland rookery later reestablishing themselves offshore. Hildebrandt and Jones suggested that canoes were necessary to procure the sea lions at Seal Rock.

In 1995, Lyman challenged many aspects of Hildebrandt and Jones' analyses, but we focus on only those related to Oregon. Lyman clarified his position that it was not that Native Americans did not have canoes, only that canoes were unnecessary to procure pinnipeds (Lyman 1995:49). In reanalyzing Hildebrandt and Jones' data and that from additional Oregon sites, Lyman did not find a statistically significant shift in migratory to resident breeder abundances in all cases. At Seaside, Ecola Point, Yaquina Head, Seal Rock, and Cape Perpetua, Lyman (1995:63) found evidence that Steller sea lion rookeries may have persisted into the mid-19th century. Lyman (1995:68–69) stressed the importance of the distinction between rookeries and haul-outs and the requisite age and sex data necessary to distinguish them. Lyman still resisted the idea that nearshore and offshore rocks were accessed by canoe, at least partly because he recognized how rocky shorelines change over time.

In the same journal issue, Jones and Hildebrandt responded to Lyman. They cast California and Oregon pinniped hunting as a "tragedy of the commons." Although they do not cite Hardin's (1968) original formulation of the tragedy of the commons, this idea suggests that shared resources will eventually be overexploited, an idea that Mirschitzka (1992) has countered with regard to Northwest Coast fishing and shell-fishing. Jones and Hildebrandt reasserted that migratory pinnipeds, which they considered the most optimal taxa, were vulnerable and overexploited. Jones and Hildebrandt reorganized their site data into four types of environmental setting. With regard to Oregon, they claimed that where mainland sites are adjacent to offshore rocks, before AD 500 migratory breeders were obtained from the mainland, but after AD 500 people were obtaining them from offshore rocks. Where mainland sites lacked offshore rocks, people switched to resident breeders. Hildebrandt and Jones (2002, 2004) recently reiterated these arguments, insisting that Native Americans overexploited optimal prey and were neither passive nor active resource managers (Hildebrandt and Jones 2002:109).

Lyman (2003) did not engage the latest Hildebrandt and Jones contributions as directly as in previous articles, but he placed the Oregon coast data within larger foraging theory debates. In contrast to his earlier positions, Lyman's latest analyses indicated that Oregon coast otariid metapopulations suffered from resource depression prehistorically. Citing Brown and Lomolino (1998), Lyman (2003:377) explains that a metapopulation "comprises all of a set of geographically isolated and/or local populations of a species." Such resource depression means "the rate at which those prey were encountered by human predators decreased" (Lyman 2003:376). Lyman maintained, however, that the local population at Seal Rock was not depressed, because humans concentrated on taking males who defended their territories "against all comers" (Lyman 2003:385). This resulted in a population in "equilibrium" with human predation, in an example of unintentional "epiphenomenal conservation." The competing Lyman and Hildebrandt and Jones models have provided a framework employed by many researchers investigating pinniped use in Oregon and California (e.g., Jones et al. 2004; Porcasi et al. 2000). No ethnographic evidence for resource ownership is considered by any of the authors who have addressed pinniped hunting on the Oregon coast. Since resource ownership or stewardship is often associated with hunter-gatherers deemed to be "complex," this warrants additional research, although this topic will not be pursued further here. Of course, "resource ownership" is a modern concept that probably does not capture the relationships Northwest Coast groups had with these various animals.

VARIABILITY IN PINNIPED AND SEA OTTER BEHAVIOR ALONG THE SOUTHERN NORTHWEST COAST TODAY • *Implications for Understanding the Past*

In their previous papers, archaeologists have compiled basic biological data about various marine mammals, information that we will not merely repeat. We believe that contemporary pinniped behavior can yield insights into precontact conditions, particularly into the behavioral flexibility and adaptability of pinniped species. Table 8.1 summarizes information on the distribution and seasonal patterns of abundance of five pinniped species in Oregon and Washington today. Primary sources of information for this table and the text that follows include Mate (1981), NOAA (1997, 2000, 2001a, 2001b, 2003a, 2003b, 2005, 2007), Oregon Coastal Atlas (2005), Brueggeman et al. (1992), U.S. Fish and Wildlife Service (2005), Sea Lions Caves (2005), and Shoreline Education for Awareness (1999).

As mentioned earlier, Hildebrandt and Jones (1992) classified the marine mammals under consideration as migratory breeders (Steller sea lion, northern fur seal, California sea lion) and resident breeders (harbor seal, sea otter). Their subsequent analyses (Hildebrandt and Jones 1992, 2002, 2004; Jones and Hildebrandt 1995) and that of Lyman (1995) are based on changing frequencies in these two groups over time. Lyman (1995:50–51) pointed out that the categories of resident breeder and migratory breeder "are founded on the historic behaviors of these species" and that these categories might not adequately characterize the past. More recently, Jones et al. (2004:70) have substituted the term "terrestrial breeders" for "migratory breeders," and the term "aquatic breeders" for "resident breeders," but they retain the same classificatory dichotomy. We prefer Etnier's (2007:198–199) approach; he uses the terms "migratory" and "nonmigratory" in the case of northern fur seals and points out their age- and sex-specific patterns of seasonal dispersals. This complexity also characterizes Steller and California sea lions today, as will be illustrated below. In the following accounts, we evaluate what contemporary behavior might suggest about the likelihood of particular places as breeding and pupping sites in the past. We use contemporary information to question some other assumptions and interpretations in previous archaeological studies.

HARBOR SEALS

Harbor seals are the most abundant and geographically widespread pinniped in Oregon today. They are year-round residents and biologists do not designate rookeries for this species because "they reproduce everywhere they occur" (Robin Brown, 2005, pers. comm. to Moss). A total of 101 haul-out sites have been documented in Oregon; these occur on the pocket beaches and rocks along and off the rocky coast but also on sandspits within bays and river mouths. These animals are very sensitive to human activity and tend to be wary and skittish today; hence, an encounter may result in the procurement of only one or two animals at a time. Some archaeologists have characterized harbor seals as one of the least optimal marine mammals because of their comparatively small size and the fact that pups can swim at birth and they and their parents are not confined to terrestrial rookeries. We disagree that harbor seals were consistently less optimal than other pinnipeds and instead suggest that prey ranking was highly conditional. Harbor seals at times may have been more optimal than other pinnipeds because (1) they frequent waters inside bays and estuaries that are less hazardous to travel than the outer coast, and (2) their small size makes them easier to transport and butcher. Furthermore, they were probably present in abundance when other pinnipeds were less accessible due to seasonal movements—in other words search time for harbor seals likely would have been comparatively low during some seasons due to the absence or rarity of other pinnipeds. Further, we reason that because of their widespread geographical distribution and the ability of pups to swim at birth, harbor seals would be highly resistant to overexploitation by humans.

CALIFORNIA SEA LIONS

Today, California sea lions breed mainly in Baja California and on the Channel Islands near Santa Barbara, California. There is no evidence for the breeding of this species in Oregon or Washington during modern times. After breeding, some adult and subadult males move north to winter in central and northern California, while others go to Oregon, Washington, and British Columbia. Those found in Oregon include animals moving to and from more northerly localities and those that winter in Oregon. California sea lions arrive in Oregon in August, with numbers peaking in September and October. Cape Arago is one of the localities where some California sea lions spend the winter. In the spring, those that had ranged north to British Columbia and Washington head southward, swelling the ranks of California sea lions in Oregon during April and May.

Biologists have found increasing numbers of California sea lions in coastal bays and up rivers as they congregate during salmon runs and herring and smelt spawning (NOAA 1997:3). California sea lions are very bold; they pester sport fishers on the Columbia River and haul out on heavily used docks in Portland and Newport, and fishers at Cape Arago report them "jumping onto their vessels and stealing bait" (NOAA 1997:16). Subadult and adult male California sea lions migrate to Oregon coastal waters and regularly haul out at 32 locations. Although California sea lions were considered migratory breeders by some investigators, the presence of California sea lion pups at Seal Rock, Whale Cove, and Umpqua-Eden suggested to Lyman that this species was a resident breeder "in the recent past" (Lyman 1991b:234; see also Lyman 1988). Whether or not California sea lions bred on the Oregon coast in the past and were year-round residents is not yet resolved.

STELLER SEA LIONS

This species occurs year-round on the Oregon and Washington coasts today, with rookeries at Rogue Reef and Orford Reef in Oregon and St. George Reef on the California border. Pups are also born at Three Arch Rocks and Sea Lion Cave, although biologists disagree over whether these sites should be considered "rookeries" (Robin Brown 2005, pers. comm. to Moss). Losey's ongoing research on the Oceanside faunal assemblage may provide evidence relevant to whether or not Three Arch Rocks was a Steller sea lion breeding area in the past. At least 10 sites on the Oregon coast are regularly used as haul-outs. The Sea Lion Caves population of about 200 animals is considered resident (non-migratory), with no mass movements. These sea lions spend fall and winter in the cave and spring and summer on rocks just outside the cave, although individuals and small groups forage substantial distances from this location (Sea Lion Caves 2005). At Sea Lion Cave, Steller sea lions breed and bear young in May and June, sometimes as late as July and August. The strongest bulls drive the bachelor bulls away during the breeding season, remaining on their territories for 3 months. This would appear to contradict Mate (1981:449) who writes

TABLE 8.1

Contemporary Pinniped Distribution Data—Oregon Coast

SPECIES	DISTRIBUTION	COUNT IN OR (YEAR OF COUNT IN PARENTHESES); FOLLOWED BY SOURCE	NOTES
PHOCIDS			
Phoca vitulina harbor seal	Harbor seals are year-round residents in OR and reproduce everywhere they occur. 101 haul-outs have been reported. Pups are born late March through April; largest pupping sites are Columbia River and Cape Arago.	5735 (1999); NOAA 2007b	Stock has reached carrying capacity
Mirounga angustirostris northern elephant seal	Shell Island is the major haul-out in OR for this species today. Most are 1–4-year-old animals born at AñoNuevo, and they haul out here to rest year-round. Since 1993, several pups have been born yearly, making this the most northerly pupping site within the range of this species.	Not available	
OTARIIDS			
Callorhinus ursinus northern fur seal	Fur seals rarely haul out in OR. Females and juveniles migrate south from Alaska's Pribilof Islands starting in November, so they may occur in waters offshore OR in the winter. They are most abundant off northern WA in January and off central and southern OR in April. Whether or not all animals too young to breed make the return trip to the Pribilofs each spring is unknown.	~7000 (1990); Brueggmann et al. 1992	

Eumetopias jubata Steller sea lion	Steller sea lions occur year-round, with rookeries at Rogue Reef, Orford Reef, and St. George Reef. Pups are also born at Three Arch Rocks and Sea Lion Caves. At least ten sites are regularly used as haul-outs. The Sea Lion Caves population is considered resident (nonmigratory), with fall and winter spent in the cave.	5305 (2002); Pitcher et al. 2007	Counts in OR have gradually increased since 1976
Zalophus californianus California sea lion	No evidence for breeding in OR in modern times; almost exclusively males found in OR today. After breeding in Baja and the Channel Islands, they migrate north en route to B.C., arriving in August, with peak numbers in September and October. The southward migration peaks in April and May in OR. The largest haul-out is Cape Arago. California sea lion are also found in rivers, in the Columbia from Astoria to Bonneville Dam, at the mouths of Lewis and Cowlitz rivers, in Multnomah Channel at mouth of the Willamette, and as far as Willamette Falls in Oregon City; also in Rogue River, Coos, Yaquina, and Tillamook bays. At least 30 haul-out sites are known.	~7000 (early 1990s); NOAA 1997	
Arctocephalus townsendi Guadalupe fur seal	OR is substantially north of the contemporary range of Guadalupe fur seal, although pups have been stranded as far north as Homer, AK. 17 Guadalupe fur seals stranded on the coast of OR and WA during the summer of 2007 (Armstrong 2007).	Not available	

that "adult males in California, Oregon, and Washington move north into British Columbia and Alaska so that by the end of October, no adult males are found along the Oregon coast." It seems possible that Mate's observations from the 1970s and 1980s might hold for those times, but that since then, the behavior of Steller sea lions appears to have changed.

Other Steller sea lion populations in Oregon are essentially nonmigratory as well; according to NOAA (2001b:10), "this species is not known to migrate," but juveniles and adult males disperse widely at times other than the breeding season. This would seem to contradict the Sea Lion Cave information above. Taken together, this conflicting information suggests to us that it is a mistake to characterize all populations of this species as migratory, when they appear to exhibit such flexibility in behavior. Whether Steller sea lions are resident or migratory depends on which group one considers as well as on the age and sex of the animals in question.

NORTHERN FUR SEALS

Today, northern fur seals are infrequently seen in Oregon and Washington except when stranded onshore. The animals that are found are females and juveniles that migrate south from Alaska's Pribilof Islands starting in November, so they occur in Oregon and Washington waters during the winter. In recent years, small numbers of fur seals have hauled out on the Shell Island – Simpson Reef area off Cape Arago (Oregon Coastal Atlas 2005). Many pups remain at sea between their first and second years before making the return trip to the Pribilof Island breeding areas as they approach their second year (NOAA 2000:1). In Oregon, this species is rarely seen of the five pinnipeds discussed in this chapter.

Despite their reputation for site fidelity (Gentry 1998), northern fur seals have established new breeding areas in the 20th century on San Miguel Island (one of the Channel Islands) by 1968, and on the Farallon Islands by 1996 (Pyle et al. 2001). Even though they originated in the Pribilofs, the San Miguel group is now considered a separate stock and is essentially nonmigratory. Although most northern fur seals in Oregon are probably part of the migratory Pribilof population, some may be from the San Miguel stock. These data would suggest that in the past, northern fur seals in Oregon and Washington could have been members of both northerly and southerly populations.

GUADALUPE FUR SEALS

Guadalupe fur seals (*Arctocephalus townsendi*) are the only members of this genus found north of the equator (NMML 2009). In the late 19th century, it was feared that Guadalupe fur seals had gone extinct, but the species rebounded in the 20th century and is concentrated at the Isle de Guadalupe off Baja California, Mexico. In the past 15 years, a few pups have been born on San Miguel Island. This species in discussed in more detail in Chapter 1.

NORTHERN ELEPHANT SEAL

Northern elephant seals (*Mirounga angustirostris*) breed and give birth on islands in California and Baja California from December to March (NOAA 2007). After breeding, males feed off Alaskan waters, whereas females feed in waters as far north as Oregon and Washington. Adults return to land to molt during the spring/summer season. Coming back from the brink of extinction, northern elephant seal populations have grown dramatically over the last 100 years. In the 1980s, they established breeding colonies at Point Reyes and Año Nuevo Point, California, places not known to have been used before the modern era. Recently, they have expanded their pupping range into Oregon, on Shell Island off Cape Arago (Hodder et al. 1998).

NORTHERN SEA OTTER

Recent efforts to reintroduce the sea otter into Oregon waters have been unsuccessful. Oregon and Washington fall within the historic range of this species, but the commercial fur trade of the 18th and 19th centuries led to extirpation of the sea otter throughout much of its range. In the 1970s, sea otters from Amchitka Island in

Alaska were introduced to southern Oregon, but by 1982, sea otters were absent from the Oregon coast (Valentine et al. 2007:1–2). Recent genetic work on living populations establishes a distinction between modern sea otters in Alaska versus those in California. Archaeological samples from the Ozette site in western Washington match the modern Alaskan otters (Larson et al. 2002), while those from Oregon are closer to modern California populations (Valentine et al. 2007).

SUMMARY OF BIOLOGICAL OVERVIEW

The biological overview shows that today harbor seals are the most abundant and most evenly distributed of the pinnipeds across the southern Northwest Coast. Next in abundance are California sea lions and Steller sea lions, with both Northern fur seals and Northern elephant seals being comparatively few in number. Steller sea lions have three well-known breeding sites in southern Oregon, while there is no evidence that California sea lions breed in Oregon. Excepting harbor seals, in the case of all the other pinnipeds, different age and sex groups are more or less migratory, nonmigratory, and/or resident. Further, all pinnipeds aggregate in places other than breeding areas, including haul-outs ranging from offshore and nearshore rocks to mainland sand beaches to ephemeral sandbars exposed only at low tide. In addition, pinnipeds follow prey concentrations into estuaries. In light of this variability, we argue that the distinction between resident and migratory breeders (or terrestrial breeders vs. aquatic breeders, as proposed by Jones et al. 2004) has outlived its usefulness. Due to the absence of the sea otter on the Oregon coast today, we lack comparable information on its behavior and movements.

RESHAPING THE CONTOURS OF PREVIOUS DEBATES

Now that we have delineated the contours of previous debates and placed them within the context of current biological information on key species, we examine the following archaeological contexts in some detail: the Netarts site (35-TI-1), Palmrose (35-CLT-47), Par-Tee (35-CLT-20) in Oregon, and Minard (45-GH-15) in southwest Washington (Figure 8.1). Our starting point will be the pinniped and sea otter remains from these sites, but we will also consider other categories of archaeological and environmental data. We (1) take a more holistic approach to faunal assemblages by assessing pinniped and sea otter remains along with other taxa, (2) consider sea mammal hunting technology in sites and its implications for understanding marine hunting off the Oregon and southern Washington coasts, and (3) assess the overall marine productivity of estuaries in the region.

NETARTS SANDSPIT (35-TI-1)

The Netarts site (35-TI-1) is located on the northern Oregon coast, approximately 100 km west of Portland. The site is located on a sandspit enclosing Netarts Bay, a shallow saltwater-dominated lagoon. About 5 km north of the site and 800 m offshore is Three Arch Rocks, a series of small rocky islands that are used today as haul-outs for Steller sea lions, California sea lions, and harbor seals. A small number of Steller sea lions are born here annually. The Netarts site was first dug during the 1940s, although work by professional archaeologists did not occur until 1952 (Cressman 1952). Extensive excavations (~500 m^2) continued in 1956, 1957, and 1958 (Losey 2002:193, 2005; Newman 1959), mostly focusing on the remains of several wood plank houses found at the site. Losey recovered additional material in his small-scale excavations using intensive recovery methods in 1999, 2000, and 2001, and he dated the site occupations to cal AD 1300 to 1800 (Losey 2002:246–259). He analyzed 5108 vertebrate specimens excavated in the 1950s (curated at the University of Oregon Museum of Natural and Cultural History) and 62,156 specimens from his 1999 to 2001 investigations. Of the total 67,264 vertebrate remains, over 90% were fish and only 2.5 % were marine mammals.

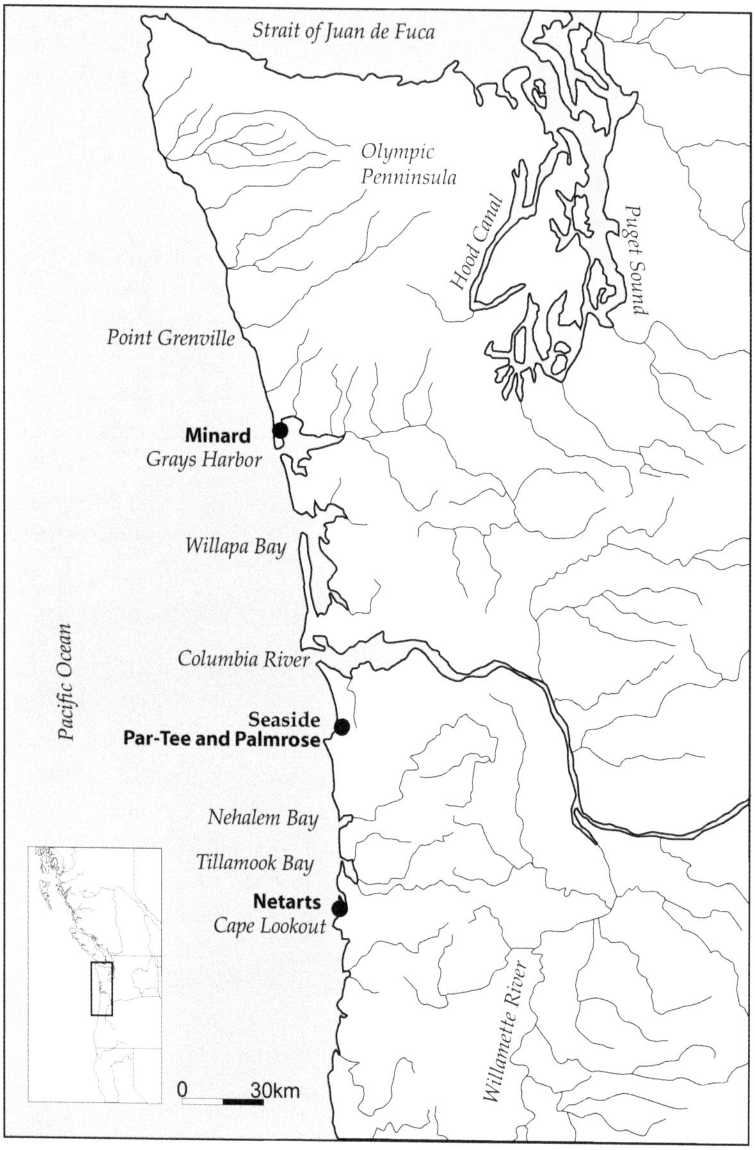

FIGURE 8.1. Map of the northern Oregon and southern Washington coasts. Prepared by Rob Losey.

Of the taxa of concern here, in order of abundance, sea otter, Steller sea lion, harbor seal, northern fur seal, California sea lion, and Guadalupe fur seal are represented (Table 8.2). Genetic work on ancient DNA indicated that a minimum of 12 northern fur seals is represented in the assemblage (Moss et al. 2006). In addition, Losey (2002) identified both whale- and porpoise-sized cetacean bones.

When the faunal assemblage from Netarts is viewed as a whole, pinnipeds and sea otters were clearly part of a meat diet that included a wide array of species of highly variable body sizes, fat contents, handling costs, and habitat preferences. Overall though, both the fish and shellfish assemblages from the site indicate a focus on the species inhabiting Netarts lagoon. Shellfish are dominated by species preferring sheltered waters, namely, bay mussel, cockle, butter clam, horse clam, and Dungeness crab. The fish assemblage also shows an estuarine focus, with shiner perch, pile perch, flatfishes,

TABLE 8.2

Marine Mammal Remains Identified from Minard, Par-Tee, Palmrose, and Netarts Archaeological Sites

		MINARD 45-GH-15	PALMROSE 35-CLT-47	PAR-TEE 35-CLT-20	NETARTS 35-TI-1
Enhydra lutris	Sea otter	780	196	422	380
Arctocephalus townsendii	Guadalupe fur seal				3
Callorhinus ursinus	Northern fur seal	255	105	32	31
Eumetopias jubatus	Steller sea lion	234	179	57	166
cf. *Mirounga angustirostris*	Northern elephant seal	1			
Phoca vitulina	Harbor seal	256	30	85	124
Zalophus californianus	California sea lion	7	15	5	2
Subtotal		1533	525	601	706
Balaenoptera acutorostrata	Minke whale			4	
Lissodelphis borealis	Northern right-whale dolphin				1
Phocoena phocoena	Harbor porpoise		22	25	1
Phocoena sp.		187			
Stenella attenuata	Pantropical spotted dolphin			1	
Tursiops truncatus	Bottlenose dolphin		50	4	
Delphinidae			92	46	
Cetacea—whale-sized			9	61	81
Cetacea—porpoise-sized				13	24
Cetacea—undiff.		57	10		
Subtotal		244	183	154	107

and herring or sardines present in abundance, but some hake and salmonids are also present. Birds show more diversity, with pelagic seabirds (albatross, fulmars), neritic seabirds (pelicans, cormorants, scoters, common murres), and waterfowl (ducks, geese) all represented. Tools from Netarts have not been described in detail, but like other sites in the lower Columbia River area dating within the last 1000 years or so (Minor 1983), the harpoon assemblage consists exclusively of three-piece toggling harpoons (Newman 1959). Contracting-stem narrow-necked chipped stone projectile points at the site indicate use of the bow and arrow.

PALMROSE (35-CLT-47)

The Palmrose site is located in the town of Seaside on the northern Oregon coast about 110 km northwest of Portland. This site is on the east side of Interstate Highway 101 about 500 m inland from the beach. Tillamook Head, a large basalt outcrop, is located a few hundred meters to the south and forms the southern boundary of a beach that extends around 22 km northward to the mouth of the Columbia River. In the mid-1960s, Robert Drucker, a local artifact collector, began excavating the site with the Oregon Archaeological Society. The excavations came to the attention of George Phebus, a collections assistant at the Smithsonian and longtime resident of the northern Oregon coast. Phebus convinced Drucker and crew to excavate in a grid system, record provenience information for artifacts, and to take notes. Phebus and Drucker excavated in 5-foot squares, using arbitrary 12-inch levels, and midden materials were screened over ¼-inch mesh. Phebus and Drucker excavated the site for over a decade and dug 240 such squares, making it one of the most extensively excavated sites on the southern Northwest Coast. The only published description of their excavations at the site consists of a short progress report written in 1977–78 (Phebus and Drucker 1979). Today the site is perhaps best known for containing one of the culture area's oldest rectangular houses, presumably a plank house similar to those used during the early historic period. Most of the artifact collection remains in private hands, but nearly all faunal remains from the site were curated by the Smithsonian Institution, a sample of which was studied by Roger Colten.

Palmrose also was test excavated in 1988 by archaeologists from the Oregon Museum of Anthropology at the University of Oregon (Connolly 1992, 1995). Although they excavated less than 4 cubic meters, their geomorphological studies combined with extensive faunal analyses led to a reconstruction of how the setting of the site evolved over time. The site was first occupied about 4000 BP, with the principal occupation dating from 2700 to 1600 BP. The presence of horse clam, butter clam, and littleneck clam in the deposit indicate that the site was located adjacent to a saltwater-dominated estuary. During this period of occupation, massive dune building created the prograding Clatsop Plains to the west. The Palmrose site apparently was abandoned ca. 800 BP, perhaps when the estuary filled in with sediment (Connolly 1992, 1995).

Of the total 6495 mammal and bird remains in the Smithsonian collection identified by Colten (2002), ~29% were marine mammals. In order of abundance, sea otter, Steller sea lion, northern fur seal, harbor seal, and California sea lion are represented (Table 8.1). Colten (2002) also identified bottle-nosed dolphin, harbor porpoise, and whale- and porpoise-sized cetaceans. The most abundant birds found at the site include ducks and geese, although seabirds including gulls, albatross, and shearwaters were also identified.

From the 1988 excavations, 5313 vertebrate specimens were recovered, with 3% as marine mammals. In order of abundance, sea otter, Steller sea lion, northern fur seal, harbor seal, and California sea lion were represented (Greenspan and Crockford 1992). Although the total pinnipeds and sea otters identified to the species level is only 60 (number of identified specimens [NISP]), the rank order in abundance is the same as that found by Colten. Greenspan and Wigen also identified cetacean bones from the site. The fish assemblage at Palmrose indicates less use of rocky shore species than seen at Par-Tee, with the most abundant fish being salmonids, followed distantly by hake and skates. Phebus and Drucker (1979) report recovering only 27 unilaterally barbed harpoons and eight bilaterally barbed harpoons from Palmrose by the end of the 1977 field season. In addition, 102 toggling harpoon valves appear to have been recovered by that time (Phebus and Drucker 1979). Only one unilaterally barbed point and four toggling harpoon valves were found in the later testing of Palmrose (Connolly 1992:86–7).

PAR-TEE (35-CLT-20)

The Par-Tee site is located in Seaside on the northern Oregon coast and is situated on the northward-flowing Necanicum River about 300 m west of the Palmrose site. Phebus and Drucker also excavated Par-Tee and employed the same procedures they used at Palmrose. Over a period of nearly 10 years, they dug 234 units at Par-Tee, making it one of the most extensively excavated sites on the southern Northwest Coast. An estimated 550 m^2 was excavated (Losey and Yang 2007:662). Their excavations were only briefly reported (~20 pages) in the same booklet describing the archaeology of Palmrose. Artifacts and faunal remains were curated at the Smithsonian Institution, and in recent years they have been studied by Roger Colten, Ruth Greenspan, and Rob Losey.

The Par-Tee site is a large shell midden, over 2 m deep in some areas. The primary occupation of the site spans between roughly 2000 and 1000 BP, but some use of the site area probably occurred as early as 3500 BP and as late as the 1800s. Losey and Power (2005) studied the curated shellfish remains and found them to be dominated by the same estuarine species found at Palmrose, indicating that the site also was on the margins of a shallow saltwater embayment during its occupation. In addition, relatively small quantities of razor clams and California mussels were identified, indicating that site residents had access to both sand beach and rocky shoreline habitats on the outer coast.

Of the total 6362 mammal and bird remains in the Smithsonian collection identified by Colten (2002), ~34% were marine mammals. In order of abundance, sea otter, harbor seal, Steller sea lion, northern fur seal, and California sea lion are represented (Table 8.1). Colten (2002) also identified harbor porpoise, minke whale, other dolphins, and whale- and porpoise-sized cetaceans. As with the other sites discussed here, the remainder of the faunal assemblage is quite diverse. The most abundant birds found at the site include common murres, sea ducks, geese, and shearwaters, but cormorants, gull, and albatross were also identified. Greenspan's (n.d.) analysis of the fish remains indicates common use of fish preferring rocky shorelines (lingcod, rockfish) as well as those more often harvested from estuaries such as flatfishes and skates. Other fish present in smaller quantities include hake, cabezon, sturgeon, salmon, sculpins, and surfperches.

Phebus and Drucker recovered nearly 6300 artifacts from Par-Tee, the largest artifact assemblage from the entire Oregon coast. This assemblage was analyzed by Losey in 2003 (Losey n.d.) and is dominated by well-preserved bone and antler tools, including a minimum of 350 modified whale bones (Figures 8.2–8.6). Bilaterally and unilaterally barbed harpoon heads dating between roughly 3000 and 1000 BP are relatively common constituents of assemblages from this region (Minor 1983), and Par-Tee is exemplary of this pattern. Single-piece harpoon heads are very abundant; at least 324 complete and fragmentary harpoon heads are present, with only one having a tip slotted for the insertion of an end blade—the rest are self-tipped. In addition, 148 valves from toggling harpoons, similar to those used historically for taking salmon, seals, sea lions, and sea otters, were also found. The most unusual tools found at the site were the remains of at least 23 whale bone atlatls. These, in combination with numerous broad-necked projectile points, clearly indicate the use of the atlatl and dart weaponry. Narrow-necked projectile points are also present and likely mark the introduction and use of the bow and arrow at the site toward the end of its primary period of occupation. These hunting weapons and faunal remains strongly suggest that Par-Tee residents were frequent and expert maritime hunters. Some archaeologists have characterized Oregon coast Natives as coastal foragers who were largely shore-bound, limited to taking sea mammals on their mainland rookeries, butchering the occasional drift porpoise or whale, or taking advantage of seabirds or hake that washed ashore. As Losey and Yang (2007) have already shown, the residents of Par-Tee appear to have struck and recovered a humpbacked whale using an elk-bone point made of local materials, indicating that they hunted whales on an opportunistic basis. They also undoubtedly scavenged whale carcasses, but the number of bone and antler weapons recovered from Par-Tee evidence systematic open water hunting.

MINARD (45-GH-15)

The Minard site is a large shell midden located on the northern spit of Grays Harbor, identified by Richard Daugherty in 1947, tested by

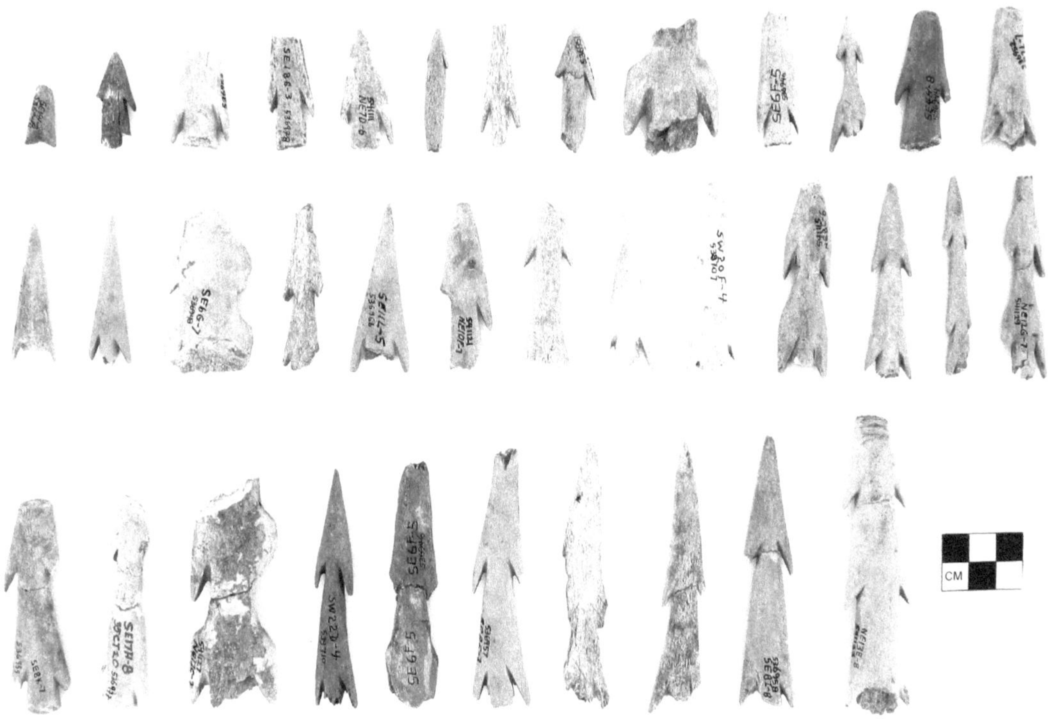

FIGURE 8.2. Midbody sections of selected bilaterally barbed points recovered from Par-Tee. Photograph by Jennifer Stringfellow, Smithsonian Institution.

him in 1969, and excavated by a Washington State University field school supervised by Tom Roll in 1970 (Bovy 2005). The site is located on the first two dune ridges from the bay side of Grays Harbor, which lies less than 100 m to the east. In 1969, it was located 1.2 km from the Pacific Ocean shoreline to the west, but because of jetty construction and littoral drift in the modern era, Roll estimated that in 1894, the site was about half as far from the Pacific (Bovy 2005).

The site was excavated in 2-meter square units, in 20 cm thick arbitrary levels. Excavated fill was passed through ¼-inch screens, and while all mammal and bird bones were retrieved, fish and shellfish remains were not systematically recovered. Bovy's (2005) study of the bird remains from Minard also included an intensive radiocarbon dating program of curated samples. She established that the site was continuously occupied from AD 1000 to AD 1600 (Bovy 2005:117). A small number of tools, most notably a carved figurine of an individual holding a flintlock rifle and powder horn, indicate some use of the site in the early historic period.

Fancher (2001) identified the mammal bones from the Minard site, curated at Washington State University. Of the 20,397 remains he studied, 11% were from marine mammals. The pinnipeds and sea otters, in order of abundance, include sea otters, followed by nearly equal numbers of harbor seal, northern fur seal, and Steller sea lions, with California sea lion following distantly. One bone was tentatively identified as representing a northern elephant seal. Harbor porpoises were relatively abundant, and other cetacean bones were also identified.

Bovy's study of the bird remains from the site showed that sooty shearwaters were the most abundant bird recovered but that their

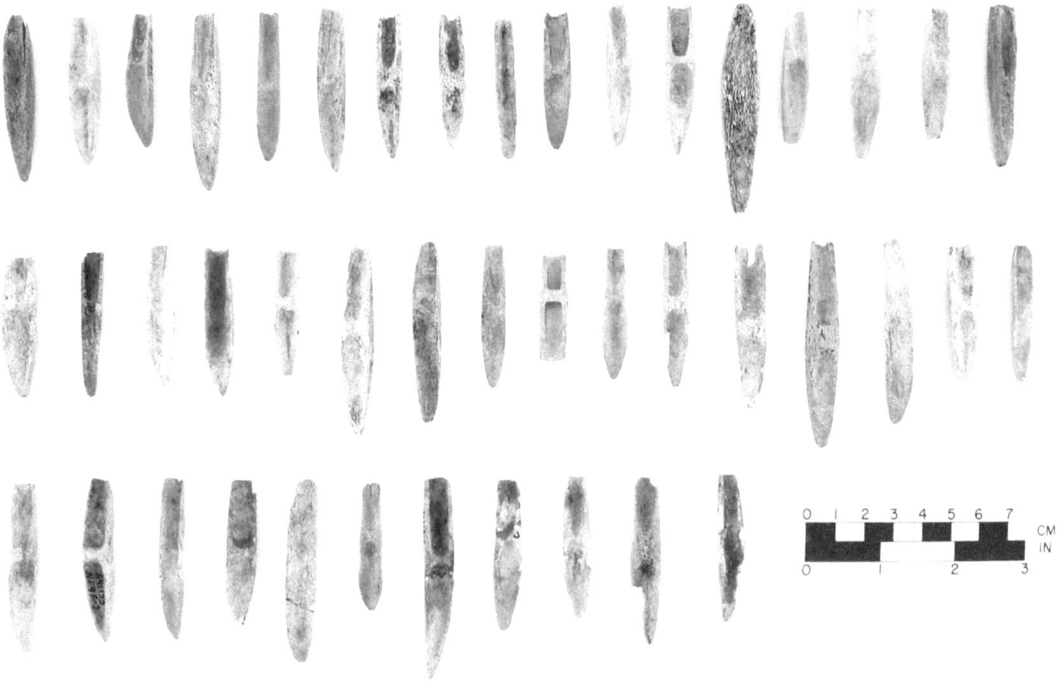

A

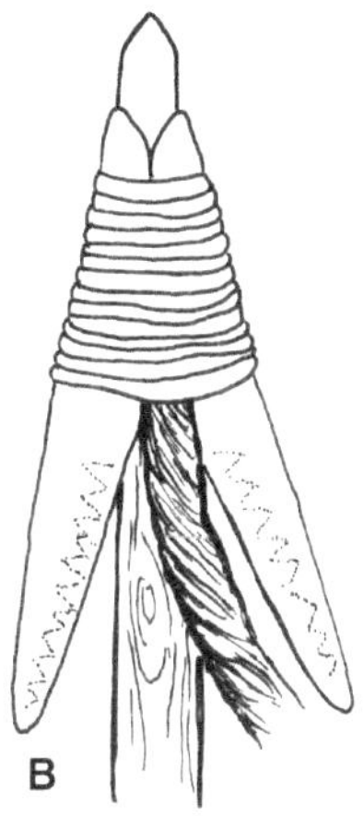

B

FIGURE 8.3. (A) Selected toggling harpoon valves recovered from Par-Tee. (B) A reconstructed toggling harpoon, drawn by Megan Wilson. Photograph by Jennifer Stringfellow, Smithsonian Institution.

numbers decreased after AD 1300. Cassin's auklet abundance increases significantly through time, and common murres were abundant in the upper levels of the site. Bovy (2005:122) convincingly argues that the site was occupied in summer and winter, and possibly year-round, even though the intensity of summer versus winter occupation may have changed over the site's history. She proposes that mutton-birding in New Zealand reduced the numbers of sooty shearwaters migrating to Grays Harbor, explaining the decline in shearwater use at Minard after AD 1300 (see also Bovy 2007).

Roberts' (2008, pers. comm.) analysis of the unsystematically collected fish remains from Minard shows the assemblage to be heavily dominated by salmon and flatfishes, which account over 80% of the identified specimens (total fish NISP identified to at least family level is 16,610). The remaining 20% is mostly composed of surfperches and sturgeon, but hake, lingcod, rockfishes, sculpins, and dogfish are also present in more than trace amounts. Shellfish remains at the site were also unsystematically collected—typically only whole valves were taken. Losey and MacKenzie's (n.d.) analysis of this sample shows the assemblage to be dominated by estuarine taxa, primarily cockle, macoma clams, and horse clams. Small quantities of razor clam and California mussel are also

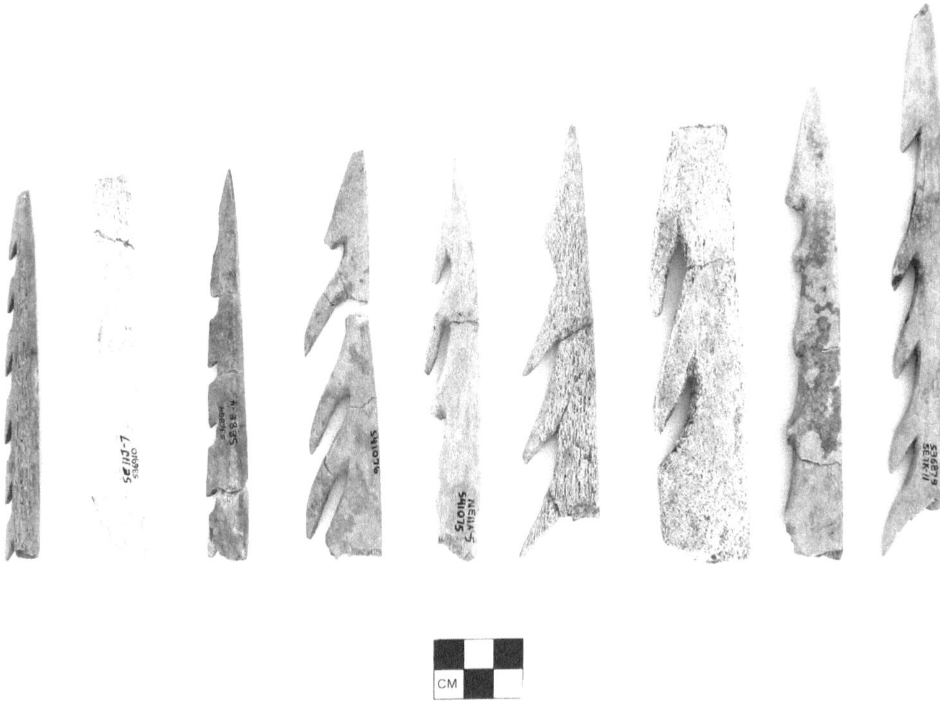

FIGURE 8.4. Selected large unilaterally barbed points recovered from Par-Tee. Photograph by Jennifer Stringfellow, Smithsonian Institution.

present and indicate some use of shellfish from the open Pacific coastline.

Roll (1974) recovered 57 toggling harpoon valves and only six single-piece barbed points from Minard, but none of these latter points have intact line attachment elements. These latter tools could be tips from arrows or spear points designed not to detach from their shafts (i.e., they are not true harpoon points). All of the stemmed bifaces at the site appear to be narrow necked and thus likely from arrows, but 10 bifaces with convex or flat bases are also present and may have functioned as end blades for implements such as harpoons.

DISCUSSION

As described in an earlier section of this paper, previous research on pinniped remains from Oregon has focused on outer coast sites, with the notable exception of Lyman's (1988, 1991b) work at Umpqua-Eden. Some workers have claimed to establish the presence of "rookeries" where they do not exist today; others interpret the remains as evidence for overexploitation and resource depression. Some have created narratives of evolutionary change in which migratory breeders were preferred prey, and when these were overhunted, people switched to resident breeders. Others have evaluated whether or not the Native American residents of the region practiced "resource conservation"—a contemporary concept that is not easily projected onto the past.

We believe trying to decipher long-term subsistence change on a regional level in the Pacific Northwest such as the Oregon and Washington coasts is not yet possible because the archaeological record is still too poorly known and environmental variability and change are not adequately appreciated. The record thus far, however, does allow us to conclude the following regarding the use of sea otters and pinnipeds.

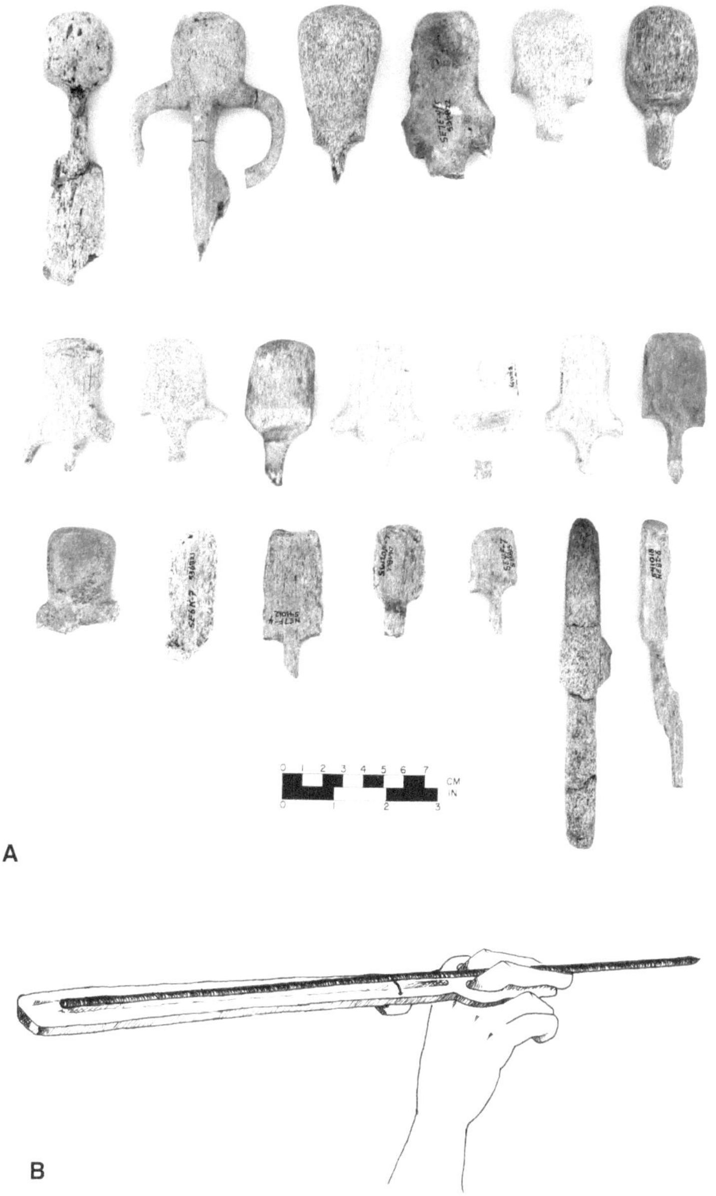

FIGURE 8.5. (A) Fragments of whale bone atlatl handles recovered from Par-Tee. (B) A reconstruction of atlatl in use, by Megan Wilson. Photograph by Jennifer Stringfellow, Smithsonian Institution.

1. Northern fur seals were clearly more abundant off the Oregon and Washington coasts in the Late Holocene than they are now. Crockford et al. (2002), Etnier (2002a) and Gifford-Gonzalez et al. (2005) have presented strong evidence that northern fur seals had breeding colonies in southern British Columbia, Washington, and California. Lyman has argued that Seal Rock was a northern fur seal rookery off the Oregon coast, although the number of northern fur seal pups from this site is relatively low. Northern fur seals are present in all four assemblages discussed

FIGURE 8.6. Embellished toggling harpoon valves recovered from Par-Tee. Photograph by Jennifer Stringfellow, Smithsonian Institution.

in this paper, but they are most abundant in Minard, the northernmost site in the sample. Perhaps these seals were affiliated with a breeding colony in Washington or southern British Colombia.

2. California sea lions are not abundant in any precontact archaeological site in Oregon or Washington. Today, biologists have documented increasing numbers of California sea lions in coastal bays and up rivers (NOAA 1997:3). If they were this abundant in the past, they should be present in consistently high numbers in archaeological assemblages from the Oregon and Washington coasts. Because they are not, we must conclude that today this species occupies a niche that belonged to one or more other pinnipeds in the past. We suggest that California sea lions did not breed in Oregon during the Late Holocene and that their numbers in archaeological sites in the study region are consistent with intermittent presence during migration through the region.
3. The record of precontact use of Steller sea lions and harbor seals in Oregon and southern Washington does not support radically different patterns of distribution and abundance of these species from those of today. Because of their behavioral flexibility (Pitcher et al. 2007), it seems likely that Steller sea lions bred at locations other than Orford, Rogue, and St. George reefs. Three Arch Rocks may have been a Steller sea lion breeding area in the past. More detailed ageing of the remains of these pinnipeds is needed to address this issue.
4. We believe the distinction between resident breeders and migratory breeders by species (or in its new version as between terrestrial breeders vs. aquatic breeders) is not helpful and should be abandoned. If species such as Steller sea lions or northern fur seals did have local breeding colonies, then they would have to be classified as "resident" rather than "migratory" breeders. The distinction also fails to consider the different seasonal movements of pinnipeds depending on age and sex. We also question the validity of considering harbor seals as consistently less optimal prey than migratory (or "terrestrial") breeders.
5. There is a diachronic shift in pinniped and sea otter use at the two Seaside sites that is similar to trends observed by other investigators working in the region. Namely, the earlier Palmrose site has relatively more Steller sea lions and fur seals than the more recent Par-tee site, which has a greater relative abundance of sea otters and harbor seals. The tendency toward hunting of these smaller species appears to occur in parallel with a shift in hunting technologies, in particular the increasing abundance of single-piece harpoon heads and the increasing abundance of these

relative to toggling harpoon heads. Interestingly, toggling harpoon heads continue to be found at the more recent Netarts and Minard sites, while single-piece harpoon heads are almost completely absent. Despite the difference in hunting implements, all previously harvested pinnipeds and sea otters continued to be taken at Minard and Netarts. A more complete understanding of the relationships between different hunting technologies and pinniped/sea otter use patterns requires additional research.

6. New genetic work by Valentine et al. (2007) indicates that sea otters from Oregon coast archaeological sites are more closely related to the California sea otter than to the Alaska sea otter. Sea otters are routinely abundant in archaeological assemblages of Oregon, suggesting their long-term use by Native Americans. Sea otters appear to have been relatively intensively harvested during the last 1000 years.

Evidence from Netarts, Par-Tee, Palmrose, and Minard can illuminate an important additional aspect of sea mammal use that has been neglected in recent years: the importance of estuaries as foraging areas for pinnipeds and sea otters. Of these four sites, only Netarts is located near (5 km) offshore rocks that might have served as a breeding colony for northern fur seals or Steller sea lions (Three Arch Rocks). It is possible that most of the pinnipeds and sea otters represented in these four archaeological sites were procured after the sea mammals had entered the estuaries while searching for prey. The three estuaries along which these sites occur are good representatives of the variability of Oregon and southern Washington outer coast estuaries.

Like many Pacific Coast estuaries, Grays Harbor is a drowned river valley formed by postglacial sea level rise. It has also been partially shaped by ocean-built bars. Grays Harbor is an order of magnitude larger than Coos Bay, the largest of the Oregon coast estuaries (Emmett et al. 2000). Grays Harbor is comprised of 10–20 m deep channels that cut through wide mud and sand flats. About half of its surface area lies within the intertidal zone, as does that of Netarts Bay (Hickey and Banas 2003:1020). Netarts has been shaped entirely by ocean-built bars (Hickey and Banas 2003:1019–1020) and is a true lagoon, receiving very little fresh water input. The Necanicum River estuary at Seaside is a remnant of a larger estuary that existed at the time the Palmrose and Par-Tee sites were occupied. It has been filled in over the last 1000 years. All three estuaries fall within the sphere of influence of the Columbia River plume.

Overall biological productivity is higher off the southwest coast of Washington than it is off most of the Oregon coast, even though upwelling-favorable wind stress is weaker off Washington. This is partly because the continental shelf is almost twice as wide off Washington than it is off Oregon (except in the area of Heceta and Stonewall banks), and this shallow nearshore region is favored by juveniles and larvae of many organisms (Hickey and Banas 2003:1015). In addition, the submarine canyons off Washington contribute to upwelling intensity, and the Columbia River plume provides micronutrients during much of the year (Hickey and Banas 2003:1017).

The Minard and Netarts sites are similar in terms of their geomorphological settings, with both sites located on the inside of an ocean-built bar or spit that encloses the estuary. Palmrose and Par-Tee may have been in similar settings. Both the Necanicum River estuary and Netarts Bay are at least an order of magnitude smaller than Grays Harbor. All four sites have evidence for occupation during both the winter and the summer—whether they were occupied continuously throughout the year is less clear but certainly possible. Together, the four sites provide a portrait of estuarine site occupation within 100 km of the Columbia River that spans the Late Holocene.

All four sites contain the remains of sea otters, Steller sea lions, northern fur seals, and harbor seals in abundance, as well as the bones of porpoises, whales, seabirds, and waterfowl.

At Par-Tee, there is good evidence that site residents hunted whales opportunistically (Losey and Yang 2007). All sites also contained abundant remains of fish, with Netarts dominated by small estuarine species, Palmrose and Minard by salmonids, and Par-Tee by rocky-shore species. The four sites each have technology likely used in hunting marine mammals, although this equipment varies significantly by site and through time. The large harpoon assemblage from Par-Tee is clearly testament to the importance of marine hunting in general. Ethnographically, we know that the people in this region—Lower Chehalis, Clatsop, and Tillamook—clearly practiced open-water hunting from canoes (Hajda 1990; Losey and Yang 2007; Seaburg and Miller 1990).

We suspect that most of the sea otters and pinnipeds represented in the archaeological assemblages were procured within estuaries as the marine mammals came into the bays to feed. Estuaries are highly productive environments, but our picture of the region's estuarine ecology undoubtedly reflects relatively recent conditions. Like other coastal areas, this region has been subject to extensive commercial fishing resulting in the near eradication of some species. Perhaps the best example of this in the study area was the devastation of chum salmon populations that spawned during the fall in the small streams entering all of the estuaries described here. Pinnipeds and sea otters likely would have congregated during the runs of these and other salmon species, as well as during herring and smelt spawning; these taxa have also been severely depleted. Intensive human competition for prey surely affected pinniped behavior.

Table 8.3 presents the seasonal abundance of selected fishes from Netarts Bay and Grays Harbor, compiled from Monaco et al. (1990), Emmett et al. (1991), and NOAA (2005). Although contemporary data cannot capture precontact conditions, these data clearly show the abundance of prey species throughout the year. The number of taxa found in Grays Harbor is larger than that from Netarts Bay, as would be expected considering the larger size of the estuary and the larger watershed it drains. Netarts Bay is also a good model for the ancient estuary of the Necanicum River. As shown in the table, even in the winter, surf smelt, staghorn sculpin, English sole, and starry flounder can be found in Netarts Bay. By February, adult herring, juvenile chum salmon, and Pacific tomcod larvae arrive. By March, winter steelhead arrive, and in April, juvenile lingcod, coho salmon, topsmelt, and shiner perch are present. In May, chinook salmon enter Netarts Bay, and northern anchovy arrive in June. Almost all of these species persist through summer and fall. We would expect pinnipeds and sea otters to follow these prey into the bay, especially those that are schooling species.

Since pinnipeds and sea otters feed on salmon, herring, hake, smelt, and other fish known to have been important to Native Americans, we suggest that they were vulnerable to hunting when these prey taxa aggregate, consistent with Monks (1987) "prey as bait" model. During times of salmon runs and herring spawning, Native Americans would have pursued these fish, but they would also have pursued the marine mammals and birds that also follow the fish. Pinnipeds and sea otters were probably also taken at places where fish were artificially concentrated, such as at the fish weirs and traps present in many southern Northwest Coast estuaries (Byram 2002; Schalk and Burtchard 2001). Following Monks and Byram, we suggest that Native Americans built these weirs not only to attract a variety of fish but also to attract their mammalian and avian predators.

CONCLUSION

Pinniped and sea otter use on the Oregon and Washington coasts is likely to be best understood when examined with as much contextual data as possible. This will involve more intensive study of sea otter and pinniped remains from sites, in particular more detailed examinations of age, sex, and season of death. Such

TABLE 8.3

Seasonal Abundance of Selected Fishes in Grays Harbor and Netarts Bay

Data on Monthly Relative Abundance Rankings Are 0 = Not Present, 2 = Rare, 3 = Common, 4 = Abundant, 5 = Highly Abundant.

ESTUARY	COMMON NAME	LIFE STAGE	JAN	FEB	MAR	APR	MAY	JUN	JUL	AUG	SEP	OCT	NOV	DEC
Grays Hbr	White sturgeon	Adults	3	3	3	3	3	3	3	3	3	3	3	3
Grays Hbr	White sturgeon	Juveniles	3	3	3	4	4	4	4	4	4	3	3	3
Netarts	No white sturgeon													
Grays Hbr	Pacific herring	Juveniles	3	3	3	3	4	4	4	4	4	4	3	3
Grays Hbr	Pacific Herring	Larvae	0	0	0	3	3	3	3	0	0	0	0	0
Netarts	Pacific herring	Adults	0	3	4	3	0	0	0	0	0	0	0	0
Netarts	Pacific herring	Eggs	0	0	4	0	0	0	0	0	0	0	0	0
Netarts	Pacific herring	Juveniles	0	0	0	0	4	4	0	0	0	0	0	0
Netarts	Pacific herring	Larvae	0	0	3	4	3	0	0	0	0	0	0	0
Netarts	Pacific herring	Spawning	0	0	4	0	0	0	0	0	0	0	0	0
Grays Hbr	Northern anchovy	Adults	3	3	3	3	4	4	5	5	5	5	3	3
Grays Hbr	Northern anchovy	Eggs	0	0	0	0	3	3	3	3	0	0	0	0
Grays Hbr	Northern anchovy	Juveniles	3	3	3	3	4	5	5	5	5	5	3	3
Grays Hbr	Northern anchovy	Larvae	0	0	0	0	3	3	3	3	0	0	0	0
Netarts	Northern anchovy	Adults	0	0	0	0	0	3	3	3	3	3	0	0
Netarts	Northern anchovy	Juveniles	0	0	0	0	0	3	3	3	3	3	0	0
Grays Hbr	Pink aalmon	Adults	0	0	0	0	0	0	0	0	0	2	0	0
Netarts	No pink salmon													
Grays Hbr	Chum salmon	Adults	0	0	0	0	0	0	0	0	3	4	4	4
Grays Hbr	Chum salmon	Juveniles	0	3	4	4	3	3	0	0	0	0	0	0
Netarts	Chum salmon	Adults	0	0	0	0	0	0	0	0	0	4	4	0
Netarts	Chum salmon	Juveniles	0	3	5	5	5	3	0	0	0	0	0	0

(continued)

TABLE 8.3 *(continued)*

ESTUARY	COMMON NAME	LIFE STAGE	JAN	FEB	MAR	APR	MAY	JUN	JUL	AUG	SEP	OCT	NOV	DEC
Grays Hbr	Coho salmon	Adults	3	0	0	0	0	0	0	3	3	4	4	3
Grays Hbr	Coho salmon	Juveniles	0	0	3	4	4	3	0	0	0	0	0	0
Netarts	Coho salmon	Adults	0	0	0	0	0	0	0	0	0	3	3	0
Netarts	Coho salmon	Juveniles	0	0	0	3	3	3	0	0	0	0	0	0
Grays Hbr	Steelhead (WINTER)	Adults	4	4	4	3	0	0	0	0	0	0	3	4
Grays Hbr	Steelhead (WINTER)	Juveniles	0	0	3	4	4	4	3	0	0	0	0	0
Netarts	Steelhead (WINTER)	Adults	0	0	0	0	0	0	0	0	0	0	2	2
Netarts	Steelhead (WINTER)	Juveniles	0	0	2	2	2	0	0	0	0	0	0	0
Grays Hbr	No sockeye													
Netarts	No sockeye													
Grays Hbr	Chinook salmon (F)	Adults	0	0	0	0	0	0	3	4	4	4	3	3
Grays Hbr	Chinook salmon (F)	Juveniles	0	3	3	4	5	5	5	4	4	3	3	0
Netarts	Chinook salmon (F)	Adults	0	0	0	0	0	0	0	0	0	2	0	0
Netarts	Chinook salmon (F)	Juveniles	0	0	0	0	2	0	0	0	0	0	0	0
Grays Hbr	Chinook salmon (SP)	Adults	0	3	4	4	3	3	0	0	0	0	0	0
Grays Hbr	Chinook salmon (SP)	Juveniles	0	3	3	4	5	5	5	4	4	3	3	0
Netarts	No spring chinook													
Grays Hbr	Surf smelt	Adults	3	3	3	3	3	3	3	3	3	3	3	3
Grays Hbr	Surf smelt	Juveniles	3	3	3	3	3	3	3	3	3	3	3	3
Grays Hbr	Surf smelt	Larvae	0	0	0	0	0	3	3	3	3	0	0	0
Netarts	Surf smelt	Adults	0	0	0	0	0	2	2	2	0	0	0	0
Netarts	Surf smelt	Juveniles	3	4	5	5	4	3	3	3	3	3	3	3
Netarts	Surf smelt	Larvae	3	3	3	3	3	0	0	3	3	3	3	3

Grays Hbr	Longfin smelt	Adults	3	3	3	3	3	3	3	3	3	3	3	3
Grays Hbr	Longfin smelt	Eggs	3	3	3	0	0	0	0	0	0	0	0	0
Grays Hbr	Longfin smelt	Juveniles	3	3	3	3	4	4	4	4	4	4	4	3
Grays Hbr	Longfin smelt	Larvae	3	4	5	4	3	0	0	0	0	0	0	0
Grays Hbr	Longfin smelt	Spawning	4	4	3	0	0	0	0	0	0	0	0	3
Netarts	No longfin smelt													
Grays Hbr	Eulachon	Adults	3	3	3	0	0	0	0	0	0	0	0	0
Grays Hbr	Eulachon	Larvae	0	3	3	3	0	0	0	0	0	0	0	0
Netarts	No eulachon													
Grays Hbr	Topsmelt	Adults	2	2	2	2	2	2	2	2	2	2	2	2
Grays Hbr	Topsmelt	Juveniles	2	2	2	2	2	2	2	2	2	2	2	2
Netarts	Topsmelt	Adults	0	0	0	3	4	4	3	3	3	3	0	0
Netarts	Topsmelt	Eggs	0	0	0	0	4	4	3	0	0	0	0	0
Netarts	Topsmelt	Juveniles	0	0	0	3	3	3	3	4	4	3	0	0
Netarts	Topsmelt	Larvae	0	0	0	0	0	3	3	0	0	0	0	0
Netarts	Topsmelt	Spawning	0	0	0	0	4	4	3	0	0	0	0	0
Grays Hbr	Pacific tomcod	Adults	3	3	3	3	3	3	3	3	3	3	3	3
Grays Hbr	Pacific tomcod	Juveniles	4	4	4	4	4	5	5	5	5	5	4	4
Grays Hbr	Pacific tomcod	Larvae	0	3	3	3	3	3	3	0	0	0	0	0
Netarts	Pacific tomcod	Juveniles	0	0	0	0	0	3	3	3	3	0	0	0
Netarts	Pacific tomcod	Larvae	0	3	3	3	3	0	0	0	0	0	0	0
Grays Hbr	Shiner perch	Adults	0	0	0	3	4	4	4	4	4	4	3	0
Grays Hbr	Shiner perch	Juveniles	0	0	0	0	3	4	5	5	5	4	3	0
Grays Hbr	Shiner perch	Spawning	0	0	0	0	3	4	4	3	0	0	0	0
Netarts	Shiner perch	Adults	0	0	0	4	5	5	5	5	4	4	0	0
Netarts	Shiner perch	Juveniles	0	0	0	0	4	5	5	5	5	4	0	0
Netarts	Shiner perch	Spawning	0	0	0	0	4	5	5	4	0	0	0	0

(continued)

TABLE 8.3 *(continued)*

ESTUARY	COMMON NAME	LIFE STAGE	JAN	FEB	MAR	APR	MAY	JUN	JUL	AUG	SEP	OCT	NOV	DEC
Grays Hbr	Lingcod	Juveniles	0	0	0	3	3	3	3	3	3	3	0	0
Grays Hbr	Lingcod	Larvae	3	3	3	3	0	0	0	0	0	0	0	0
Netarts	Lingcod	Juveniles	0	0	0	3	3	3	3	3	3	0	0	0
Grays Hbr	Staghorn sculpin	Adults	3	4	4	4	4	4	4	4	4	4	3	3
Grays Hbr	Staghorn sculpin	Eggs	3	3	0	0	0	0	0	0	0	0	3	3
Grays Hbr	Staghorn sculpin	Juveniles	3	4	4	4	4	4	4	4	4	4	3	3
Grays Hbr	Staghorn sculpin	Larvae	3	3	3	3	3	3	0	0	0	0	0	0
Grays Hbr	Staghorn sculpin	Spawning	3	3	0	0	0	0	0	0	0	0	3	3
Netarts	Staghorn sculpin	Adults	3	3	4	4	4	4	4	4	4	4	3	3
Netarts	Staghorn sculpin	Eggs	3	3	0	0	0	0	0	0	0	0	0	3
Netarts	Staghorn sculpin	Juveniles	3	4	5	5	5	5	5	5	4	4	3	3
Netarts	Staghorn sculpin	Larvae	3	3	3	0	0	0	0	0	3	3	3	3
Netarts	Staghorn sculpin	Spawning	3	3	0	0	0	0	0	0	0	0	0	3
Grays Hbr	English sole	Juveniles	3	3	3	3	4	4	4	4	4	4	3	3
Grays Hbr	English sole	Larvae	3	3	3	3	0	0	0	0	0	0	0	0
Netarts	English sole	Juveniles	3	3	3	4	4	4	4	4	4	3	3	3
Netarts	English sole	Larvae	3	3	3	0	0	0	0	0	0	0	0	0
Grays Hbr	Starry flounder	Adults	3	3	3	3	3	3	3	3	3	3	3	3
Grays Hbr	Starry flounder	Juveniles	3	3	3	3	4	4	4	4	4	4	3	3
Grays Hbr	Starry flounder	Larvae	0	3	3	3	3	0	0	0	0	0	0	0
Netarts	Starry flounder	Juveniles	3	3	3	3	3	3	3	3	3	3	3	3
Netarts	Starry flounder	Larvae	0	0	0	0	3	3	0	0	0	0	0	0

SOURCE: Data from Monaco et al. 1990.

data should be considered along with analyses of other classes of fauna used by site inhabitants, hunting and transportation technologies, site location, and changing climatic and geomorphological conditions. Over 10 years ago, Erlandson et al. (1998) attempted such a holistic review; they argued that the importance of marine mammal hunting on the southern Northwest Coast had been overemphasized, and they instead favored greater recognition of the cultural significance of estuarine fishing, elk hunting, and other activities. This was an appropriate corrective to the habit of zooarchaeologists focusing more attention on mammals than to fish, birds, or shellfish.

Broad cultural historical processes also need to be considered when evaluating sea otter and pinniped use, but unfortunately these are yet to be fully refined or understood in our study area. In contrast, on the west coast of Vancouver Island, marine mammal hunting could be addressed with regard to the timing of Wakashan expansion. In the Gulf of Georgia, where the cultural historical sequence is well defined, pinniped hunting can be assessed in light of technological and sociopolitical changes. On the southern Northwest Coast, discussion of population movements (other than the Athapaskan expansion) has been largely taboo. Clearly, patterns of use of sea otters and pinnipeds are likely part of broader historical processes, but variability has to be approached at multiple scales.

The site assemblages discussed here demonstrate that pinnipeds and sea otters were regularly procured by Native Americans of Oregon and Washington who lived along estuaries of widely varying characteristics. We suggest pinnipeds and sea otters were harvested when they entered estuaries following prey such as schooling and spawning fishes. Probable sea mammal hunting technologies varied by site and through time, but clearly the abundance of harpoons in sites such as Par-Tee indicates the ability to hunt marine mammals in the water. Whether this involved harpooning animals as they fled from haul-outs to the water or the harpooning of animals already in the water cannot be assessed with the data at hand. Given that Native people were commonly traveling through estuaries to fish, to gather clams, crabs, and fuel wood, to hunt waterfowl, and to meet and trade with neighboring groups, etc., it seems probable that they also regularly hunted pinnipeds and sea otters in these same settings. Because many of the estuaries of the southern Northwest Coast have undergone substantial changes in the modern era due to jetty building, dredging, diking and farming, understanding the dynamics of precontact Native American hunting provides crucial data for managing pinnipeds and sea otters today.

We maintain that attempts to provide regional narratives of long-term subsistence change are premature. The archaeological records of the North Pacific must be analyzed against independent data on environmental variability that are mediated by local settings. Only then can data be compiled in a way in which differences between contemporary, historical, and precontact patterns of pinniped and sea otter abundance become meaningful. Ideally, this will allow us to reconstruct a series of shifting baselines (Jackson et al. 2001; Pauly 1995; Rick and Erlandson 2008) that can better inform contemporary wildlife, fisheries, and ocean management.

ACKNOWLEDGEMENTS

We extend our thanks to Todd Braje and Torrey Rick for inviting us to contribute a paper to their effort. The following colleagues have generously provided us source materials: Robin Brown, Kris Bovy, Ken Pitcher, Jay Fancher, and Tanya Roberts.

REFERENCES CITED

Armstrong; Michael

2007 Rare Fur Seal Found in Bay Now Back Home. *Homer News*, Homer, AK. Electronic document, http://homernews.com/stories/103107/news_1_002.shtml, accessed August 25, 2008.

Bovy, K. M.

2005 *Effects of Human Hunting, Climate Change and Tectonic Events on Waterbirds along the Pacific Northwest Coast during the Late Holocene*. Ph.D. dissertation, University of Washington, Seattle.

2007 Global Human Impacts or Climate Change? Explaining the Sooty Shearwater Decline at the Minard Site, Washington State, USA. *Journal of Archaeological Science* 34:1087–1097.

Brown, J. H., and M. V. Lomolino
1998 *Biogeography*, 2nd ed. Sinauer Associates, Sunderland, MA.

Brueggeman, J. J., M. L. Bonnell, C. E. Bowlby, and G. A. Green
1992 *Oregon and Washington Marine Mammal and Seabird Surveys—Pinniped Distribution and Abundance Off Oregon and Washington, 1989–1990*. Pacific OCS Region, Minerals Management Service, U.S. Department of the Interior, Los Angeles, CA.

Burton, R. K., D. Gifford-Gonzalez, J. J. Snodgrass, and P. L. Koch
2002 Isotopic Tracking of Prehistoric Pinniped Foraging and Distribution along the Central California Coast: Preliminary Results. *International Journal of Osteoarchaeology* 12:4–11.

Burton, R. K., J. J. Snodgrass, D. Gifford-Gonzalez, T. Guilderson, T. Brown, and P. L. Koch
2001 Holocene Changes in the Ecology of Northern Fur Seals: Insights from Stable Isotopes and Archaeofauna. *Oecologia* 128(1):1–16.

Byram, R. S.
2002 *Brush Fences and Basket Traps: The Archaeology and Ethnohistory of Tidewater Weir Fishing on the Oregon Coast*. Ph.D. dissertation, University of Oregon, Eugene.

Colten, R. H.
2002 Prehistoric Marine Mammal Hunting in Context: Two Western North American Examples. *International Journal of Osteoarchaeology* 12:11–22.

Connolly, T. J.
1992 *Human Responses to Change in Coastal Geomorphology and Fauna on the Southern Northwest Coast: Archaeological Investigations at Seaside, Oregon*. University of Oregon Anthropological Paper 45, Eugene.
1995 Archaeological Evidence for a Former Bay at Seaside, Oregon. *Quaternary Research* 43: 362–369.

Cooper, C. F., and B. S. Stewart
1983 Demography of Northern Elephant Seals, 1911–1982. *Science* 219:969–971.

Cressman, L. S.
1952 Report on Research Project, Oregon Coast Prehistory. Letter on file at the State Museum of Anthropology, University of Oregon, Eugene.

Crockford, S. J., S. G. Frederick, and R. J. Wigen
2002 The Cape Flattery Fur Seal: An Extinct Species of *Callorhinus* in the Eastern North Pacific? *Canadian Journal of Archaeology* 26(2): 152–174.

Emmett, R., R. Llanso, J. Newton, R. Thom, M. Hornberger, C. Morgan, C. Levings, A Copping, and P. Fishman
2000 Geographic Signatures of North American West Coast Estuaries. *Estuaries* 23:765–792.

Emmett, R. L., S. L. Stone, S. A. Hinton, and M. E. Monaco
1991 *Distribution and Abundance of Fishes and Invertebrates in West Coast Estuaries Volume II: Species Life History Summaries*. ELMR Report No. 8. U. S. NOAA/NOS Strategic Environmental Assessments Division, Rockville, MD.

Erlandson, J. M., M. A. Tveskov, and R. S. Byram
1998 The Development of Maritime Adaptations on the Southern Northwest Coast of North America. *Arctic Anthropology* 35(1):6–22.

Etnier, M. A
2002a *The Effect of Human Hunting on Northern Fur Seal (*Callorhinus ursinus*) Migration and Breeding Distributions in the Late Holocene*. Ph.D. dissertation, Department of Anthropology, University of Washington, Seattle.
2002b Occurrences of Guadalupe Fur Seals (*Arctocephalus townsendi*) on the Washington Coast over the Past 500 years. *Marine Mammal Science* 18(2):551–557.
2007 Defining and Identifying Sustainable Harvests of Resources: Archaeological Examples of Pinniped Harvests in the Northeastern Pacific. *Journal for Nature Conservation* 15:196–207.

Fancher, J. M.
2001 *An Analysis of the Minard (45GH15) Faunal Assemblage*. Master's thesis, Department of Anthropology, Washington State University, Pullman.

Gentry, R. L.
1998 *Behavior and Ecology of the Northern Fur Seal*. Princeton University Press, Princeton, NJ.

Gifford-Gonzalez, D., S. D. Newsome, P. L. Koch, T. P. Guilderson, J. J. Snodgrass, and R. K. Burton
2005 Archaeofaunal Insights on Pinniped-Human Interactions in the Northeastern Pacific. In *The Exploitation and Cultural Importance of Sea Mammals*, edited by G. G. Monks, pp. 19–38. Proceedings of the 9th Conference of the International Council of Archaeozoology. Oxbow Books, Oxford, U.K.

Greenspan, R. L.
n.d. Vertebrate Remains. In *Archaeology of the Par-Tee Site, Northern Oregon Coast*, edited by R. J. Losey. Manuscript in preparation.

Greenspan, R. L., and S. Crockford
1992 Vertebrate Faunal Remains from the Palmrose and Avenue Q Sites. In *Human Responses to*

Change in Coastal Geomorphology and Fauna on the Southern Northwest Coast: Archaeological Investigations at Seaside, Oregon, edited by T. J. Connolly, pp. 123–165. University of Oregon Anthropological Paper 45, Eugene.

Gustafson, C. E.
1968a Prehistoric Use of Fur Seals: Evidence from the Olympic Coast of Washington. *Science* 161:49–51.
1968b Historic Change in Migratory Fur Seal Populations: Result of Pelagic Sealing? *Journal of Mammalogy* 49:842.

Hajda, Y.
1990 Southwestern Coast Salish. In *Handbook of North American Indians*, vol. 7, *Northwest Coast*, edited by Wayne Suttles, pp. 503–517. Smithsonian Institution, Washington D.C.

Hardin, G.
1968 The Tragedy of the Commons. *Science* 162:1243–1248.

Hickey, B. M., and N. S. Banas
2003 Oceanography of the U.S. Pacific Northwest Coastal Ocean and Estuaries with Application to Coastal Ecology. *Estuaries* 26(4B):1010–1031.

Hildebrandt, W. R.
1981 *Native Hunting Adaptations of the North Coast of California*. Ph.D. dissertation, Department of Anthropology, University of California, Davis.
1984a Archaeological Presence of the Northern Fur Seal (*Callorhinus ursinus*) along the Coast of Northern California. *The Murrelet* 65:28–29.
1984b Late Period Hunting Adaptations on the North Coast of California. *Journal of California and Great Basin Anthropology* 6:189–206.

Hildebrandt, W. R., and T. L. Jones
1992 Evolution of Marine Mammal Hunting: A View from the California and Oregon Coast. *Journal of Anthropological Archaeology* 11:360–401.
2002 Depletion of Prehistoric Pinniped Populations along the California and Oregon Coasts: Were Humans the Cause? In *Wilderness and Political Ecology: Aboriginal Influences on the Original State of Nature*, edited by C. E. Kay and R. T. Simmons, pp. 72–110. University of Utah Press, Salt Lake City.
2004 Evolution of Marine Mammal Hunting: A View from the California and Oregon Coasts. In *Prehistoric California: Archaeology and the Myth of Paradise*, edited by M. L. Raab and T. L. Jones, pp. 53–72. University of Utah Press, Salt Lake City.

Hodder, J., R. Brown, and C. Cziesla
1998 The Northern Elephant Seal in Oregon: A Pupping Range Extension and Onshore Occurrence. *Marine Mammal Science* 14:873–881.

Howorth, P. C.
1993 Commercial Collection of Pinnipeds in the California Channel Islands, 1877–1981. In *Third California Islands Symposium: Recent Advances in Research in the California Islands*, edited by H. G. Hochberg, pp. 579–590. Santa Barbara Museum of Natural History, Santa Barbara, CA.

Jackson, J. B. C., M. Kirby, W. Berger, K. Bjorndal, L. Botsford, B. Bourque, R. Bradbury, R. Cooke, J. Erlandson, A. James, J. A. Estes, T. Hughes, S. Kidwell, C. Lange, H. Lenihan, J. Pandolfi, C. Peterson, R. Steneck, M. Tegner, and R. Warner
2001 Historical Overfishing and the Recent Collapse of Coastal Ecosystems. *Science* 293: 629–637.

Jones, T. L., and W. R. Hildebrandt
1995 Reasserting a Prehistoric Tragedy of the Commons: Reply to Lyman. *Journal of Anthropological Archaeology* 14:78–98.

Jones, T. J., W. R. Hildebrandt, D. J. Kennett, and J. F. Porcasi
2004 Prehistoric Marine Mammal Overkill in the Northeastern Pacific: A Review of New Evidence. *Journal of California and Great Basin Anthropology* 24(1):69–80.

Larson S., R. Jameson, M. Etnier, M. Flemings, and P. Bentzen
2002 Loss of Genetic Diversity in Sea Otters (*Enhydra lutris*) Associated with the Fur Trade of the 18th and 19th Centuries. *Molecular Ecology* 11:1899–1903.

LeBoeuf, B. J.
1996 *Northern Elephant Seals at Año Nuevo during the Breeding Season of 1992–1996*. Administrative Report LJ-96-12C. Southwest Fisheries Science Center, National Marine Fisheries Service, La Jolla, CA.

Losey, R. J.
2002 *Communities and Catastrophe: Tillamook Response to the AD 1700 Earthquake and Tsunami, Northern Oregon Coast*. Ph.D. dissertation, Department of Anthropology, University of Oregon, Eugene.
2005 House Remains at the Netarts Sandspit Village, Oregon Coast. *Journal of Field Archaeology* 30(4):401–417.
n.d. *Archaeology of the Par-Tee Site, Northern Oregon Coast*. Manuscript in preparation.

Losey, R. J., and A. MacKenzie
n.d. Shellfish Remains. In *Archaeology of the Par-Tee Site, Northern Oregon Coast*, edited by R. J. Losey. Manuscript in preparation.

Losey, R. J., and E. Power
2005 Shellfish Remains from the Par-Tee Site (35-CLT-20), Seaside, Oregon: Making Sense of a

Biased Sample. *Journal of Northwest Anthropology* 39(1):1–20.
Losey, R. J., and D. Yang
2007 Opportunistic Whale Hunting on the Southern Northwest Coast: Ancient DNA, Artifact, and Ethnographic Evidence. *American Antiquity* 72(4):657–676.
Lyman, R. L.
1988 Zoogeography of Oregon Coast Marine Mammals: The Last 3000 Years. *Marine Mammal Science* 4:247–264.
1989 Seal and Sea Lion Hunting: A Zooarchaeological Study from the Southern Northwest Coast of North America. *Journal of Anthropological Archaeology* 8:68–99.
1991a Subsistence Change and Pinniped Hunting. In *Human Predators and Prey Mortality*, edited by M. C. Stiner, pp. 189–199. Westview Press, Boulder, CO.
1991b *Prehistory of the Oregon Coast: The Effects of Excavation Strategies and Assemblage Size on Archaeological Inquiry*. Academic Press, San Diego.
1995 On the Evolution of Marine Mammal Hunting on the West Coast of North America. *Journal of Anthropological Archaeology* 14:45–77.
2003 Pinniped Behavior, Foraging Theory, and the Depression of Metapopulations and Nondepression of a Local Population on the Southern Northwest Coast of North America. *Journal of Anthropological Archaeology* 22:376–388.
Lyon, G. M.
1937 Pinnipeds and a Sea Otter from the Point Mugu Shell Mound of California. *University of California Publications in Biological Sciences* 1(8):133–168.
Mate, B. R.
1981 Marine Mammals. In *Natural History of Oregon Coast Mammals*, edited by C. Maser, B. Mate, J. F. Franklin, and C. T. Dyrness, pp. 372–458. USDA Forest Service Technical Report, PNW-133.
McClenachan, L.
2002 *Abundance and Blame: The Case of the California Sea Lion*. Master's thesis, Environmental Studies Program, University of Oregon, Eugene.
Minor, R.
1983 *Aboriginal Settlement and Subsistence at the Mouth of the Columbia River*. Ph.D. dissertation, Department of Anthropology, University of Oregon, Eugene.
Mirschitzka, S.
1992 Usage and Management of Marine Resources on the Northwest Coast. *European Review of Native American Studies* 6(1):9–16.
Monaco, M. E., R. L. Emmett, D. M. Nelson, and S. A. Hinton
1990 *Distribution and Abundance of Fishes and Invertebrates in West Coast Estuaries Volume I: Data Summaries*. ELMR Report No. 4. U. S. NOAA/NOS Strategic Environmental Assessments Division, Silver Spring, MD.
Monks, G. G.
1987 Prey as Bait: The Deep Bay Example. *Canadian Journal of Archaeology* 11:119–142.
Moss, M. L., D. Y. Yang, S. D. Newsome, C. F. Speller, I. McKechnie, A. D. McMillan, R. J. Losey, and P. L. Koch.
2006 Historical Ecology and Biogeography of North Pacific Pinnipeds: Isotopes and Ancient DNA from Three Archaeological Assemblages. *Journal of Island and Coastal Archaeology* 1(2):165–190.
National Marine Mammal Laboratory (NMML)
2009 Marine Mammal Education Web: Guadalupe Fur Seals. Electronic Document, http://www.afsc.noaa.gov/nmml/education/pinnipeds/guadalupefs.php, accessed December 3, 2009.
Newman, T. M.
1959 *Tillamook Prehistory and its Relation to the Northwest Coast Culture Area*. Ph.D. dissertation, Department of Anthropology, University of Oregon, Eugene.
NOAA—Center for Coastal Monitoring and Assessment (CCMA)
2005 Estuarine Living Marine Resources Database. Online http://ccma.nos.noaa.gov/ecosystems/estuaries/elmr.html, accessed June 24, 2008.
NOAA Fisheries (Protected Resources)
1997 Stock Assessment Report, California Sea Lion (*Zalophus californianus californianus*): U.S. Stock. Electronic document, http://www.nmfs.noaa.gov/prot_res/PR2/Stock_Assessment_Program/Pinnipeds/CA_Sea_Lion_(U.S.)/PO97casealion.pdf, accessed February 3, 2006.
2000 Stock Assessment Report, Northern Fur Seal (*Callorhinus ursinus*): San Miguel Stock. Electronic document, http://www.nmfs.noaa.gov/pr/pdfs/sars/po2000senf-sm.pdf, accessed August 25, 2008.
2001a Stock Assessment Report, Northern Fur Seal (*Callorhinus ursinus*): Eastern Pacific Stock. Electronic document, http://www.nmfs.noaa.gov/pr/pdfs/sars/ak2001senf-pe.pdf, accessed August 25, 2008.
2001b Stock Assessment Report, Steller Sea Lion—Eastern U.S. Stock. Electronic document, http://www.nmfs.noaa.gov/prot_res/PR2/Stock_Assessment_Program/Pinnipeds/Steller_(East)/AK01stellersealion_East.pdf, accessed February 3, 2006.
2003a Stock Assessment Report, California Sea Lion (*Zalophus californianus californianus*): U.S.

Stock. Electronic document, http://www.nmfs.noaa.gov/prot_res/PR2/Stock_Assessment_Program/Pinnipeds/CA_Sea_Lion_(U.S.)/PO03casealion.pdf, accessed February 3, 2006.

2003b Stock Assessment Report, Northern Fur Seal (*Callorhinus ursinus*): San Miguel Island Stock. Electronic document, http://www.nmfs.noaa.gov/pr/pdfs/sars/po03northernfurseal-sanmiguel.pdf, accessed June 14, 2006.

2005 Stock Assessment Report, Northern Fur Seal (*Callorhinus ursinus*): Eastern Pacific Stock. Electronic document, http://www.nmfs.noaa.gov/pr/pdfs/sars/ak2005senf-pe.pdf, accessed June 14, 2006.

2007 Stock Assessment Report, Northern Elephant Seal (*Mirounga angustirostris*) California Breeding Stock. Electronic document, http://www.nmfs.noaa.gov/pr/pdfs/sars/po2007sene-ca.pdf, accessed June 26, 2008.

Oregon Coastal Atlas

2005 Electronic document, http://coastalatlas.net/, accessed Feb. 9, 2005.

Pauly, D.

1995 Anecdotes and the Shifting Baselines Syndrome of Fisheries. *Trends in Ecology and Evolution* 10(10):430.

Pearson, J. P.

1968 *The Abundance and Distribution of Harbor Seals and Steller Sea Lions in Oregon*. Master's thesis, Oregon State University, Corvallis.

Phebus, G. E., Jr., and R. M. Drucker

1979 *Archaeological Investigations at Seaside*. Seaside Museum and Historical Society, Seaside, OR.

Pitcher, K. W., P. F. Olesiuk, R. F. Brown, M. S. Lowry, S. J. Jeffries, J. L. Sease, W. L. Perryman, C. E. Stinchcomb, and L. F. Lowry

2007 Status and Trends in Abundance and Distribution of the Eastern Population of Steller Sea Lions (*Eumetopias jubatus*). *Fishery Bulletin* 107:102–115.

Porcasi, J. F., T. L. Jones, and M. L. Raab

2000 Trans-Holocene Marine Mammal Exploitation on San Clemente Island, California: A Tragedy of the Commons Revisited. *Journal of Anthropological Archaeology* 19:200–220.

Pyle, P., D. J. Long, J. Schonewald, R. E. Jones, and J. Rolleto

2001 Historical and Recent Colonization of the South Farallon Islands, California, by Northern Fur Seals (*Callorhinus ursinus*). *Marine Mammal Science* 17(2):397–402.

Rick, T. and J. M. Erlandson (editors)

2008 *Human Impacts on Ancient Marine Ecosystems: A Global Perspective*. University of California Press, Berkeley.

Roll, T. E.

1974 *The Archaeology of Minard: A Case Study of a Late Prehistoric Northwest Coast Procurement System*. Unpublished Ph.D. dissertation, Department of Anthropology, Washington State University, Pullman, WA.

Schalk, R. and G.C. Burtchard

2001 *The Newskah Creek Fish Trap Complex, Grays Harbon, Washington*. Report prepared for U.S. Army Corps of Engineers, Seattle District. Cascadia Archaeology, Seattle, WA.

Sea Lion Caves

2005 Sea Lion Caves website. Electronic document, http://sealioncaves.com/, accessed, Jan. 31, 2005.

Seaburg, W. R. and J. Miller

1990 Tillamook. In *Handbook of North American Indians* Vol. 7, *Northwest Coast*, edited by W. Suttles, pp. 560–567. Smithsonian Institution, Washington D.C.

Shoreline Education for Awareness

1999 Simpson Reef and Shell Island, Cape Arago. Bandon, OR.

Stewart, B. S., P. K. Yochem, R. L. DeLong, and G. A. Antonelis

1993 Trends in Abundance and Status of Pinnipeds on the Southern California Channel Islands. In *Third California Islands Symposium: Recent Advances in Research in the California Islands*, edited by H. G Hochberg, pp. 501–516. Santa Barbara Museum of Natural History, Santa Barbara, CA.

U.S. Fish and Wildlife Service

2005 Oregon Islands National Wildlife Refuge. Electronic document, http://oregoncoast.fws.gov/oregonislands/index.htm, accessed Feb. 9, 2005.

Valentine, K., D. A. Duffield, L. E. Patrick, D. R. Hatch, V. L. Butler, R. L. Hall, and N. Lehman

2007 Ancient DNA Reveals Genotypic Relationships among Oregon Populations of the Sea Otter (*Enhydra lutris*). *Conservation Genetics* 9(4):933–938.

Walker, P. L., and S. Craig

1979 Archaeological Evidence Concerning the Prehistoric Occurrence of Sea Mammals at Point Bennett, San Miguel Island. *California Fish and Game* 65:50–54.

Walker, P. L., D. J. Kennett, T. L. Jones, and R. DeLong

1999 Archaeological Investigations at the Point Bennett Pinniped Rookery on San Miguel Island. In *Proceedings of the 5th California Islands Symposium*, edited by D. R. Browne, H. S. Chaney, and K. L. Mitchell, pp. 628–632. Santa Barbara Museum of Natural History, Santa Barbara, CA.

9

Why Were Northern Fur Seals Spared in Northern California? A Cultural and Archaeological Explanation

Adrian R. Whitaker and William R. Hildebrandt

SINCE THE INITIATION OF systematic archaeological research in northwestern California and southern Oregon 40 years ago, seals and sea lions have held a central place in archaeological analyses. The topic was brought to a broader audience in a series of papers by Hildebrandt and Jones, and Lyman, that examined trends in sea lion hunting through time and first posited the possibility that prehistoric humans had adversely affected breeding seals and sea lions (Hildebrandt and Jones 1992, 2002; Jones and Hildebrandt 1995; Lyman 1988, 1995, 2003). Since then, a group of eastern Pacific Rim archaeologists have intensively examined the role of humans in shaping pinniped breeding patterns, population trends, and biogeography. The biogeography of the northern fur seal, in particular, has been of interest throughout its prehistoric range, which stretched from southern California to the Bering Sea.

In this chapter, we provide a history of pinniped hunting in Northern California and examine the record from Stone Lagoon (CA-HUM-129). We describe the evidence for continuous hunting of large sea lions up to European contact and offer an explanation for the persistence of otariid hunting in northern California despite the fact that some of these pinnipeds were no longer hunted in central California and parts of Oregon (Gifford-Gonzales et al. 2005, Hildebrandt and Jones 1992, 2002, Jones and Hildebrandt 1995, Lyman 1995, 2003).

Northern California sits at the intersection of the diverse acorn-based economies of California and the truly maritime cultures of the Northwest Coast culture area proper. The northernmost tip of California consisted of a number of groups of Algic and Athabaskan language speakers who also shared a number of cultural traits with people to the North (Kroeber 1939). Some of these traits revolved around the pursuit of status and may have created a setting in which sea lions were hunted sustainably as a component of prestige economies. Prior to a discussion of these practices, we set the stage by detailing the history of archaeological and ethnographic research in the region.

OTHER STUDIES OF PINNIPED PROCUREMENT

Over the past 25 years, several California archaeologists have identified pinniped skeletal remains that are consistent with the modern distribution of some species (Gifford-Gonzalez et al. 2005; Hildebrandt 1984; Hildebrandt and Jones 1992, 2002; Jones and Hildebrandt 1995; Lyman 1988, 1991, 1995, 2003). The discovery of northern fur seal (*Calorhinus ursinus*) remains from multiple archaeological sites was the most perplexing of these findings, as this species does not presently come to shore along the mainland coasts of California and Oregon, but instead forages off the edge of the continental shelf (Gentry 1998). Hildebrandt and Jones (1992) hypothesized that this might reflect the prehistoric presence of mainland northern fur seal rookeries along the California coast that had been eliminated by prehistoric hunters prior to 1500 cal BP. This hypothesis was particularly controversial for the biological community, as the primary breeding area for this animal is currently on the Pribiloff Islands within the Bering Sea (Gentry 1998). It was also questioned by some archaeologists because the proposal relied on the presence of northern fur seal bones in the archaeological record and did not present the age and sex data (e.g., evidence of preweaned infant remains) necessary to clearly document the presence of a local, mainland rookery (Etnier 2002; Lyman 1995).

This problem ultimately led to the formation of a research group of northern Pacific zooarchaeologists led by Diane Gifford-Gonzales. Three alternative hypotheses were offered to explain the high numbers of northern fur seal bones in California archaeological assemblages.

1. Northern fur seals were exploited pelagically off the continental shelf of California.
2. The assemblages were accretions of "pinniped driftwood," the random stranding of ill or starving individuals, accumulated over long periods of time.
3. Prehistoric pinniped rookeries existed on the mainland or on nearshore rocks, sea stacks, and islands but were extirpated by human predation.

Archaeological and ethnographic evidence do not support the first hypothesis, because there is little evidence that Native hunters along the central and northern California coast had the technological capability to locate, harpoon, and dispatch pelagic pinnipeds while on the open ocean. There is no ethnographic evidence for the use of oceangoing canoes among central California groups—between Cape Mendocino in the north and southern San Luis Obispo in the south—where prehistoric remains of northern fur seals have been found. The archaeological record also shows little indirect evidence for oceangoing craft (i.e., sophisticated harpoon technology), and fish bone assemblages from these sites have failed to produce a significant number of deep water species. A recent analysis of a faunal assemblage dating between 1300 and 1200 cal BP along the central California coast at Año Nuevo (CA-SMA-18), for example, produced a large number of northern fur seal bones but a fish assemblage dominated by surf perches and rockfishes that can be captured nearshore or in kelp forest habitats (Gifford-Gonzales et al. 2006).

In Northern California—north of Cape Mendocino, within Tolowa, Yurok, and Wiyot territories, there are many more offshore rocks and islands than elsewhere in central and northern California. These rocks are located up to 11 kilometers out to sea and support large pinniped rookeries and haul-outs today. The ethnographic record demonstrates that these rocks were accessed by large oceangoing canoes (Kroeber and Barrett 1960). Although the local archaeological record contains composite harpoons (ethnographically associated with canoes), some evidence for deep water fishes, and large numbers of pinniped remains, the ethnographic record suggests that seals and sea lions were typically clubbed while sleeping on offshore rocks and were rarely taken on the open ocean. If the animals were alerted and attempted to escape, hunters in canoes were ready to harpoon them before they could get away. Sometimes a clubbed animal would fall off the rock, but they still needed to be harpooned as they are negatively buoyant and would sink. If a large animal was harpooned and took to the water relatively unharmed, many hours of dangerous work was often required before it could be pulled close enough to the boat to be killed with a club or additional projectiles (Hildebrandt 1981; Jobson and Hildebrandt 1980; Kroeber and Barrett 1960).

Under the second ("driftwood") hypothesis, northern fur seals likely originated in the Pribilofs, the modern breeding location of all but a small fraction of the population. Stable isotope geochemistry has demonstrated that this was not the case for archaeological specimens in central California. Stable isotopes of archaeological northern fur seal bones match middle latitude foraging signatures rather than the mixed northern and middle latitude signature of modern northern fur seals (Burton et al. 2002, 2001; Gifford-Gonzales et al. 2005; Newsome et al. 2006). These data indicate that, whether or not mainland rookeries were present, archaeological specimens belonged to a mid-latitude breeding population.

The best test of the final hypothesis in California comes from the work of Gifford-Gonzales and her colleagues at the Moss Landing site (CA-MNT-234). Analysis of stable isotope ratios from archaeological specimens indicates that the animals were feeding in mid-latitudes, and it appears that the bones of pups have also been recovered from the site. Age determinations of the sample were derived from a series of growth curves developed by Etnier (2002) for long bones and mandibles based on known-age skeletons collected from modern breeding sites on the Pribiloff Islands. Infant individuals were also identified through the analysis of nitrogen isotope ratios from their bones, which show that they were feeding one trophic level above the adults—that is, on their mother's milk. Gifford-Gonzales et al. (2005) also note that northern fur seal remains drop out of the central California archaeological record after 1000 BP. While this finding provides some support for Hildebrandt and Jones's (1992) original hypothesis, Gifford-Gonzales et al. (2005) argue that we need to consider a variety of nonhuman factors such as changing sea water temperatures that may have led to their demise.

Moss et al. (2006) have conducted similar analyses using data from three sites in southern Alaska, British Columbia, and northern Oregon. Stable isotope signatures on fur seal elements from these locations are indistinguishable from those of modern fur seals belonging to the Pribilof population. Furthermore, the age and sex data from the sites were inconclusive as to the use of these locations as breeding areas. Moss et al. (2006:184) conclude that "for archaeological sites located along the Pacific coast from [northern] Oregon north to Alaska, it appears that available isotopic data are hard to use as direct evidence to support a case for or against the presence of local northern fur seal breeding areas."

Lyman's (2003) analysis of a large data set from multiple sites along the Oregon coast shows an overall decline in terrestrial breeding otariids (northern fur seals, California sea lions, Steller sea lions) and an increase in the more elusive aquatic breeding harbor seals and sea otters. These findings were largely consistent with the original proposals of Hildebrandt and Jones (1992, 2002; Jones and Hildebrandt 1995), but Lyman (2003) found no evidence for a northern fur seal rookery at any of his sites. Following Hildebrandt and Jones (1992), he argued for a Steller sea lion rookery at offshore rocks near the Seal Rock archaeological site (35LNC14). Unlike Hildebrandt and Jones (1992), however, Lyman (2003) found that the sex demographics were dominated by males. He concluded that the near-exclusive harvest of males allowed hunters to sustainably harvest Steller sea lions from the rookery at Seal Point throughout its occupation. This is an important conclusion as it would allow long-term sustainable hunting of the population. Lyman (2003) cautions, however, that this does not imply that inhabitants of Seal Rock were intentionally conserving the pinniped population.

In summary, northern fur seal remains appear in archaeological sites from California to Alaska until historic times with some degree of regularity. The exception appears to be in central California, where populations may have been extirpated around 1000 cal BP. The overall population of otariids also appears to have declined relative to harbor seals and sea otters at about this time throughout mainland California and Oregon (Hildebrandt and Jones 1992; Jones and Hildebrandt 1995). Detailed studies of specific sites have demonstrated the possibility of mainland fur seal rookeries in British Columbia, Washington, and central California. No evidence of mainland northern fur seal rookeries in northern California or Oregon has been found.

THE RECORD OF MARINE MAMMAL HUNTING IN NORTHERN CALIFORNIA

Hildebrandt and Jones (1992) provide a comprehensive overview of pinniped hunting in northwestern California. Summarized in Figure 9.1, these data document a high percentage of terrestrial breeding pinnipeds (sea lions and fur seals that breed on land, also called "migratory breeders") found in late prehistoric archaeological assemblages. Unlike regions to the north and south, there is scant data available in the form of multicomponent assemblages that can be readily compared. This is likely due to the history of coastal occupation (see below).

In the one location at which there are multicomponent data—MacKerricher State Park (CA-MEN-828, CA-MEN-428)—the ratio of terrestrial to aquatic breeders actually increases, rather than decreases through time (Whitaker 2008). In contrast to the record of sea lion hunting at Moss Landing and of other supposed instances of resource depression in prehistoric California (e.g. Broughton 1997, 1999), sea lions remained a key meat source for peoples of California's northwestern coast. This record appears to be the result of offshore procurement of otariids (especially northern fur seals and Steller sea lions) relatively late in time at sea stacks and offshore rocks (Hildebrandt and Jones 1992).

OCEANGOING CANOES AND OFFSHORE HUNTING

The ethnographically recorded offshore hunting of sea lions by the Tolowa and Yurok are cited as evidence that these groups exhibited true maritime cultures (Hildebrandt and Levulett 2002; Jobson and Hildebrandt 1980). Gould (1968) was the first to synthesize ethnographic reports of oceangoing boats on the California coast and to assert their presence in prehistoric northern California. Boats in the region are described by Gould (1968) as larger versions of the redwood dugouts used on rivers and in the estuaries of Lake Earl and Humboldt Bay. These large boats required a great deal of labor to construct and, therefore, were owned by only the wealthiest individuals, who had to provide their assistants with gifts of food during the long building process (Gould 1968). Historically, the Yurok adopted sails from European settlers and ran boats up the coast from Humboldt Bay to Crescent City. Prehistorically, however, boats were likely only used for subsistence pursuits at offshore rocks.

Gould's (1968) study stopped short of examining the distribution of oceangoing dugouts, limiting analysis to the Tolowa, Yurok, Wiyot, and Tututni. Jobson and Hildebrandt (1980) expanded the scope of analysis to include the entire north coast. They also formalized the debate by defining *oceangoing* as "the act of safely traveling out substantial distances on the ocean; 'substantial distances' being a qualitative judgment in this instance" (Jobson and Hildebrandt 1980:167). Following this definition, they ex-

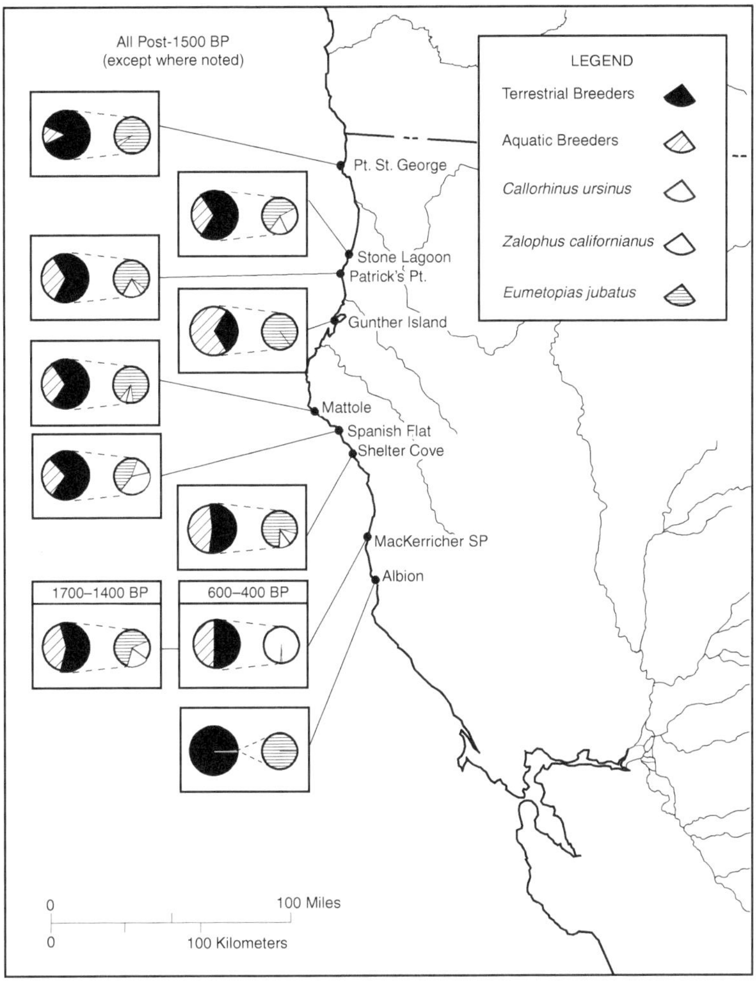

FIGURE 9.1. Proportional representation of terrestrial (otariids and northern elephant seals) vs. aquatic breeders (harbor seals and sea otters; larger circles) and distribution of otariids within the terrestrial breeders (small circles).

amined the ethnographic record for the use of dugouts and found that only groups north of Cape Mendocino had boats that could truly be considered *oceangoing*.

South of Cape Mendocino, Jobson and Hildebrandt (1980) cite the fact that marine mammal kills were always towed to shore rather than carried in boats, suggesting the watercraft used were not truly *oceangoing*. Summarizing the ethnographic description of canoes south of Cape Mendocino, Jobson and Hildebrandt (1980:168) state: "Canoes were neither large nor stable enough to be taken out onto the ocean for any appreciable distance." In lieu of oceangoing boats, peoples south of Cape Mendocino were reported to have swum to offshore rocks in pursuit of sea lions (Nomland 1938).

Other possible evidence of oceangoing craft is the distribution of offshore rocks with marine mammal haul-outs that are beyond the range of swimmers (defined as 500 m from shore). Evidence of ownership of these rocks and large quantities of pinniped bone in nearby coastal sites indicate that these rocks were exploited ethnographically.

Hudson (1981) refutes the sea-worthiness of large redwood dugouts. He believes that the rough seas and the design of the redwood dugout would leave it prone to capsizing in the choppy surf. Instead of being adapted for seaworthiness, he argues, the redwood dugout's size served to increase the cargo capacity and, therefore, reduce the risk that multiple trips with lower cargo capacity would entail. Regardless of whether these boats were truly *oceangoing* or not, it is difficult to refute their use to access offshore rocks by the Tolowa, Yurok, and Tututni. Furthermore, a lack of comparable offshore rocks and any mention of offshore hunting indicate that dugouts were not successfully used to pursue marine mammals by peoples south of Cape Mendocino.

When boats were used, hunters would paddle to offshore rocks where they would hunt sea lions, which were not pursued on the open ocean (Kroeber and Barrett 1960, Gould 1966). Once on offshore rocks, individual hunters attempted to sneak up on sleeping sea lions. Often they wore sea lion skin decoys and imitated sea lion calls and movements. If possible, sea lions were clubbed on the forehead or nose either killing or stunning them. Other hunters in canoes and on the rocks were ready to harpoon fleeing animals once the hunt began. On some rocks, sea lions were harpooned as they swam. Once an animal was harpooned, hunters took to the boats and held the line, allowing the sea lion to tow them until it tired and they were able to dispatch it. It often took several hours for the sea lion to tire, especially when larger individuals were harpooned. Because of the possibility of a long pursuit, sea lion hunts almost always took place close to dawn. The Tolowa and Yurok preferred young sea lions because they were easier to catch and the meat was more tender (Kroeber and Barrett 1960).

THE RECORD FROM NORTHERN CALIFORNIA • *The Timing of Northern California Coastal Occupation and Pinniped Hunting*

The ethnographic record enhances our understanding of the history of pinniped exploitation in northern California. This is because, unlike the 8000-to-10,000-year record of human occupation along California's central and southern coasts (Erlandson and Colten 1991), the time depth of the coastal archaeological record in northern California is fairly shallow (<2500 years). Some believe that this incongruity is a problem of archaeological visibility related to the disproportionate inundation of the shallow northern California continental shelf by rising Holocene sea levels (Arnold 1991; Erlandson and Colten 1991; Jones 1991; Lyman 1991; Moss 1998). They argue that the paucity of archaeological work on the north coast, coupled with the problem of sea level rise and subduction along the shores of Humboldt Bay, have conspired to hide archaeological sites that ought to be present.

Hildebrandt and Levulett (2002) argue that intensive survey and excavation in the King Range National Conservation Area (KRNCA), an area where tectonic uplift of marine terraces has kept pace with sea level rise, has failed to find a site older than 2500 cal BP. Furthermore, sea level rise appears to have slowed dramatically around 6000 years ago (Barron et al. 2003; Bickel 1978; Milliman and Emery 1968). If this is the case and older sites do exist, it seems unlikely that archaeologists on the north coast of California would have failed to uncover a single coastal site dating to earlier than 2500 cal BP. In fact, we should find a number of sites dating to ca. 6000 cal BP when sea levels stabilized (Bickel 1978).

Hildebrandt and colleagues (Hildebrandt and Carpenter 2007; Hildebrandt and Levulett 2002) offer an alternative explanation for the late coastal record. They propose that terrestrial resources in northern California were superior to those in southern California. Because precipitation decreases on a north-south and west-east gradient in California, the northwest coast contains the most lush vegetation and supports more terrestrial biomass than the central coast and, even more so, than the south coast. Hildebrandt and colleagues predict that early populations of hunter-gatherers, living at low population densities, would have been able to make a living in the interior of northern California, subsisting on high-return resources such as deer, elk, and anadromous fishes. This supposition is based on a number of Early Holocene interior sites found on Pilot Ridge in Northern Humboldt County (CA-HUM-573, CA-HUM-367) and recent evidence of early interior riverine occupation at Jed Smith State Park (DNO-26) on the Smith River (Hildebrandt 2007; Tushingham et al. 2008). In southern California, where these resources were not as readily available, hunter-gatherers were forced to subsist on coastal resources, resulting in early coastal settlement.

The priority of interior settlement is an important notion in the construction of culture history on the north coast. Developed by Hildebrandt and Hayes (1993) and later refined by Hildebrandt and Levulett (1997, 2002), the culture history of northwestern California is broken into three general periods: the Borax Pattern (7000–3500 cal BP), the Mendocino Pattern (3500–1500 cal BP), and the Gunther Pattern (1500–150 cal BP). With the exception of one site dating to the Borax Lake Period, all coastal sites in the northwestern portion of the state date to either the Mendocino or Gunther Patterns (Hildebrandt 2007; Jones 1992). These two patterns are described briefly to allow us to establish the timing of the initial hunting of sea lions in the region.

COASTAL MANIFESTATIONS OF THE MENDOCINO PATTERN (3000–1500 CAL BP)

The Mendocino Pattern contains the earliest evidence of prolonged coastal occupation in Mendocino, Humboldt, and Del Norte counties. The pattern is identified by the use of side-notched, corner-notched, and concave-based dart points of the Mendocino, Willits, and Trinity series. Handslabs and millingslabs as well as various flake and cobble tools and, rarely, mortars and pestles are associated with Mendocino Pattern sites. Leaf-shaped, McKee unifaces are found north of Sonoma County. Mendocino Pattern sites on the Del Norte and Humboldt coasts include Point St. George (CA-DNO-11; Gould 1966), CA-HUM-277 (Levulett 1985; Whitaker 2006), and CA-HUM-351 (Eidsness 1993). These sites all date no earlier than 2500 cal BP and demonstrate, primarily, terrestrial-based subsistence strategies. They are thought to have been used as seasonal camps by coastal hunters (Hildebrandt and Levulett 1997, 2002). Farther south, coastal sites dating to the Mendocino Pattern, between 3000 and 1500 cal BP, also represent, mainly, temporary camps or seasonal occupations with a terrestrial subsistence focus.

COASTAL MANIFESTATIONS OF THE GUNTHER PATTERN (1500–150 CAL BP)

The Gunther Pattern marks the initiation of truly coastal and estuarine adaptations in northern California. A fluorescence of sites along the coastal margin at this time is marked by thick middens and extremely diverse artifact assemblages relative to earlier occupations. The use of the bow and arrow is inferred from the prevalence of Gunther-series projectile points, presumably arrowheads. A strong littoral focus is indicated by a myriad of fishing and marine mammal hunting technologies, including harpoon tips and evidence of composite harpoons, net sinkers, bone and antler spears and harpoons, and bone and shell fishhooks (Bennyhoff 1950). Seed and shellfish processing is implied by the elaborate, flanged pestles, groundstone bowls, and hopper mortars. Well-made mauls along with antler wedges are thought to indicate woodworking, including the manufacture of redwood dugouts north of Cape Mendocino. Faunal assemblages, dominated by marine mammals, fish, and shellfish, demonstrate a shift to a marine-focused subsistence economy. Artifact assemblages and faunal remains at these sites substantiate the ethnographic record of hunting and fishing at offshore rocks and on the open ocean north of Cape Mendocino (Gould 1968; Hildebrandt and Jones 1992; Jobson and Hildebrandt 1980).

Nonsubsistence aspects of Gunther Period sites demonstrate a great deal of cultural elaboration. The remains of several redwood-plank houses similar to those in British Columbia have been found in large village sites. These houses are often identified by their stone patios and prepared clay floors. Ceremonial items such as zoöform clubs, large obsidian blades, and dentalium beads have been recovered from a limited number of burial contexts. These associations are likely indicative of competition for social prestige during the Gunther Pattern (Hildebrandt and Levulett 2002).

If the Gunther Pattern marks the inception of truly littoral and/or maritime foci of subsistence economies on the Northwest Coast, it is unlikely that sea lions have been hunted for longer than approximately 1500 years in northwestern California. In comparison with central or southern California or the Northwest Coast Culture area, this is a relatively recent development. However, modeling of pinniped population dynamics indicates that sea lions are prone to overhunting even under low levels of predation (Barlow and Boveng 1991; Whitaker 2008).

Hildebrandt and Jones initially proposed that the hunting of otariids during the late period resulted from the development of seafaring redwood dugouts that were used to pursue otariids at offshore rocks following the extirpation of mainland populations. However, there is little evidence that mainland rookeries ever existed in northern California. Regardless, otariids appear to have been an important part of coastal subsistence in northern California, and, as large, slowly reproducing mammals, they would be prone to overhunting, even at offshore rocks. Why, then, were northern Californian foragers able to continuously hunt otariids for over 1500 years of occupation on the coast? In an attempt to answer this question, we examine the record from one site in northwestern California: the Stone Lagoon site (CA-HUM-129).

STONE LAGOON (CA-HUM-129)

The Stone Lagoon site (HUM-129) was identified by Waterman (1920) as the ethnographic village of "*Tsahpek*", the third-largest ethnographic Yurok village on the coast, located at the southern end of Stone Lagoon in northern Humboldt County (see Figure 9.1). Situated on a bluff overlooking Stone Lagoon and the Pacific Ocean, the site is 150 m south of the breach point that forms during most, but not all, winters. This breach point would have been ideal for hunting marine mammals that might enter the lagoon to feed on anadromous or surf fish funneling into the small channel, which would also have been exploited by humans. Furthermore, the site is located at the closest point on the coast to Redding Rock, a large sea stack on which sea lions regularly haul out.

Sonoma State University began work at HUM-129 in the summer of 1976 to mitigate damage from erosion and looters. Their initial excavation was designed to stabilize the bluff on which the site sits, to recover human remains encountered during the stabilization effort for later reburial, and to hide the surface deposits from looters (Milburn et al. 1979). Stabilization was completed in 1978, and the faunal assemblage was partially analyzed by Hildebrandt and curated at the California State Parks Archaeological Laboratory facility in West Sacramento. A detailed identification of bird and mammal remains, however, was not undertaken until a recent study by Whitaker (2008).

Excavation of the site was concentrated on the eroding bluff faces on the northern and eastern borders and on several slump areas. Large-scale excavation was undertaken in 1.0×2.0 m or 1.0×1.0 m blocks placed adjacent to one another to form trenches. The deposits were screened through ¼-inch mesh. Basic cleaning, sorting, and cataloguing of all artifacts was completed at Sonoma State University, and a detailed analysis of bone tools, antler wedges, projectile points, and net sinkers was also undertaken (Milburn et al. 1979).

HUM-129 is a quintessential Gunther Pattern site. In fact, it is one of the sites used to create the cultural chronology of northern California (Hildebrandt 2007; Hildebrandt and Levulett 1997, 2002). Radiocarbon dates substantiate a Gunther Pattern occupation between A.D. 1240 and 1730 (Table 9.1). However, two dates indicate a pre-Gunther occupation of

TABLE 9.1

Radiocarbon dates from Hum-129

SAMPLE ID#	PROVENIENCE	DEPTH	^{14}C AGE	MATERIAL	2-SIGMA CAL BC/AD[a]	MEDIAN[a]
OS-63547	Unit N2 W16	0–10 cmbs	635 ± 35	*Callorhinus* bone	Off curve[b]	Off curve[b]
OS-57904	Unit NW2 SW 9	20–30 cmbs	1150 ± 25	*Mytilus* shell	AD 1406–1586	AD 1479
OS-63552	Unit N2 W16	30–40 cmbs	1270 ± 35	*Callorhinus* bone	AD 1307–1463	AD 1390
OS-63553	Unit NW2 SW7	30–40 cmbs	950 ± 30	*Callorhinus* bone	AD 1536–1806	AD 1660
OS-57902	Slump B	40–50 cmbs	1070 ± 25	*Mytilus* shell	AD 1465–1651	AD 1550
OS-63548	Unit NW2 SW10	40–50 cmbs	1470 ± 40	*Callorhinus* bone	AD 1098–1328	AD 1238
OS-63549	Unit NW1 SW6	50–60 cmbs	1050 ± 30	*Callorhinus* bone	AD 1475–1664	AD 1569
OS-63550	Unit N4 W16	60–70 cmbs	1040 ± 35	*Callorhinus* bone	AD 1477–1672	AD 1577
UCR-0884	Unit NW 1 SW 8	60–70 cmbs	1490 ± 100	Charcoal	AD 340–710	AD 552
UCR-0885	Unit N2 W18	75–80 cmbs	1860 ± 120	Charcoal	117 BC–AD 424	AD 158
OS-63551	Unit N3 W16	80–90 cmbs	1010 ± 30	*Callorhinus* bone	AD 1497–1687	AD 1598
UCR-0883	Unit NW 1 SW 10	95 cmbs	215 ± 100	Charcoal	AD 1610–1953	AD 1734
OS-57903	Slump B	90–100 cmbs	1250 ± 30	*Mytilus* shell	AD 1318–1474	AD 1406
OS-57905	Unit NW2 SW 9	100–110 cmbs	1370 ± 25	*Mytilus* shell	AD 1246–1414	AD 1323

[a] All dates calibrated using Calib 5.01 calibration software. Assays on *Mytilus* shell were corrected for marine reservoir effect with a correction of $\Delta R = 308 \pm 40$.

[b] Falls into modern range on curve with reservoir correction, possible modern or historic.

TABLE 9.2A

Number of Identified Marine Mammal Specimens at CA-HUM-129

TAXA	COMMON NAME	NISP
Pinnipedia	Seals and sea lions	251
Otariidae	Fur seals and sea lions	896
Arctocephiline	Fur seals	1
Callorhinus ursinus	Northern fur seal	147
Eumetopias jubatus	Steller sea lion	438
Zalophus californianus	California sea lion	192
Phoca vitulina	Harbor seal	18
Enhydra lutris	Sea otter	334

TABLE 9.2B

Index Values for Marine Mammal Assemblage at CA-HUM-129 by Stratum

STRATUM	LOWER	UPPER	ALL
Marine Index	0.64	0.64	0.63
Otariid Index	0.71	0.69	0.74

NOTE: Marine index = Σ Marine Mammals/(Σ Marine Mammals + Nonburrowing Terrestrial Mammals), Otariid index = Σ Otariids/(Σ Pinnipeds + Sea Otter).

AD 160 to 550. The site stratigraphy is divided into two major strata: a lower stratum of dark reddish-brown clay-loam and an upper stratum of intermixed midden and sand deposits rich in faunal material (Milburn et al. 1979). Radiocarbon assays demonstrate considerable stratigraphic mixing, with no correlation between sample depth and age.

Artifacts recovered from HUM-129 are indicative of intensive littoral or maritime resource exploitation including evidence of woodworking, fishing (use of nets), and sea mammal hunting. A total of 72 antler and bone wedges, used ethnographically to split wood, were recovered. Woodworking may also be indicated by the presence of 139 choppers, which are easily made or resharpened, and a single groundstone maul. Stone Lagoon offers some of the strongest evidence for the use of composite harpoons in California. Twelve harpoon fragments were recovered comprising six simple harpoons and six toggling varieties. Three of the simple harpoons are classified as Category I, large unilaterally barbed, and the others are classified as Category II, small unilaterally barbed (Bennyhoff 1950). Two tips and four spur fragments comprise the toggling harpoon collection.

With 1674 otariid bones in the faunal assemblage, otariids make up 43% (1674/3870) of the mammalian elements identified to family or better (Table 9.2). The overwhelming presence of these animals in a Late Period site dem-

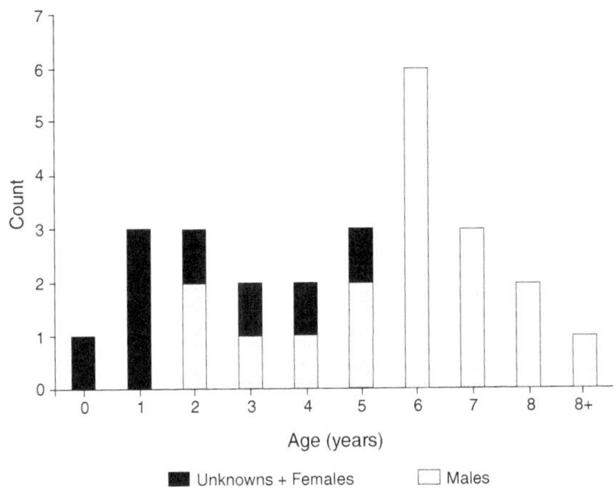

FIGURE 9.2. Age and sex data for northern fur seal remains from Stone Lagoon.

onstrates that they were not overhunted at the site, and we offer three possible explanations of this pattern.

OTARIID DEMOGRAPHICS

Both sex and age profiles are important in understanding how otariid populations at Stone Lagoon were hunted without detriment to their overall population. In general, size overlaps between otariid species often make the differentiation of males of one species from females of another species difficult. Because extensive work has been done to differentiate the sex of northern fur seal bones, the identification of both males and females was possible. Etnier (2002) used extensive modern, known-age osteological collections from northern fur seals to construct growth curves for a number of skeletal elements. He has since developed growth curves for all long bones and the mandible. These equations are based on best-fit curves that follow Von Bertalannfy's (1938) theoretical growth curve formulas. The element specific formulas and the measurements of individuals in the study can be found elsewhere (Etnier 2002; Whitaker 2008).

Once fur seals reach maturity, sexual dimorphism allows for the differentiation of sex in elements based upon osteometrics. For this study, metric sex identifications were confirmed with visual identifications, based on discrete osteological landmarks. These visual identifications were made with the assistance of Diane Gifford-Gonzales and comparative collections from the zoological collections from the University of California, Santa Cruz, University of California, Davis, and the Museum of Vertebrate Zoology at U.C. Berkeley.

Age estimates based on Etnier's (2002) growth curves are shown in Figure 9.2 for all individuals for which age could be determined. Males between the ages of 3 and 8 are considered subadult because they are generally too small to successfully defend territory on a rookery (Gentry 1998). Seventeen of 18 male elements produced age estimates within this subadult range. Furthermore, all female and unsexed elements have age estimates less than 5 years of age—too young to be members of a breeding population. Age-sex demographic data from the site indicate that the exploited individuals were not part of a breeding population and, therefore, their deaths would have less effect on the overall health of the local population at the site. Interestingly, radiocarbon dates on fur seal bones from Stone Lagoon (Table 9.1) indicate that the hunting of fur seals at the lagoon began after the extirpation of fur seal rookeries from central California. This raises questions about where the fur seal hunted at Stone Lagoon originated.

STABLE ISOTOPE ANALYSIS

As discussed earlier, stable isotope analysis provides a method with which to identify the foraging strategies of pinnipeds (among other animals). Stable isotope analysis in plants and animals is based on the assumption that "you are what you eat." In other words, the isotopic composition of an animal's tissues reflects the composition of its diet with some predictable differences. Carbon-13 (^{13}C) is measured in relation to a standard—fossils from the belamnite limestone of the Pee Dee formation in South Carolina—which is given a ^{13}C value of 0 (PDB). Because the standard has a higher $^{13}C/^{12}C$ ratio than most natural substances, $\delta^{13}C$ ("delta C 13") values are reported as a negative per-mil (‰) statistic.

In marine environments, the isotopic contribution of carbon to high-level foragers (such as pinnipeds) is determined by a number of things including the isotopic composition and concentration of dissolved CO_2, phytoplankton growth and species contribution, and carbon from benthic macrophytes, which are significantly enriched in ^{13}C relative to phytoplankton (Burton and Koch 1999:579; Rau et al. 1992). In areas of coastal upwelling—such as the nearshore environments along the California coast—all of these factors lead to a relatively ^{13}C-rich environment, signaled by relatively less negative $\delta^{13}C$ values. Further offshore, nutrients are more limited, and organic carbon at the bottom of the food web is more ^{12}C enriched. In general, this should lead to an average 2 to 3‰ negative difference between nearshore and offshore diets. Furthermore, high-latitude environments are ^{13}C depleted due to lower ^{13}C particulate organic carbon in high-latitude surface water (Rau et al. 1982). This should also lead to more negative $\delta^{13}C$ values for high-latitude foragers.

Nitrogen is often used to track the trophic level of a forager. The standard for nitrogen is atmospheric N_2 (AIR), which, by definition, has a $\delta^{15}N$ value of 0. Natural concentrations of $^{15}N/^{14}N$ are higher than N_2 and are therefore reported as a positive $\delta^{15}N$ ‰ statistic. In general, nitrogen values enrich by an average of 3‰ from diet to animal (Michener and Schell 1994; Schoeninger and DeNiro 1984). This has been useful in examining the contribution of meat to the diet of human foragers (e.g., Schoeninger and Moore 1992) and weaning patterns of humans (e.g., Clayton et al. 2006; Fuller et al. 2006) and pinnipeds (Newsome et al. 2006). Weaning studies are possible because the infant is, in essence, feeding on its mother and its tissue will have nitrogen isotope values that are enriched by close to 3‰.

Burton and Koch (1999) used carbon and nitrogen isotopes to examine the isotopic signatures of terrestrial (otariids) and aquatic breeding (phocids) pinnipeds in the eastern Pacific. They found that both offshore/nearshore and latitudinal patterns of carbon contribution to bone tissue matched expectations. These data represent a modern baseline for archaeological stable isotope studies discussed below. By comparing archaeological samples with these baselines, it is possible to examine changes in the foraging patterns of otariids through time.

To evaluate the origins of the northern fur seals at Stone Lagoon, samples were taken from 66 individual archaeological specimens. These samples were prepared following methods outlined by Burton and Koch (1999) and analyzed using a Carlo-Erba elemental analyzer interfaced with a Micromass Optima gas source mass spectrometer in the Department of Earth and Ocean Sciences, University of California, Santa Cruz. Results by sex are plotted in Figure 9.3.

The isotopic signature of $\delta^{13}C$ from modern and archaeological northern fur seal bone has been used to identify two prehistoric breeding populations, one located in the northern Pacific, likely the Pribilof Islands—the current home to the majority of northern fur seal rookeries—and one in central and southern California.

Females in both northern and southern breeding populations likely foraged off the nutrient-rich continental shelf of California similar to modern populations. The difference inherent in the stable isotope data, however, is that the California-breeding population appears

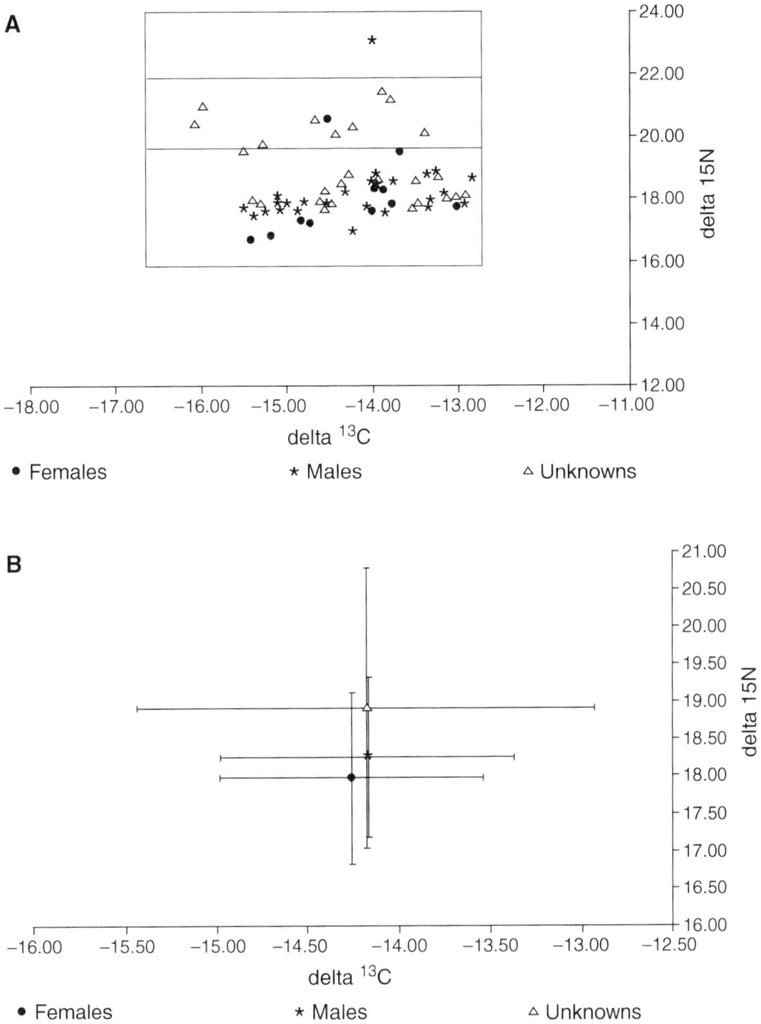

FIGURE 9.3. Stable isotope values for northern fur seal: (A) individual samples, (B) mean with error bars at 2 s.d.

to have remained in mid-latitudes year round whereas northern breeding populations undertook southward migrations during the winter to feed in mid-latitude zones.

Interestingly, the $\delta^{13}C$ signatures for samples from Stone Lagoon are more similar to those from both modern and archaeological northern breeding populations. Table 9.3 summarizes isotope data from modern and archaeological samples. T-test comparisons of means and standard deviations between the Stone Lagoon sample and both modern and archaeological samples were calculated. The null hypothesis of the t-test is that the two means being compared are similar. Therefore, a low p-value indicates that the samples are drawn from unique populations. Thus, high p-values imply a greater likelihood that samples were drawn from the same population. In Table 9.3 the sample from Stone Lagoon compares most favorably with modern samples from the Pribilof Islands ($p = .07–.27$) and with archaeological samples from Netarts in Oregon ($p = .08–.32$; Moss et al. 2006). In all likelihood the Stone Lagoon fur seals are individuals from a northern breeding population.

TABLE 9.3A

$\delta^{13}C$ and $\delta^{15}N$ Data from Modern and Archaeological Data

LOCATION	$\delta^{13}C$		$\delta^{15}N$		N	REFERENCE
	MEAN	SD	MEAN	SD		
MODERN SAMPLES						
San Miguel Island	–13.3	0.4	18.4	0.4	6	Burton et al. 2001
Pribilof (male)	–14.5	0.7	16.5	1.4	9	Burton et al. 2001
Pribilof (female)	–13.9	0.7	16.3	1.0	9	Burton et al. 2001
Northern Pacific (female)	–13.2	0.6	17.3	1.0	101	Newsome et al. 2007
California (female)	–12.1	0.4	17.9	0.7	72	Newsome et al. 2007
ARCHAEOLOGICAL SAMPLES						
Stone Lagoon	–14.2	0.9	18.5	1.2	66	This study
Moss Landing	–13.2	0.5	18.5	1.4	61	Burton et al. 2001
MacKerricher	–13.5	0.7	18.3	1.1	10	Burton et al. 2001
Cape addington (pups)	–14.8	0.5	19.0	0.6	6	Moss et al. 2006
Netarts (adult male)	–14.6	0.5	16.7	0.8	8	Moss et al. 2006
Netarts (immature male)	–14.6	0.9	19.5	1.3	6	Moss et al. 2006

TABLE 9.3B

Statistical Comparison of Previous Stable Isotope Studies to Stone Lagoon Data

	T-VALUE	D-F	P-VALUE
San Miguel Island	3.558	9	0.005
Pribilof (female)	–1.981	11	0.067
Pribilof (male)	1.141	11	0.268
Northern Pacific	8.166	105	0.000
California	18.057	90	0.000
Moss Landing	8.022	105	0.000
MacKerricher	2.822	13	0.012
Cape addington (pups)	–2.642	8	0.030
Netarts (male)	–1.978	12	0.079
Netarts (male)	–1.066	5	0.322

The northern origin of northern fur seal individuals hunted at Stone Lagoon explains how northern fur seals could continue to be hunted even after the extirpation of the southern breeding population along California's central coast. The northern origin of these individuals is slightly perplexing, however, since modern adult male fur seals from Alaskan populations rarely forage so far south (Gentry 1998). It is possible that, previous to the extirpation of central California rookeries, males from Alaskan populations were able to establish breeding grounds with females from California populations. Even after the loss of these females, males may have continued—though unsuccessfully—to establish territories at offshore rocks in northern California.

Regardless of the sex of individuals from other otariid species, the lack of infant or fetal elements in the assemblage indicates that hunters at Stone Lagoon were not targeting rookeries. Only 15 of 1646 (<1%) of all otariid elements recovered at the site were from infants.

Stable isotope data for northern fur seals further demonstrate a lack of infants at the site. Only one individual has a nitrogen signature that is enriched by 3‰, but this element has fused epiphyses and, therefore, could not come from an infant. A number of individuals have slightly enriched nitrogen signatures indicating that bone remodeling had not fully occurred between weaning and death (Figure 9.3). In a study of weaning signatures for nitrogen isotopes, Newsome et al. (2006) found that northern fur seal individuals between 6 to 8 months old had $\delta^{15}N$ signatures between 18 and 20‰. Therefore, it appears that, while there were some younger individuals, including a few infants, there is not a strong "rookery" signal in either the otariid faunal assemblage or the $\delta^{15}N$ stable isotope signature. Instead, it appears that the hunted population consisted of a "bachelor herd" of juvenile and subadult males with a small proportion of females. If this is the case, it is not surprising that northern fur seals were not extirpated at the site since the hunting of males has almost no effect on the maintenance of a healthy breeding population and the hunting of juveniles likewise has little effect given high levels of natural juvenile mortality.

PRESTIGE HUNTING

It is apparent that males were preferentially hunted by the Yurok at Stone Lagoon. This sex bias in the targeting of male sea lions may stem from a trait unique to Native California. With the exception of groups along the Santa Barbara Channel, the majority of Native Californians lived under loose political entities known as tribelets (Kroeber 1925). In contrast, a high degree of socioeconomic stratification existed among the Tolowa, Yurok, and Wiyot, where, rather than allegiance to the broader tribelet group, the individual or immediate family took precedence over the village or the tribe (Goldschmidt 1951). There was a universal concept of privately owned property, including money (e.g., dentalium shell), which was linked to differential wealth and power within the population (Gould 1975; Kroeber 1925). Individual households possessing superior pools of labor could generate substantial food surpluses and other items of wealth, ultimately separating themselves from less successful family units. According to Gould, the system was supported by polygynous marriage because:

> . . . the amount of food a family could accumulate and store was directly dependent upon the number of women in the household. A man with several hard-working wives and daughters could store up large reserves, and not only could he use these as security against possible scarcities, but he could also commission "feasts" when he wished to recruit people to construct a large canoe or house. (1975:166)

Construction of the large oceangoing canoe required a great deal of labor that could only be commissioned by the wealthiest members of society. There was usually only one canoe per village, reaching sizes in excess of 40 feet long and 10 feet wide. The diary of Bodega y Quadra (1865:285) indicates that his ship was once greeted near Trinidad by a single dugout canoe containing 28 to 30 individuals. The boats were typically manned by five or six men during hunting forays, and, according to Gould (1968) and Powers (1925), the largest vessels could transport several sea lions of various sizes during a single trip.

While large loads could be carried by these dugouts, trips to distant offshore rocks were quite dangerous, sometimes resulting in loss of life among the participants (Spott and Kroeber 1942). When the boat returned to the village, there was often a feast, where the meat and oil were divided and shared with the larger group in accordance with the owner of the canoe. According to Gould (1968), the owner would give the first cuts to himself and the members of the hunting party, which were often his patrilineal relatives, but would share the remainder with the larger group.

Recent analyses of feasting from other parts of the world show that many of these events involve the conspicuous consumption of foods that go well beyond basic subsistence needs and often enhance the prestige of the producer (Brumfiel and Fox 1994; Dietler and Hayden 2001). The ability to generate large amounts of food for a feasting event is thought to improve the symbolic capital and prestige of the producer, thereby signaling their wealth and other desirable characteristics to the large audiences attending these events (Bliege Bird and Smith 2005). According to Brumfiel (1994), the items used in these contexts are usually highly valued due to their nonlocal origin and the quantity of labor required to produce them.

The benefits of feasting behavior are many, even given the high costs of the activity and the absence of any obvious forms of material reciprocity from members of the larger group. These relationships are usefully explained through the application of costly signaling theory, which seeks to determine how seemingly inefficient types of behavior can evolve, as long as they communicate a series of underlying qualities that can benefit both the signaler and the observer. Following the lead of Bliege Bird and Smith (2005:237), the ability of a boat owner to organize the labor necessary to construct and maintain the vessel, organize the hunting expedition, and share the meat and oil signals the cohesion of his group, which could potentially enhance his social standing through attracting high-quality marriages and other forms of alliance building.

The nonsubsistence importance of pinniped hunting is also reflected in the ceremonial aspects of the archaeological record. Spott and Kroeber (1942) noted that a hunter who presented the owner of beach with the flipper of any sea lion taken there,could not possibly be economically driven in his distribution:

> Another point which the story illustrates is the high prestige values of certain intangible rights the utilitarian value of which must have been insignificant. The flippers were evidently a prized delicacy, but the amount of food received in this way cannot have been considerable. (Spott and Kroeber 1942:190)

Similar to the ideological prominence of high-prestige animals like big horn sheep and swordfish in other parts of California, which are expressed in prehistoric rock art and grave offerings (Hildebrandt and McGuire 2002), the importance of pinniped hunting is reflected by a ceremonial site located on an offshore rock near Patrick's Point (Heizer 1951). This rock (CA-HUM-174) has a deposit containing roughly 1000 sea lion skulls and no other cultural material. The foramen magnum was purposefully enlarged on many specimens, probably to remove the brain, and all the observed crania were apparently cleaned before being placed on the rock because no mandibles were observed at the site. Although no ethnographic accounts mention this site or the activities associated with it, it likely represents some type of sea lion shrine created by people with a deep respect for the animal and the skill required in its capture.

The ethnographic record also provides ample evidence for the prestigious nature of pinniped hunting. Elaborate sea lion clubs were an

FIGURE 9.4. Sea lion tusk headband worn as regalia for the white deer skin dance. (Courtesy of Charles Deering McCormick Library of Special Collections, Northwestern University Library.)

outward, visible sign of an individual's participation in the hunt, and some of these clubs were passed down through generations as heirlooms (Waterman 1920). Perhaps the best example of the ritual importance of pinniped remains is the use of canine-studded headbands in the white deer skin dance (Figure 9.4; Goddard 1903; Kroeber and Gifford 1949). The Yurok focused on canine teeth from the Steller sea lion and northern fur seals, with particular status given to the latter species (Kroeber and Barrett 1960:116–117). If hunters were selecting for large males to enhance their prestige through the acquisition of the large canines from the most dangerous animals for their head bands, they would have been culling the most superfluous age/sex class from the population, as other males would quickly fill any spots left vacant by the activity. This practice might explain not only the sustained hunting of sea lions and fur seals through time but also the dominance of male elements in the Stone Lagoon assemblage.

OWNERSHIP OF RESOURCES

In addition to the direct effects of prestige economies on sea lion hunting, the concept of individual ownership that resulted from the system may have further conditioned the sex bias toward hunting of male sea lions. The best ethnographic evidence of ownership is documented for the Yurok, but mention is made of ownership at the family level of fishing, hunting, and acorn-gathering locations for other groups in northwestern California. Table 9.4 summarizes the nature of property rights for certain key coastal resources among the Yurok.

More widely noted were hunting or snaring tracts owned by families or wealthy individuals. Whereas squatters' rights might apply to vacant fishing holes, snaring or hunting areas were always private property for the Yurok (Waterman 1920:223), and, while a wounded animal could be pursued onto neighboring property, outright poachers were shot when found encroaching on another's tract (Kroeber 1925:34). Important to resource management, these tracts did not, Kroeber notes, supply total control over game populations, simply access to the game trails and drinking holes held within them. Thus, animals moved freely between numerous owned resource patches. All sea stacks, offshore rocks, and beaches were owned by "definite and limited groups" (Waterman 1920:220). Similar to fishing rights, beach rights may have been shared by more than one group. The rights to beaches extended to any food procured on the beach or immediately offshore, including sea lions, seals, shellfish, seaweed, and beached whales. Ownership of offshore rocks was denoted by sea lion clubs that were inherited along with sealing rights. Property rights extended to the ability to will, sell, barter, or exchange any of these spots. Inheritance of property did not simply follow descent lines but could be given or received by any member of the group.

TABLE 9.4

Ownership of Select Natural Resources among the Ethnographic Yurok

	COMMON-POOL RESOURCE	VILLAGE OWNERSHIP	HOUSE OR FAMILY OWNERSHIP	INDIVIDUAL OWNERSHIP	FRACTIONAL INDIVIDUAL OWNERSHIP
Oak groves	X	X	X	X	X
Artiodactyl snaring areas		X	X	X	
Stranded whales on beaches	X	X			
Whale cuts			X	X	
Whale flippers				X	
Sea lion hunting grounds	X		X		
Sea lion flipper			X	X	
Shellfish beds	X		X	X	

SOURCE: Pilling 1978, table 3.

The ownership of resources is necessary, though not sufficient, for them to be conserved. Once resources are owned, the owner may have incentive to judiciously procure them in a manner that maximizes long-term returns. The surer that an owner can be of passing a resource patch on to a relative, the more incentive there will be for him to conserve that resource (Rogers 1991). Thus, the Yurok at Stone Lagoon may have had incentive to carefully manage the harvest of pinnipeds at offshore rocks, conserving their population for future generations and leading to the continued exploitation of otariids up to contact.

DISCUSSION AND CONCLUSIONS

Unlike the original formulation of the model of marine mammal hunting in northern California, we conclude that age/sex ratios of northern fur seal at Stone Lagoon are not indicative of rookery exploitation. Instead, we find that subadult and adult males were harvested disproportionately. This harvest could have been the result of any of the following scenarios, all of which would have allowed for the sustained harvest up until contact:

1. So-called bachelor herds may have been the only individuals present at offshore rocks. These individuals, hailing from the Alaskan breeding population of northern fur seals, for instance, could have been harvested with little effect on the overall population.
2. If females were present, males may have been preferentially hunted because they garnered more prestige than females or juveniles. Evidence for prestige hunting includes the use of sea lion canines for dance regalia and the presence of the ceremonial offshore site adjacent to Patrick's Point. In this case, prestige hunting would have allowed for the long-term sustainable harvest of sea lions and fur seals as the culling of males

generally has little effect on overall population structure.

3. The individual and family ownership of resource tracts, unique to north coast tribes within California, may have provided the necessary social situation in which conservation could occur. If this were the case, the culling of males may have been the result of careful management of resources by the inhabitants of Stone Lagoon.

It is apparent that these conclusions are neither mutually exclusive nor easily distinguishable in the archaeological record. In fact, it is quite possible that sea lion hunting was sustainable throughout the use of the site due to a combination of all three. One thing is certain at Stone Lagoon, the northern fur seals that were hunted were predominantly subadult males and spent portions of the year off both the Alaskan and Californian coasts. This does not preclude the presence of a rookery at the site, as these males may have been attempting to establish breeding colonies with females from a Californian or Alaskan population.

We also conclude that hunting of sea lions by the Yurok began approximately 1500 years ago and was sustainable in the sense that otariid hunting was still a large part of the subsistence economy at contact. This would indicate that hunting practices were not detrimental to otariid populations at the site. A great deal of previous research indicates that instances of conscious indigenous conservation are rare (Alvard 1994, 1998, 2002). More often, what appears to be conservation is, in fact, the result of too-small indigenous populations hunting in large patches that are not easily depleted. Specific tests of conservation related to hunting have failed to demonstrate restraint on the part of hunters. Hames (1987), for instance found that hunters routinely pursued prey which they encountered in areas which they acknowledged were depleted. Hames (1987, 2007), Alvard (1998), and others (Broughton 1999; Kay 2002; Krech 1999) have asserted that modern environmentalists have romanticized hunter-gatherers as naturally conserving stewards of the environment, when, in fact, this is rarely documented in the ethnographic, historic, or archaeological records. Furthermore, countless reviews of the archaeological and ethnographic records have found evidence of adverse environmental impacts as the result of human foraging practices (e.g., reviews in Grayson 2001; Hames 2007; Rick and Erlandson 2009). This is not to say that hunters at Stone Lagoon were not practicing conscious conservation, it is just harder to believe in light of other evidence from California and North America.

Given this caution, we favor the second option listed above (prestige economies). The ethnographic record of socioeconomic complexity and the role of sea lion hunting within the prestige economy coupled with the dominance of male elements in the archaeological assemblage are persuasive evidence for prestige-motivated hunting. This option is also appealing because it does not require foresight (though it also does not rule it out) or conscious conservation but rather might represent an instance of epiphenomenal sustainable hunting practices. Furthermore, this explanation takes into account human culture as a factor mediating the human/prey relationship, rather than considering humans no different than any other predator.

ACKNOWLEDGEMENTS

We thank Robert Bettinger, Christyann Darwent, the editors of this volume, and two anonymous reviewers for insightful comments and suggestions on the original manuscript. Age/sex determinations on northern fur seal bone were made with assistance from Diane Gifford-Gonzales and Charlotte Sunseri. Stable isotope methods and analysis were performed following excellent protocols shared by Paul Koch and Seth Newsome. The Department of Wildlife and Fish Conservation Biology allowed us to use their facilities in preparing isotope samples. Whitaker was funded in part by a National Science Foundation Pre-Doctoral Fellowship.

REFERENCES CITED

Alvard, M. S.

1994 Conservation by Native Peoples: Prey Choice in a Depleted Environment. *Human Nature* 5:127–154.

1998 Evolutionary Ecology and Resource Conservation. *Evolutionary Anthropology* 7:62–74.

2002 Evolutionary Theory, Conservation and Human Environmental Impact. In *Wilderness and Political Ecology: Aboriginal Influences and the Original State of Nature*, edited by Charles Kay and Randy T. Simmons, pp. xix, 342. University of Utah Press, Salt Lake City.

Arnold, J. E.

1991 Foreword. In *Hunter-Gatherers of Early Holocene Coastal California*, edited by Jon M. Erlandson and R. H. Colten. Perspectives in California Archaeology. vol. 1. UCLA Institute of Archaeology, Los Angeles.

Barlow, J., and P. Boveng

1991 Modeling Age-Specific Mortality for Marine Mammal Populations. *Marine Mammal Science* 7(1):50–65.

Barron, J. A., L. Heusser, T. Herbert, and M. Lyle

2003 High-Resolution Climatic Evolution of Coastal Northern California. *Paleoceanography* 18(1):1020–1029.

Bennyhoff, J. A.

1950 California Fish Spears and Harpoons. *University of California Anthropological Records* 9(4):295–337.

Bickel, P.McW.

1978 Changing Sea Levels along the California Coast: Anthropological Implications. *Journal of California Anthropology* 5(1):5–20.

Bliege Bird, R., and E. A. Smith

2005 Signaling Theory, Strategic Interaction, and Symbolic Capital. *Current Anthropology* 46(2):221–248.

Bodega y Quadra

1865 Primer viaje de . . . Bodega y Quadra . . . año de 1775. Annuario de la direccion de hidrografia, año III, 1964:279–294. Madrid.

Broughton, Jack M.

1997 Widening Diet Breadth, Declining Foraging Efficiency, and Prehistoric Harvest Pressure: Ichthyofaunal Evidence from the Emeryville Shellmound, California. *Antiquity* 71:845–862.

1999 Resource Depression and Intensification during the Late Holocene, San Francisco Bay. *University of California Press Anthropological Records* 35.

Brumfiel, E. M.

1994 Part I: Introduction. In *Factional Competition and Political Development in the New World*, edited by E. M. Brumfiel and J. W. Fox. Cambridge University Press, Cambridge.

Brumfield, E. M., and J. W. Fox

1994 *Factional Competition and Political Development in the New World.* Cambridge University Press, Cambridge.

Burton, R. K., D. Gifford-Gonzalez, J. J. Snodgrass, and P. L. Koch

2002 Isotopic Tracking of Prehistoric Pinniped Foraging and Distribution along the Central California Coast: Preliminary Results. *International Journal of Osteoarchaeology* 12(1):4–11.

Burton, R. K., and P. L. Koch

1999 Isotopic Tracking of Foraging and Long-Distance Migration in Northeastern Pacific Pinnipeds. *Oecologia* 119:578–585.

Burton, R. K., J. J. Snodgrass, D. Gifford-Gonzalez, T. Guilderson, T. Brown, and P. L. Koch

2001 Holocene Changes in the Ecology of Northern Fur Seals: Insights from Stable Isotopes and Archaeofauna. *Oecologia* 128:107–115.

Clayton, F., J. Sealy, and S. Pfeiffer

2006 Weaning Age among Foragers at Matjes River Rock Shelter, South Africa, from Stable Nitrogen and Carbon Isotope Analyses. *American Journal of Physical Anthropology* 129(2): 311–317.

Dietler, M., and B. Hayden

2001 *Feasts: Archaeological and Ethnographic Perspectives on Food, Politics, and Power.* Smithsonian Institution Press, Washington, D.C.

Eidsness, J. P.

1993 *Archaeological Investigations at Ca-Hum--351/H on Humboldt Bay, California.* Report on File, Northwest Information Center, Sonoma State University, Rohnert Park, California.

Erlandson, J. M., and R. H. Colten

1991 An Archaeological Context for Early Holocene Studies on the California Coast. In *Hunter-Gatherers of Early Holocene Coastal California*, edited by J.M. Erlandson and R. H. Colten, pp. 1–10. Perspectives in California Archaeology. vol. 1. UCLA Institute of Archaeology, Los Angeles.

Etnier, M. A.

2002 *The Effects of Human Hunting on Northern Fur Seal (*Callorhinus ursinus*) Migration and Breeding Distributions in the Late Holocene.* Unpublished Ph.D. dissertation, Department of Anthropology, University of Washington, Seattle.

Fuller, B. T., J. L. Fuller, D. A. Harris, and R. E. M. Hedges

2006 Detection of Breastfeeding and Weaning in Modern Human Infants with Carbon and Nitrogen Stable Isotope Ratios. *American Journal of Physical Anthropology* 129(2):279–293.

Gentry, R. L.

1998 *Behavior and Ecology of the Northern Fur Seal.* Princeton University Press, Princeton, New Jersey.

Gifford-Gonzalez, D., K. W. Gobalet, and J. C. Geary

2006 Vertebrate Faunal Remains: Environment, Subsistence, Taphonomy, and Historical Ecology. In *Archaeological Investigations at CA-SMA-18:A Study of Prehistoric Adaptations at Año Nuevo State Reserve,* by W. Hildebrandt, J. Farquhar, and M. Hylkema. Report on File at California Department of Parks and Recreation, Sacramento, CA.

Gifford-Gonzalez, D., S. D. Newsome, P. L. Koch, T. P. Guilderson, J. J. Snodgrass, and R. Burton

2005 Archaeofaunal Insights on Pinniped-Human Interactions in the Northeastern Pacific. In *The Exploitation and Cultural Importance of Sea Mammals,* edited by G. G. Monks, vol. 19–38. Oxbow Books, Oxford.

Goddard, P. E.

1903 Life and Culture of the Hupa. *University of California Publications in American Archaeology and Ethnology* 1(1):1–88.

Goldschmidt, W. R.

1951 Nomlaki Ethnography. *University of California Publications in American Archaeology and Ethnology.* 42: 303–443

Gould, R.

1966 Archaeology of Point St. George Site and Tolowa Prehistory. *University of California Publications in Anthropology* 4:1–141.

1968 Seagoing Canoe Use among the Indians of the Northwestern California Coast. *Ethnohistory* 15(1):11–42.

1975 Ecology and Adaptive Response among the Tolowa Indians of Northwest California. *Journal of California Anthropology* 2:148–170.

Grayson, D.

1999 The Archaeological Record of Human Impacts on Animal Populations. *Journal of World Prehistory* 15(1):1–68.

Hames,R.

1987 Game Conservation or Efficient Hunting. In *The Question of the Commons: The Culture and Ecology of Communal Resources,* edited by B.J. McCay and J. Acheson, pp. 92–107. University of Arizona Press, Tucson, Az.

2007 The Ecologically Noble Savage Debate. *Annual Review of Anthropology* 37:177–190.

Heizer, R. F.

1951 A Prehistoric Yurok Ceremonial Site (Hum-174). *University of California Archaeological Survey Reports* 11:1–5.

Hildebrandt, W. R.

1981 *Native Hunting Adaptations on the North Coast of California.* Unpublished Ph.D. dissertation, Department of Anthropology, University of California, Davis.

1984 Late Period Hunting Adaptations on the North Coast of California. *Journal of California and Great Basin Anthropology* 6(2):189–296.

2007 Northwest California: Ancient Lifeways among Forested Mountains, Flowing Rivers, and Rocky Ocean Shores. In *California Prehistory: Colonization, Culture and Complexity,* edited by Terry L. Jones and Kathryn A. Klar. Altamira Press, Blue Ridge Summit, PA.

Hildebrandt, W. R., and K. Carpenter

2007 Native Hunting Adaptations in California from Prehistory to European Contact, In *Handbook of North American Indians, Vol. 3: Environment, Population and Origins* edited by Bruce Smith. William C. Sturtevant, general editor. Smithsonian Press, Washington D.C.

Hildebrandt, W. R., and T. L. Jones

1992 Evolution of Marine Mammal Hunting: A View from the California and Oregon Coasts. *Journal of Anthropological Archaeology* 11(4): 360–401.

2002 Depletion of Prehistoric Pinniped Populations along the California and Oregon Coasts: Were Humans the Cause? In *Wilderness and political ecology: aboriginal influences and the original state of nature,* edited by C. Kay and R. T. Simmons, pp. 72–110. University of Utah Press, Salt Lake City.

Hildebrandt, W. R., and V. Levulett

1997 Middle Holocene Adaptations on the Northern California Coast. In *Archaeology of the California Coast during the Middle Holocene,* edited by Jon M. Erlandson and Michael A. Glassow, pp. 143–150. Perspectives in California Archaeology. vol. 4. Institute of Archaeology, University of California, Los Angeles, Los Angeles.

2002 Late Holocene Emergence of Marine-Focused Economies in Northwest California. In *Catalysts to Complexity: Late Holocene Societies on the California Coast,* edited by Jon M. Erlandson and Terry Jones, pp. 303–319. Cotsen Institute of Archaeology, University of California, Los Angeles.

Hildebrandt, W. R., and K. R. McGuire

2002 The Ascendance of Hunting during the California Middle Archaic: An Evolutionary Perspective. *American Antiquity* 67(2):231–256.

Hudson, T.

1981 To Sea or Not to Sea: Further Notes on the "Oceangoing" Dugouts of North Coastal Cali-

fornia. *Journal of California and Great Basin Anthropology* 3(2):269–282.
Jobson, R. W., and W. R. Hildebrandt
1980 The Distribution of Oceangoing Canoes on the North Coast of California. *Journal of California and Great Basin Anthropology* 2(2): 165–174.
Jones, T. L.
1991 Marine-Resource Value and the Priority of Coastal Settlement—a California Perspective. *American Antiquity* 56(3):419–443.
1992 Settlement Trends along the California Coast. In *Essays on the Prehistory of Maritime California*, Terry L. Jones, editor, pp. 1–38. Center for Archaeological Research at Davis Publication 10, University of California.
Jones, T. L., and W. R. Hildebrandt
1995 Reasserting a Prehistoric Tragedy of the Commons—Reply. *Journal of Anthropological Archaeology* 14(1):78–98.
Kay, Charles E.
2002 False Gods, Ecological Myths, and Biological Reality. In *Wilderness and Political Ecology: Aboriginal Land Management—Myths and Reality*, C.E. Kay and R.T. Simmons, University of Utah Press, Salt Lake City.
Krech III, S.
1999 *The Ecological Indian: Myth and History.* W.W. Norton and Company, New York.
Kroeber, A. L.
1925 *Handbook of the Indians of California.* Bureau of American Ethnology, Washington.
1939 *Cultural and Natural Areas of Native North America.* University of California Publications in American Archaeology and Ethnology 38:1–242, Berkeley.
Kroeber, A. L., and S. A. Barrett
1960 Fishing among the Indians of Northwestern California. *University of California Anthropological Records* 21(1):1–210.
Kroeber, A. L., and E. W. Gifford
1949 The World Renewal: A Cult System of Native Northwest California. *University of California Anthropological Records.* 13(1):1–156.
Levulett, V.
1985 *The Prehistory of Southwestern Humboldt County: A Study of Coastal Archaeological Sites in the King Range National Conservation Area.* Unpublished Ph.D. dissertation, Department of Anthropology, University of California, Davis.
Lyman, R. L.
1988 Zoogeography of Oregon Coast Marine Mammals: the Last 3,000 Years. *Marine Mammal Science* 4(3):247–264.
1991 Subsistence Change and Pinniped Hunting. In *Human Predators and Prey Mortality*, edited by M. C. Stiner, pp. xv, 276. Westview Press, Boulder.
1995 On the Evolution of Marine Mammal Hunting on the West Coast of North America. *Journal of Anthropological Archaeology* 14(1):45–77.
2003 Pinniped Behavior, Foraging Theory, and the Depression of Metapopulations and Nondepression of a Local Population on the Southern Northwest Coast of North America. *Journal of Anthropological Archaeology* 22:376–388.
Michener, R. H., and D. M. Schell
1994 Stable Isotope Ratios as Tracers in Marine Aquatic Food Webs. In *Stable Isotopes in Ecology and Environmental Science*, edited by K Lathja and R.H. Michener, pp. 138–157. Blackwell, Boston.
Milburn, J. W., D. A. Fredrickson, M. Dreiss, L. Demichael, and W. Van Dusen
1979 *A Preliminary Report on the Archaeology of Ca-Hum-129: Volume 1.* Report on file with the California Department of Parks and Recreation, Sacramento, CA.
Milliman, J. D., and K. O. Emery
1968 Sea Level during the Past 35,000 Years. *Science* 162:1121–1123.
Moss, M. L.
1998 Early Holocene Adaptation on the Southern Northwest Coast. *Journal of California and Great Basin Anthropology* 20(1):13–25.
Moss, M. L., D. Y. Yang, S. D. Newsome, C. F. Speller, I. McKechnie, A. D. McMillan, R. J. Losey, and P. L. Koch
2006 Historical Ecology and Biogeography of North Pacific Pinnipeds: Isotopes and Ancient DNA from Three Archaeological Assemblages. *Journal of Island and Coastal Archaeology* 1:165–190.
Newsome, S. D., P. L. Koch, M. A. Etnier, and D. Aurioles-Gamboa
2006 Using Carbon and Nitrogen Isotope Values to Investigate Maternal Strategies in Northeast Pacific Otariids. *Marine Mammal Science* 22(3):556–572.
Newsome, Seth D., Michael A. Etnier, Diane Gifford-Gonzalez, Donald L. Phillips, Marcel van Tuinen, Elizabeth A. Hadly, Daniel P. Costa, Douglas J. Kennett, Tom P. Guilderson, and Paul L. Koch
2007 *The Shifting Baseline of Fur Seal Ecology in the Northeast Pacific Ocean.* Proceedings of the National Academy of Sciences 104:9709–9714.
Nomland, G.
1938 Bear River Ethnography. *University of California Anthropological Records* 2(2):91–126.

Powers, S. J.
1925 *Tribes of California*. University of California Press, Berkeley.
Rau, G. H., R. E. Sweeney, and I. R. Kaplan
1982 Plankton Δ13C:12C Ratio Changes with Latitude: Differences between Northern and Southern Oceans. *Deep Sea Research* 29:1035–1039.
Rick, Torben C., and Jon M. Erlandson
2009 Coastal Exploitation. *Science* 325:952–953.
Rogers, A.
1991 Conserving Resources for Children. *Human Nature* 2:73–82.
Schoeninger, Margaret J., and M. J. DeNiro
1984 Nitrogen and Carbon Isotopic Composition of Bone Collagen from Terrestrial and Marine Mammals. *Geochimica et Cosmochimica Acta* 48:62
Schoeninger, Margaret J., and K. Moore
1992 Bone Stable Isotope Studies in Archaeology. *Journal of World Prehistory* 6(2):247–296.
Spott, R., and A. L. Kroeber
1942 Yurok Narratives. *California. University of California Publications in American Archaeology and Ethnology* 35:143.
Tushingham, S., W. Hildebrandt, J. Garabaldi, and A. Ruby
2008 *Archaeological Test Excavations and Sensitivity Assessment for Jedediah Smith State Campground, Redwood National and State Parks Del Norte County, California*. Report on File with Redwood National and State Parks, Orick , California.
von Bertalannfy, L.
1938 A Quantitative Theory of Organic Growth. *Human Biology* 10:181–213.
Waterman, T. T.
1920 Yurok Geography. *University of California Publications in American Archaeology and Ethnology* 16(5):177–314.
Whitaker, A. R.
2006 2005 *Archaeological Investigations in the King Range National Conservation Area*. Report on file at the Arcata Field Office, Bureau of Land Management, Arcata, CA.
2008 *The Role of Human Predation in the Structuring of Prehistoric Prey Populations in Northwestern California*. Unpublished Ph.D. dissertation. Department of Anthropology, University of California, Davis.

10

Holocene Monterey Bay Fur Seals

DISTRIBUTION, DATES, AND ECOLOGICAL IMPLICATIONS

Diane Gifford-Gonzalez

HISTORICAL ECOLOGISTS HAVE recognized for some time the relevance of paleontological and archaeological data for understanding longer-term ecological dynamics than could be apprehended from relatively short-term historical records (Jackson et al. 2001). The research reported here has proceeded on the assumption that zooarchaeological, stable isotopic, and ancient DNA (aDNA) analyses can, in combination, elucidate the longer-term histories of northern fur seals (*Callorhinus ursinus*) in the North Pacific. After a decade of collaborative research by investigators from several institutions and agencies, this expectation has proved to be justified (e.g., Newsome, Etnier, Gifford-Gonzalez et al. 2007). Nonetheless, while these results have established some findings reasonably well, they have led to new research questions. These are best solved by closer collaboration with marine mammalogists working with the present-day *Callorhinus* stock and its management.

The first part of this chapter presents our current state of knowledge about the distribution of northern fur seal remains in the Greater Monterey Bay region, commenting on emerging geographic and temporal patterns. The second part of this chapter presents detailed data on the ecology of present-day northern fur seals as a prelude to a third section discussing possible differences in ecological parameters between the ancient California *Callorhinus* population and its contemporaneous cousins north of Oregon. It then considers how these factors, in concert with human predation, may have contributed to the Middle to Late Holocene demise of near-coastal rookeries in California. The final section outlines some ways in which collaborative approaches can shed light on emerging questions and problems in northern fur seal historical ecology.

This chapter focuses more on the ecology of *Callorhinus*, rather than on more archaeologically oriented questions addressed by others in this volume. I take as a given that human predation can depress species metapopulations in one or more of several ways (see Lyman 2003) and that, under certain circumstances, humans can drive species to extinction. At the same time, the contrasting fates of *Callorhinus* populations in Holocene California and those farther north imply that humans and fur seal interactions produced different outcomes throughout their range, before the advent of industrial-scale sealing (Etnier 2007). With the much richer archaeofaunal, temporal, and contextual records available to us now, archaeologists would do well to reframe the tenor of their investigations to one that parallels that of other historical sciences. That is, rather than make grand, supraregional generalizations about human effects on marine mammals, it is time to use general ecological principles to ask focused questions about how and why specific taxa fared as they did in specific regional settings. This chapter attempts to begin such an inquiry, focusing on processes that may have made California otariid populations more vulnerable to extirpation than were those farther north. A regional focus is not a return to archaeological parochialism; rather, it considers both regional and supra-regional processes that may affect a taxon possessing a given set of biological parameters. It asks, first and foremost, whether other factors than human predation could affect the stability of a population at a regional scale and likewise at a supra-regional scale. This approach is especially important with reference to marine mammal species, such as eared seals, with long-range migration and flexible colonization strategies. It finally asks whether the nature of age- and sex-specific human offtake could have destabilized a local or regional population.

GREATER MONTEREY BAY FUR SEALS

• *Zooarchaeological, Isotopic, and aDNA Findings*

Since 2003, my laboratory team at UC Santa Cruz has undertaken analysis or reanalysis of vertebrate archaeofaunas from Point Año Nuevo to the Monterey Peninsula. A central aim is documenting the spatiotemporal distribution of northern fur seals, their age/sex class representation, and evidence for human handling and other taphonomic modifications. This research is part of whole-assemblage analyses of all mammal and bird specimens from as many Monterey Bay assemblages as accessible and, depending upon sample size, whole-assemblage or sampling-to-redundancy subsets of fish remains. We have completed or are finalizing analyses of ten such assemblages and plan to analyze ten more regional samples reported to contain remains of *Callorhinus* or Guadalupe fur seal (*Arctocephalus townsendi*).

Figures 10.1 and 10.2 present a rough distribution of fur seal remains, known either from our direct analyses or from earlier faunal reports in the literature (see also Table 10.1). Some generalizations can be made about spatiotemporal distributions and densities. First, the occurrences span a considerable amount of time, from the so-called Millingstone Period in the 7th/8th millennia BP to the late 2nd millennium BP. The early use of marine mammals in discrete Millingstone occurrences at the Moss Landing Hill Site (CA-MNT-234) is supported by roughly contemporaneous human bone isotope assays from Harkins Slough (CA-SCR-60), just north of the Pajaro River and only about 15 km from Moss Landing. Burials dating to the 8th millennium BP reflect a diet high in marine mammals and marine fish (Burton et al. 2002).

However, the long temporal span over which *Callorhinus* remains are found in the Greater Monterey Bay should not necessarily be interpreted as *continuous* presence of the species in the region for 9000 years. Combinations of environmental effects and impacts of human predation may have produced a discontinuous presence of breeding colonies in the region. This theme will be taken up in more detail later in this chapter.

Second, although we have not yet submitted these data to spatial statistical analysis, they do suggest the existence of two "point sources," one

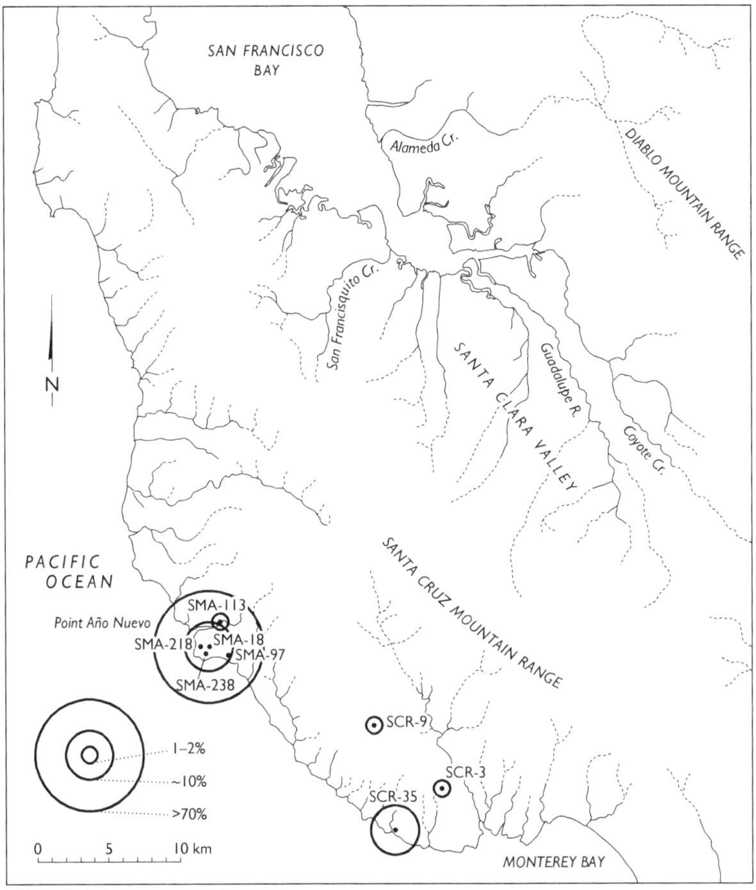

FIGURE 10.1. Map of San Mateo and north Santa Cruz County coastlines, showing present knowledge of proportions of *Callorhinus ursinus* in analyzed or reanalyzed archaeological sites. These proportions and additional site inventories may be added as the balance of SMA and SCR assemblages are analyzed. Map from Mark Hylkema (2002:234, figure 13), used with permission. For detailed data and sources, see Table 10.1.

at Point Año Nuevo and one around Moss Landing/Monterey/Carmel, with high proportions of northern fur seals in sites at these localities and lower proportions in areas inland (Figures 10.1 and 10.2). The densest concentrations of archaeofaunal *Callorhinus* remains along the Greater Monterey Bay (Figures 10.1 and 10.2) are thus near loci of strong upwelling, even in El Niño Southern Oscillation (ENSO) years: Point Año Nuevo, Point Lobos, and Point Sur (Trainer et al. 2000).

We have verified *Callorhinus* remains at CA-SCR-44, formerly considered to be occupied only in the last 500 years, and obtained new radiocarbon dates suggesting a more complex occupational history for the site (Table 10.1). Three northern fur seal elements and about 12 other pinniped specimens have been found by Charlotte Cooper Sunseri at CA-SCL-119, in southern Santa Clara County at the eastern side of Pacheco Pass, roughly 40 km due east of the mouth of the Pajaro River. This multicomponent site contains materials dated to the Middle (2500–950 cal BP), Middle-Late Transition (950–700 cal BP) to Late (700–180 cal BP) periods (Jones et al. 2007), and we are currently directly dating several of the pinniped remains.

Third, reanalyses have consistently distinguished more *Callorhinus* specimens than reported in the original analyses. These include, in any substantial sample ($n \geq 50$), remains of breeding age (≥5–6 years) males and young-of-the-

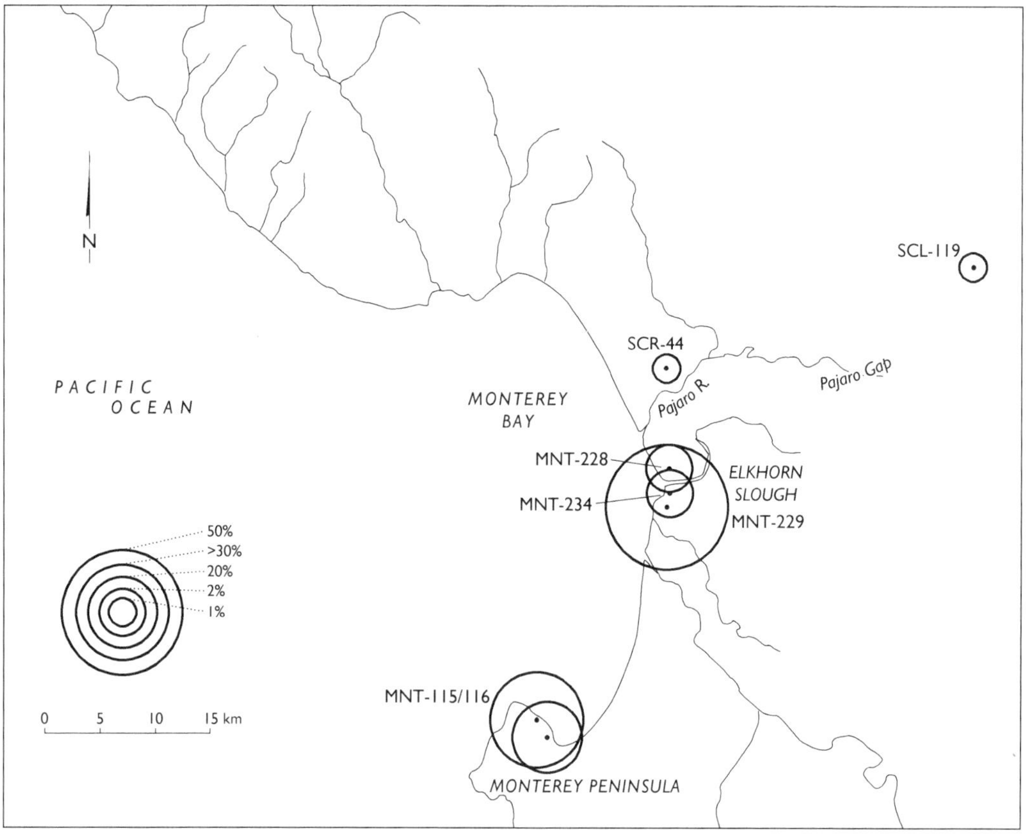

FIGURE 10.2. Map of south Santa Cruz and Monterey County coastlines inland to southern Santa Clara Valley, showing present knowledge of proportions of *Callorhinus ursinus* in analyzed or reanalyzed archaeological sites. These proportions and additional site inventories may be added as the balance of SCR, MNT, and SCL assemblages are analyzed. Map from Hildebrandt and Mikkelson 1993: map 9; original map drawn by Tammara Ekness, used with permission. For detailed data and sources, see Table 10.1.

year (≥5 months, YOY). Elements from breeding-age males are either those of the extremities, especially the succulent flippers, or axial parts often referred to an "otariid indet" category in many analyses (including the author's earlier ones). This element representation pattern probably reflects "schlepp effect" transport decisions, as reported by Savelle et al. (1996) for larger-bodied members of the Otariidae family, as adult male fur seals can weigh up to 275 kg.

In assemblages dominated by female *Callorhinus* remains, rare elements attributed by some analysts to California or Steller sea lions may instead be from adult male fur seals, and we are pursuing this possibility in such assemblages. *Callorhinus* male skeletal elements are morphologically distinguishable from those of like-sized male California sea lions (*Zalophus californianus*) and female Steller sea lions (*Eumetopias jubatus*), but they are often overlooked for lack of fur seal comparative specimens or the will to sort through pinniped metapodia and phalanges.

OSTEOMETRICS AND AGE ESTIMATION: APPLICATIONS OF ETNIER'S LOGISTICAL EQUATIONS

In our research, we have relied upon osteometric indices of age at death, developed by Michael Etnier with modern comparative samples of known-age-at-death *Callorhinus* specimens. This approach, based on von Bertalanffy's logistical curves, was developed in Etnier's (2002) dissertation research. Etnier and biostatistician

TABLE 10.1

Occurrence of Northern Fur seals in Monterey Bay Sites

SITE	NISP	%ΣNISP	%AD. F	%AD. M	%≤5 M	^{14}C YBP 2σ	REFERENCE/NOTES
SMA-18	111	10	58	9	17	**1070–1480**[a]	Hylkema et al. 2006; Newsome, S. D., M. A., Etnier, D. Gifford-Gonzalez et al. 2007
SMA-218	142	48	unknown	unknown	unknown	**2811–2871**[a]	Hylkema 1991
SMA-113	3	1	2	1	0	880–940[a]	Beta Analytic #238125, 238126, 238127
SMA-97	5	2	unknown	unknown	unknown	277–533	Hylkema 1991, calibrated by author
SMA-238	4	3	3	0	1	352–651	Hylkema 1991, calibrated by author
SCR-9	71	3	66	9	14	2095–2357	Hylkema 1991, pooled mean of ^{14}C ages on *Mytilus* calibrated by author, *C. u.* reanalysis and identifications by author, Potenzone 2007
SCR-35	19	11	19	0	0	**2870–2970**	Newsome, S. D., M. A. Etnier, D. Gifford-Gonzalez et al. 2007
SCR-3	3	0.7	67	33	0	2040–2200	Beta #239732
SCR-44	6	0.5	50	0	33	2531–2623[a] 1606–1682[a] 785–1062 390–525	Breschini and Haversat 1989, 2000; Sunseri 2009:269[b]
MNT-228	4	2	Unknown	Unknown	1	1140–1850 7000–7550	Dates approximate from Jones 2002, Jones et al. 1996
MNT-229	114	10	57	6	16	900–2700 6000–8200	Dietz et al. 1988; Jones 2002, faunal reanalysis by author, no direct dates on *Callorhinus*
MNT-234-PM	2334	50	39	2	8	**2470–2438**[a]	Newsome, S. D., M. A. Etnier, D. Gifford-Gonzalez et al. 2007
MNT-234-MS	4	20	2	0	2	7250–8100	Milliken, et al. 1999:80
MNT-115	45	32	Unknown	Unknown	Unknown	1670–1980	Dietz and Jackson 1981
MNT-116	2	20	Unknown	Unknown	Unknown	1670–1980	Dietz and Jackson 1981
SCL-119	2	0.5	100	0	0	1538–1702	Hildebrandt and Mikkelson 1993; Sunseri 2009

NOTE: Dates in bold: direct dates on *Callorhinus* bone. *Callorhinus ursinus* NISP, as a percentage of nonrodent mammal bone, and age/sex distribution, if known, with dates for site. Percentages of age/sex classes are calculated on all *Callorhinus* elements, including those of juveniles and nonageable specimens. KEY: MNT-234-PM:MNT-234 Primary Midden; MNT-234-MS: MNT-234 Millingstone Phase, based on summaries and examination of Area C fauna. Ad. F: adult female; Ad. M: breeding age males. All dates calibrated. A marine reservoir correction of 250 ± 35 years to δ^{13}C-corrected, ^{14}C ages of marine mammal and molluscan specimens (Newsome et al. 2007). Dates given as 2 σ ranges BP.

[a] Pooled averages of ^{14}C ages calculated by Calib®, then calibrated as above.

[b] Dates for *Callorhinus* ambiguous because of salvage of disturbed deposits; recent dates by Sunseri (2009) suggest multiple occupations.

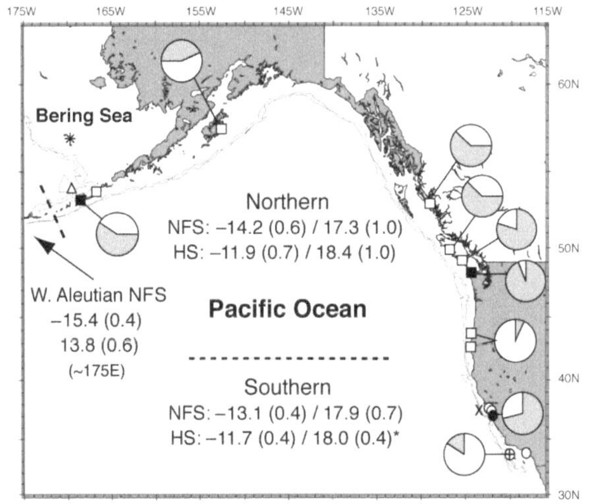

FIGURE 10.3. Mean bone collagen isotope values of ancient northern fur seal (NFS) and harbor seal, *Phoca vitulina* (HS), plus abundance estimates from selected archaeological sites. Pie diagrams show the relative abundance of NFS remains (shaded) versus other pinniped remains based on the number of identifiable specimens. Mean $\delta^{13}C/\delta^{15}N$ values (SD) for adult female NFS and (HS) are reported beneath each labeled group of NFS or HS; the asterisk denotes previously published HS data from the southern region (18). NFS cluster into northeastern Pacific (squares), California (circles), and western Aleutian populations based on significant differences in isotopic values. Filled symbols denote sites with harvest profile data (see Fig. 2). Locations of islands mentioned in text: Pribilof Islands, eastern Bering Sea; Δ, Bogoslof Island, eastern Aleutians; X, Farallon Islands off San Francisco Bay; Q, SMI off southern California. From Newsome, Etnier, Gifford-Gonzalez et al. 2007.

Phillips rechecked and refined the method with modern specimens with other indices of age such as counts of dental annuli in the same skeletons (Newsome, Etnier, Gifford-Gonzalez et al. 2007). We thus have a reasonable and conservative estimator of age at death.

This chapter does not present detailed age/sex data on *Callorhinus* from Monterey Bay sites. However, the largest sample, Moss Landing Hill Primary Midden, is dominated by YOY with modal ages of 2 to 4 months. Even given ranges of error in the age estimates, the majority of YOY thus far aged (NISP=23) are between 1 and 5 months. Females, represented by over 1000 specimens, range from slightly under 3 to over 8 years. The assemblage contains at least seven males of breeding age, ranging from 5 or 6 years to 8.3 years, according to Etnier's age determination methods.

BONE STABLE ISOTOPE EVIDENCE FOR ANCIENT NORTHERN FUR SEAL ECOLOGY

As our lab identified and aged fur seal specimens, our geochemist colleague Koch and his students Burton and Newsome developed bone stable isotope studies of several northern Pacific pinniped species relevant to fur seal ecology. They proceeded stepwise from ground truthing bone stable isotope values for several northern Pacific pinniped species of known provenience (Burton et al. 2002; Burton and Koch 1999). Ground truth stable carbon and nitrogen isotope assays revealed consistent isotopic signals for nearshore versus offshore foragers and for latitudinal zonation of foraging ranges in a single species. Turning to archaeofaunal *Callorhinus*, Koch's research team concluded that those recovered in western North

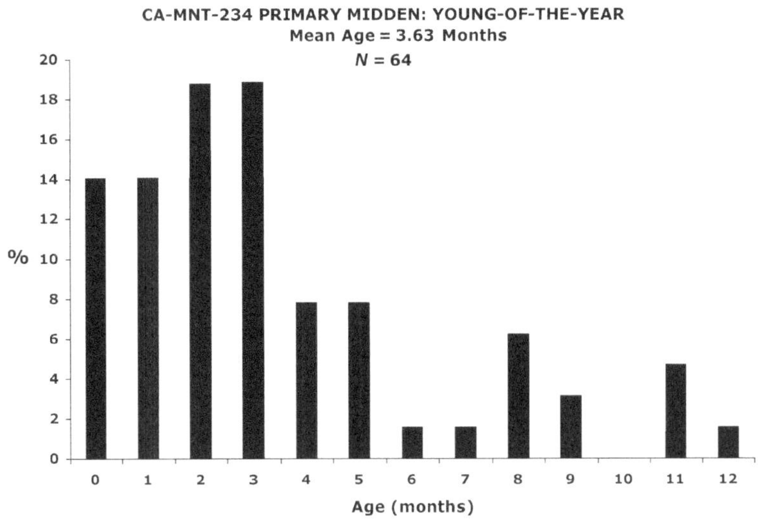

FIGURE 10.4. MNT-234, Moss Landing Hill Site Primary Midden: age structure of *Callorhinus* young of the year, with modern Alaska-Siberian populations' modal weaning age indicated. All specimens aged using Etnier's method (see text). Negative ages are fetal bone. Sample size is anticipated to increase when all specimens are recorded.

America predominantly foraged epipelagically, as do modern members of the species. However, in contrast to females and males breeding on the Pribilof and Siberian islands, California archaeofaunal *Callorhinus* specimens lack the isotopic signals of foraging in far northern latitudes, instead displaying stable carbon and nitrogen isotope ratios similar to the modern San Miguel Island (SMI) population (Newsome, Etnier, Gifford-Gonzalez et al. 2007).

Koch's research group and collaborators further distinguished two geographically distinct subgroupings of isotopic foraging "signatures" in the Holocene northeastern Pacific. A northern subpopulation, comprising animals from Oregon, Washington, British Columbia, and Alaska (Moss et al. 2006), and a southern subpopulation, from California (Newsome, Etnier, Gifford-Gonzalez et al. 2007), showed statistically significant differences in carbon and nitrogen values. Such interpopulational differences probably reflect the documented latitudinal differences in nitrogen ratios and differing levels of photosynthetic productivity in the respective foraging ranges of the two populations.

Recent satellite tracking of far North Pacific *Callorhinus* and long-term records of the location of fur seals sighted or taken at sea show them foraging in the Bering Sea, Gulf of Alaska, and clustering around "hotspots" along the eastern and western Pacific coastlines to about the 35th parallel (National Marine Fisheries Service 2007). The same records suggest that some North Pacific fur seals feed across the Pacific at the transition zone chlorophyll front (TZCF), the basin-wide zone of convergence of the subtropical and subarctic gyres (National Marine Fisheries Service 2007). The meeting of warm and cold waters creates a chlorophyll-dense "front" exploited by many invertebrate and vertebrate species, including cephalopods, juvenile albacore tuna, and smaller fishes, attracting such predators as loggerhead turtles (Polovina et al. 2001) and fur seals. The TZCF moves between the 35th and 40th parallels seasonally and on longer-term cycles.

By contrast, members of the SMI *Callorhinus* population, as reflected by telemetry and other forms of data collection, appear to remain in the California Current during their months foraging at sea, ranging as far north as Oregon and Washington. The recent *Callorhinus* recolonization of South Farallon Island (Pyle et al. 2001), largely by subadult males and females from SMI, reflects that northward movement of many SMI animals along the continental shelf break. Seasonal upwelling creates localized "hotspots" of anchovy, spawning herring, squid, and other fur seal prey along the California coast (Arntz et al. 1991; York 1991).

Whitaker and Hildebrandt (Chapter 8) have discerned a northern subpopulation signature in the Stone Lagoon archaeofaunal in Humboldt County, north of Arcata, California. Thus, zooarchaeologists may be picking up the geographic boundary between prehistoric representatives of the two foraging groups around Humboldt County. As discussed in detail below, preliminary aDNA results suggest that the existence of distinct prehistoric *foraging* subpopulations does not necessarily imply the existence of distinct *genetic* subpopulations.

BONE $\delta^{15}N$ ISOTOPE ENRICHMENT, LACTATION DURATION, AND MATERNAL ATTENDANCE

For some time, an isotopic signal of suckling, in the form of a >3‰ ^{15}N-enrichment in relation to maternal $d^{15}N$ levels, has been known from other mammalian species. Hobson et al. (1997) previously demonstrated this relationship in hair and muscle samples from modern otariids, including *Callorhinus*. Burton et al. (2001) demonstrated the presence of this 3‰ $\delta^{15}N$ lactation signal in YOY bones of archaeofaunal *Callorhinus* from the Moss Landing Hill Site. Newsome et al. (2006) investigated details of the bone turnover rate for this signature using known-age modern specimens of northern fur seal and California sea lion. They found a consistent post-weaning bone turnover rate and shift to adult signature in both species, which permits discernment of the end of suckling and beginning of independent foraging. The Newsome et al. (2006) ground-truth study served as the basis for research by Newsome and Etnier on two large prehistoric samples of juveniles, Chaluka (Umnak Island), Alaska, and Ozette, Washington. Assaying $\delta^{15}N$ values in dentary bones aged by Etnier's osteometric approach, they estimated modal weaning ages. These ages, 9 to 12 months and 12 to 15 months, respectively, are substantially later than the 4 to 5 months weaning age documented for populations in present-day Alaskan and Siberian islands (Newsome, S. D., M. A. Etnier, D. Gifford-Gonzalez et al. 2007:figure 3).

Such findings have implications for the Holocene historical ecology of the species. They suggest that a more diverse range of trade-offs in maternal attendance and foraging tactics existed for earlier populations of *Callorhinus* than seen in modern representatives. Later modal ages of weaning in more southerly fur seal populations would resemble those of other otariids feeding in temperate, subtropical, or tropical latitudes (Costa 2008; Trillmich 1990). SMI females, descendents of far North Pacific females that colonized the island in the 1960s, have slightly extended their span of attendance in normal years but display nothing approaching the extended lactation periods diagnosed for Ozette and Chaluka. It is possible that they may lack a genetic potential that permits trade-offs of attendance strategies seen in some other otariids. National Marine Mammal Laboratory head Robert DeLong has noted (personal communication 2008) that it would be impossible for female *Callorhinus* to sustain nursing much past the present SMI duration, as maternal condition reaches a very low point by that time. The Ozette profile (Figure 10.5) may suggest how maternal attendance might be prolonged. Between 6 to 9 and 9 to12 months, $\delta^{15}N$ values decline from their maximum of 3‰ above the adult female value but do not go into the steep decline that is characteristic of complete weaning. This suggests that, between ages 6 to 15 months, YOY may be provisioning themselves part of the time, while still taking mothers' milk. Ozette lies near areas where herring aggregate to spawn from January to April (Washington Department of Fish and Wildlife Fish Management Program 1997). At this time pups born June of the preceding year would be in the beginning of the 6 to 15 month age span shown in Figure 10.5. Herring spawning grounds, including those around the tip of the Olympic Peninsula, historically have

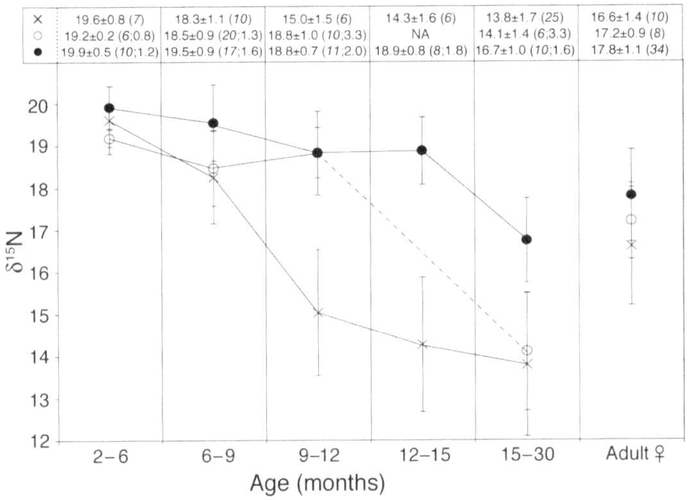

FIGURE 10.5. Bone collagen $\delta^{15}N$ values for one modern and two Holocene NFS ontogenetic series (2–30 months of age): X, modern Pribilof Islands (Bering Sea); °, prehistoric Umnak Island (eastern Aleutians, Alaska); •, prehistoric Ozette (Olympic Peninsula, Washington). Mean $\delta^{15}N$ values ±SD for age group. Numbers in parentheses: sample size median SD of age error estimates. Source: Newsome, Etnier, Gifford-Gonzalez et al. 2007.

formed a zone of concentration for fully weaned YOY from the Pribilofs (York 1991:figure 10.5). Thus, concentrations of this fat-rich food source known to be preyed upon by YOY of this age range, as well by adults, might have enabled extended maternal attendance as pups relied more on their own foraging efforts while still taking some milk from their mothers. The herring spawning season does not last the entire span of elevated $\delta^{15}N$ values, ending by March, but it could have formed a "weaning food" that supported limited but extended nursing.

The historical ecology of maternal effort in *Callorhinus* could further be elucidated by analysis of other large archaeofaunal samples from Canada and farther south. Another possible window into Late Holocene maternal attendance in central California would be to recover more specimens from early 19th-century industrial sealing middens on the Farallon Islands, already sampled in the 20th century (Pyle et al. 2001), with special attention to $\delta^{15}N$ ratios in bones of YOY and slightly older juveniles. This could provide an interesting "baseline" for *Callorhinus* life history in the greater San Francisco Bay and Monterey Bay regions.

ADNA EVIDENCE FOR ANCIENT NORTHERN FUR SEAL MOBILITY AND POPULATION SIZE

Ancient fur seal bone samples have been analyzed for aDNA in two laboratories so far (Moss et al. 2006; Newsome, Etnier, Gifford-Gonzalez et al. 2007), with the most extensive study by Elizabeth Hadly's Stanford laboratory. Hadly and coworkers have compared aDNA to modern DNA from the far North Pacific population (Ream 2002). From haplotypic diversity in aDNA dating 1000 to 2000 BP, they estimate northern fur seal metapopulation size throughout its ancient range. Preliminary results of the Hadly lab research are presented here with permission of the investigators, with the caution that all results are currently undergoing experimental replication in another laboratory and thus subject to modification.

First, prehistoric metapopulation size appears about the same as historic 20th-century levels, with a population reduction between the 2nd millennium BP comparative baseline and then recovery in recent times. These results suggests that the population levels of the early- to middle-20th-century northern fur seal stocks were similar to their maxima 1 to 2 millennia earlier, despite late-18th- and 19th-century industrial-scale sealing.

Second, past migration rates across the fur seal range were high enough to forestall development of genetically isolated regional populations (cf. Crockford et al. 2002). One way to express this is through mapping distinctive clades in relation to their provenience, as shown in Figure 10.6 (from Newsome, Etnier, Gifford-Gonzalez

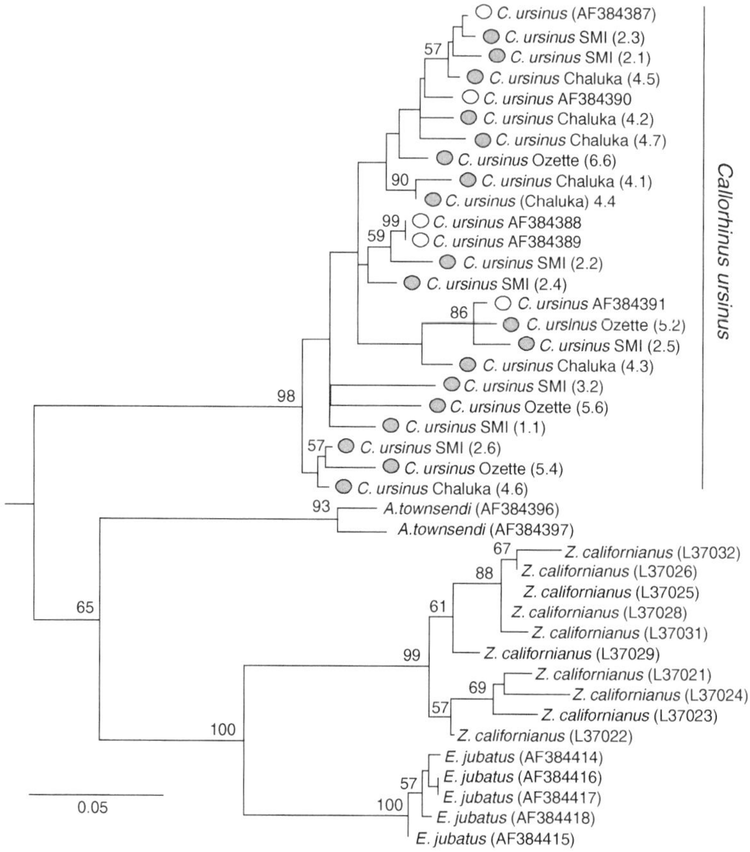

FIGURE 10.6. Cladogram showing initial aDNA data (gray circles) from northeast Pacific *Callorhinus*, 1000–2000 BP, in relation to modern *Callorhinus* (open circles) and other eared seals of the North Pacific. Based on a 50% consensus bootstrap neighbor-joining tree inferred from 5 modern and 19 ancient northern fur seal DNA sequences, also including other sympatric otariids: Guadalupe fur seals (*Arctocephalus townsendi*), California sea lions (*Zalophus californianus*), and Steller sea lions (*Eumetopias jubatus*). Jukes-Cantor corrected distances were estimated from a 156-bp mitochondrial control region fragment. Values above branches show bootstrap support based on 1000 bootstrap iterations. Note the significant monophyly of the four otariid species and the clustering of the aDNA sequences (solid ovals) among representative modern NFS (open ovals). Specimens from different archaeological sites highlighted in different fonts to show the essential genetic homogeneity of the ancient population from San Miguel Island, California, (SMI) through the Olympic Peninsula (Ozette), to Umnak Island, Aleutian archipelago, Alaska (Chaluka). Source: fig. 6. Newsome, Etnier, Gifford-Gonzalez et al. 2007. Data preliminary, courtesy of Elizabeth Hadly, Stanford University (see text).

et al. 2007). This contrasts with the isotopic signatures of distinct foraging ranges discussed above, one involving the far North Pacific to the TZCF, and the other more tied to the California Current. However, it is consistent with documented movements of tagged individuals from the far North Pacific rookeries to SMI or vice versa (DeLong and Antonelis 1991; Peterson et al. 1968).

Third, a longer-term signal of rapid population growth over the last 250,000 years exists, from Marine Isotope Stage 7, an interglacial span, through the Last Glacial and into the Holocene in the northern Pacific (Malin Pinsky et al. 2010). This reflects the species' long-term success in coping with the major climatic and oceanographic changes over that time span.

If fur seal aDNA samples dating from the 2nd to 3rd mellenium BP represent a similarly sized but differently distributed metapopulation in comparison to the modern one, climatic conditions in the North Pacific during the immediately preceding time span may be implicated. Crockford and Frederick (2007) argue that faunal evidence, specifically remains of bearded seals, ring seals, and northern fur seals, from Unalaska Island, eastern Aleutians, documents the expansion of sea ice as far south as that island in the interval of the Holocene Neoglacial (~4700 to 2500 BP). The southern extension of sea ice would effectively close off the more northern Pribilof Islands to fur seal breeding, which requires ready access to the sea for female foraging trips. If these authors are correct, the occurrence of fur seal elements in sites along the coasts of southern Alaska, Canada, and the western continental United States during this span may reflect a southward displacement of the entire metapopulation. The "waning" of fur seal populations in their southern range may be the result of a complex set of interactions among large-scale climate change, shifts in northern ice distribution, alterations in oceanic currents and upwelling patterns, their effects up the food chain, and, finally, human offtake. This topic will be taken up again in the last section of this chapter.

LATITUDINAL PATTERNS OF AGE/SEX CLASS REPRESENTATION

From Stone Lagoon (Humboldt County, California) to the north, a male-dominant pattern in aboriginal offtake of *Callorhinus* often emerges in larger samples. In sites such as Stone Lagoon, Ozette (Washington), and Chaluka (Umnak Island, Alaska), the majority of individuals are younger males. The Oregon coastal sites of Umpqua/Eden and Seal Rock yielded small *Callorhinus* samples that included all age/sex classes but proportionately more subadult males and adult males than seen farther south in California. As noted by Etnier (2002, 2007) and Lyman (2003), the sexually segregated aggregation habits of otariids during breeding season lend themselves to sustainable harvests of nonbreeding males. This does seem to be the pattern of offtake reflected in archaeological sites from far northern California into British Columbia and Alaska.

By contrast, in California south of the 40th parallel, another ubiquitous harvest pattern dominates that of females and YOY. This is typical of sites from San Francisco to Ventura and San Miguel Island. It is widespread in sites of widely varying sample sizes and extends across several prehistoric culture areas. If juvenile-male-dominant harvests are the best path to sustainable offtake, then the California pattern, which focuses on the age/sex classes most likely to destabilize rookery populations, calls for an explanation. Before we knew much about the prehistoric northern subpopulation and human offtake patterns in that region, it might have been easy to attribute this pattern to a "typically human" propensity to overcrop high-ranked resources. However, we are now more knowledgeable about the offtake patterns at multiple sites in the far North Pacific, where long-term sustainable use of juvenile male otariids is evidenced (cf. Etnier 2007). I will return to this problem in the last section of this chapter.

ACCOUNTING FOR NORTHERN FUR SEAL OCCURRENCES IN THE GREATER MONTEREY BAY

Robert DeLong (personal communication, 2005) has informally suggested that the existence of fur seal remains in the Monterey Bay might reflect younger pups that were swept away from the Farallons during their early experimentation with swimming, which begins preweaning (cf., Baker and Donohue 2000). According to this scenario, YOY and adults in distress would be carried on southerly currents toward the Monterey Bay. At present, the beaches north and south of Moss Landing do act as a "catcher's mitt" for oiled birds and stranded marine mammals during some seasons. To assess the plausibility of this scenario for the many fur seal remains at Moss Landing, Charlotte Cooper Sunseri, a member of my lab, compared ages of YOY expected to be carried there by the currents dominant during the fur seal breeding season (roughly June to November) with the age classes actually found at the Moss Landing Hill Site (CA-MNT-234). A substantial lack of fit exists between the expected and the observed age structures, which does not support the likelihood of this interpretation (Gifford-Gonzalez and Sunseri 2008). It is also notable that the 6-month-to-2-year age class of *Callorhinus*, which York (1991) reports as likely to follow spawning herring or other prey into estuaries or straits, are absent from the Moss Landing sample. Perhaps, with the growth of the *Callorhinus* colony on South Farallon Island, we will be able to test this hypothesis with concrete stranding data.

Another possibility raised by DeLong is that mainland or near-mainland breeding colonies around Point Año Nuevo or the dunes west of the old Salinas River course at Moss Landing could have been less preferred "overflow" colonies spawned by the Farallon Island rookeries in times of peak fur seal density on the offshore islands. According to this scenario, males unable to stake a claim to breeding territories on the Farallons would colonize along the coast, as California sea lion males occasionally do, since populations have burgeoned in recent years.

This is a reasonable explanation, implying a noncontinuous history of mainland colonization and hence accessibility to human predation along the central California coast. However, it leads to new questions. In the absence of such colonizing events along the central coast today, how can we understand more about these processes? Is it possible to argue from first principles and data from SMI? For example, might occurrences of fur seals in mainland or near-mainland colonies follow on with optimal years in the ENSO cycle? Can we predict *Callorhinus* breeding site locations along the California coast, that is, the "ecological space" required by a middle-latitude population of epipelagic foragers in the California Current that requires specific terrestrial and offshore conditions for their rookeries?

The next section outlines the costs and benefits to fur seals of living year-round in middle latitudes. It first summarizes some key parameters of northern fur seals as a species, especially the articulation of their foraging behavior with reproductive effort in females and pup survival. It then reviews the effects of ENSO-associated fluctuations in the California Current and associated upwelling on the contemporary *Callorhinus* population on SMI. The SMI population can serve as an indicator of the costs to fur seals of living at middle latitudes. It then touches on other benefits and costs to otariids associated with ENSO.

NORTHERN FUR SEAL LIFE HISTORY AND ECOLOGY: NORTH AND SOUTH

Modern arctic *Callorhinus* populations spend close to 8 months in dispersed foraging at sea, largely on prey in the water column rising nocturnally from deeper water. During the day, they rest at the surface. Young of the year in the Alaskan colonies go to sea independent of their mothers at around 4 months of age, many leaving before their mothers do (Gentry 1998:299). Pups head south through passes in the Aleutian Islands and, until about 2 years of age, forage on smaller prey in offshore waters off Canada and the United States, diving at night and resting during the day (Baker 1978; Baker 2007; Baker and Donohue 2000; Ragen et al. 1995). This age class is the only one encountered close to land pursuing herring and other schooling fish (York 1991). Until their third year, immature *Callorhinus* do not return to their natal rookery, nor do they normally haul out in other terrestrial locations.

Young females appear on their natal breeding grounds c. 3 to 4 years of age, but most do not bear their first young until their fifth or sixth years, reflecting successful ovulation and fertilization in the fourth to fifth years (York 1983). Female fur seals have a maximum lifespan of 23 years, though most die earlier. Parous females enter estrus within a week of giving birth and swiftly become pregnant. Like all otariids, they display embryonic diapause, or delayed development and implantation of the embryo, such that their subsequent birth is nearly exactly 12 months after their previous one (Gentry 1998).

As Alaskan male juveniles mature, they begin to forage in the cooler waters of the Gulf of Alaska and the Bering Sea. Juvenile males usually return to haul-out grounds near their natal rookeries in their third year of life, remaining there for much of the breeding season (Trites and Larkin 1989). Although spermatogenesis may begin around 5 to 6 years of age, male *Callorhinus* on well established colonies are not large enough nor socially experienced enough until 8 to 9 years to succeed in competing for breeding territories (Gentry 1998). In arctic rookeries, immature males aggregate densely, although individuals leave land to forage or visit other all-male herds during their sojourn (Gentry 1998). Like breeding-age members of their species, this age/sex class shows high site fidelity in North Pacific islands, despite yearly subsistence harvests focused on their haul-outs in July (Gentry 1998). As territorial males' defense behaviors wane and they leave the rookeries in August, juvenile males may enter rookeries and contact females; however, their chances of impregnating a female at this time are very low. It should be noted that some older juvenile males do "scout" other haul-out locations in their overall foraging range, and these appear to be the founders of new colonies, along with females that come ashore to pup. This was the case with the colonization of SMI and South Farallon. Tagged juveniles and females have also transferred from the eastern Pribilofs to Bogoslof Island in recent times (National Marine Fisheries Service 2007).

Experienced males occupy breeding sites and compete for territories in May or early June, and territoriality reaches its peak before females arrive in June and July (Gentry 1998). Males return to these territories annually and hold them as long as they can, even under conditions where females and smaller males may be abandoning a location (Gentry 1998:97). Because territories are gained and held through male-male threat displays and sparring, territorial tenure depends on male size and condition and size, since territory-holding males fast for up to 2 months. Up to 30% of mature males are estimated never to win a breeding territory (Vladimirov 1987).

Like the rest of the otariids, northern fur seals are "income" breeders. They do not store energy for an entire lactation season in thick fat deposits, as do most phocids; instead, females must subsidize the lactation span by intermittent foraging trips. In documented Alaskan cases, foraging trips average 5.5 to 10 days, with relatively shorter spans on the rookery (Robson et al. 2004). The pups' nursing span is thus a set of discrete episodes punctuated by maternal absences and fasting. The longer the trip length, the more nutritional stress is placed on the growing pup. Arctic-breeding females appear to be able to offset short- and long-term variations in prey type, location, and density (Costa 2008) with little effect on their reproductive success. By contrast, SMI females have historically showed that they are unable to cope successfully with variability conditioned by extreme ENSO events. To appreciate this contrast, one must understand a bit more of female *Callorhinus* foraging habits.

Callorhinus females are flexible foragers, capable of using different tactics to suit available prey types and locations. Telemetry of breeding females foraging off various Pribilof colonies has shown that breeding females employ different diving tactics according to their foraging grounds (Robson et al. 2004; Zeppelin and Ream 2006). They may forage over the continental shelf itself or at its margins, and they may vary their diving tactics from relatively shallow to deep, or a combination of the two. Telemetry, scat, and isotope analyses show that Pribilof females partition foraging ranges according to their home breeding colony (Kurle and Worthy 2001; Robson et al. 2004; Zeppelin and Ream 2006). Individuals can switch prey species, or age classes within a single species, according to their availability (Costa 2008; DeLong and Antonelis 1991).

Observational, telemetric, and experimental data likewise indicate that *Callorhinus* females are efficient foragers, spending on average less than 30% of their time at sea actually seeking prey (Costa et al. 2006). Costa (2008) presents evidence that Pribilof females can intensify their foraging effort and raise their field metabolic rates (FMR) while still adding body mass and not lengthening their foraging trip duration. He argues that the Pribilof data imply that, in contrast to benthic foraging species of *Arctocephalus*, Pribilof breeding *Callorhinus* females are so well within their energy budgets that they can compensate for lower prey availabilities without compromising the duration of their foraging forays and, consistently, can provide pups with predictable, high-quality nursing sessions. Costa et al. (2006) have argued that epipelagic foragers like *Callorhinus* can more swiftly recover from cyclical marine productivity lows or other events than can benthic foraging species.

In contrast to the northern subpopulation, observations on SMI *Callorhinus* female condition, foraging behavior, and attendance and on pup mortality during the 1982–83 and 1998–99 El Niño events show that there is a point past which females' efficient foraging strategies cannot cope. The low prey availability typical of El Niño events in the middle latitudes imposes cyclical challenges to population maintenance in all pinnipeds, as will be discussed in the next section.

EFFECTS OF ENSO OSCILLATIONS ON CALIFORNIA NORTHERN FUR SEAL POPULATIONS

From the Middle Holocene to the present, coastal California and the northern tier of contiguous states have been affected by ENSO climatic fluctuations, in 4-to-7-year cycles (Liu et al. 2000). During El Niño events, sea surface temperatures (SST) rise substantially off the northeastern Pacific coast, stalling the seasonal upwelling of cold benthic waters. In the absence of such nutrient-rich upwelling, die-offs occur through trophic levels, beginning with plankton and primary consumers and up to apex vertebrate predators. The Pacific jet stream shifts, resulting in displaced storm tracks and changes in terrestrial weather patterns. Historic data for El Niño years indicate that, in November and December central to northern California receives an average amount of precipitation, while far southern California receives rainfall 20 to 30% above average. In January to March, El Niño precipitation in northern California runs 20 to 30% higher than average, and far southern California sees precipitation 40 to 45% above normal (Service 2008).

Close monitoring of adult *Callorhinus* condition and pup mortality on SMI during two strong El Niño events (1982–1983 and 1998–99) has shown the devastating effect such conditions have on the species at middle latitudes. In 1983, pregnant females arrived on SMI in poor condition, and pups were born later and smaller (DeLong and Antonelis 1991). In 1983 and 1984, mean pup weights at 3 months of age were significantly less than in years before or after, and maternal foraging trips increased in duration, further stressing underweight pups (DeLong and Antonelis 1991). The vast majority of tagged pups in the 1983 and 1998 cohorts did not survive to the next year. Notably, adult female and juvenile male arrivals at SMI diminished in 1983, suggesting death or emigration to better foraging grounds (DeLong and Antonelis 1991).

By contrast, SMI *Callorhinus* adult males showed little deviation from the norm in body mass or length of stay on rookery. DeLong and Antonelis argue that this is because males can leave breeding colonies and resume foraging appreciably earlier than can lactating females. Moreover, they speculate that, with many prey species remaining in very deep water during El Niño years (cf., Arntz et al. 1991), *Callorhinus* males' ability to dive to greater depths may allow them to obtain prey where females cannot.

During the same 1982–1983 El Niño event that affected the SMI fur seals so negatively,

Pribilof Island breeding-age females and males showed no appreciable changes in their time on rookery, in other indices of condition, or in female foraging trip duration (Gentry 1991). York (1991) in fact suggests that the increase in SSTs characteristic of El Niño years off the coasts of western North America may favor higher survival of juveniles from the northern rookeries. Unlike adults of both sexes, weaned pups 4 months to 2 years old travel down the Pacific coast closer to land, exploiting spawning herring. York argues that, since herring cohorts actually thrive in El Niño years, this and localized increases in other prey species may be the reason for higher rates of younger juvenile survival during those times.

Subadult and adult *Callorhinus* foraging in California waters suffer another hazard during El Niño years: domoic acid (DA), a neurotoxin produced by *Pseudo-nitzschia* species, transmits up the food chain through filter feeders and smaller finfish. Filter-feeding mollusks and such finfish as sardines and anchovies are unaffected by DA but concentrate it in their tissues. Seabirds and pinnipeds are vulnerable to DA poisoning if they consume such prey. Domoic acid was implicated in deaths of 70 *Zalophus* and 1 *Callorhinus* along the central California coast in May to June 1998 (Trainer et al. 2000). During *Pseudo-nitzschia* blooms, DA concentrates in high levels in sardines and anchovies, favored prey of *Callorhinus* along the California coast.

Pseudo-nitzschia blooms have been hypothesized to result from the influx of nutrients into coastal waters with heavy El Niño rains and stream discharge. However, Trainer et al. (2000:1826–27) note that some of the highest collected DA levels were in areas of lower stream inputs but consistent upwelling, even during El Niño. These included upwelling centers near Point Año Nuevo, Point Sur/Point Lobos, and Point Conception. When they are nutrient depleted, *Pseudo-nitzschia* cells lose their buoyancy and sink to deeper waters, where finfish species foraging at shallower depths may be less likely to accumulate it and its by-products in their tissues and pass them up the food chain.

California sea lions were the most affected pinniped species in the 1998 DA poisoning, probably because they are now the ubiquitous otariid of the central California coast. However, they are not so common in the archaeofaunal samples, accounting for 23%, while they comprise 57% percent of recent documented strandings (Burton et al. 2002). Were *Callorhinus* as numerous and widespread as their osteological traces suggest they were in the greater Monterey Bay region, they would have been major victims of DA poisoning. The waters off Point Año Nuevo, Point Lobos, and Point Conception, where upwelling cells exist even in El Niño years, would have been magnets for foraging fur seals then as they are today for *Zalophus*. Thus, prey search would have drawn foraging seals into precisely the zones of highest risk of DA poisoning.

In sum, in addition to having to cope with the substantial thermal stresses of hot dry weather on YOY while on their rookeries (Trites 1990), *Callorhinus* breeding along the California coast would have been affected by multiple effects of El Niño. Prey scarcity and its cascading effects on female foraging success and pup survival can effectively eliminate entire birth cohorts from the metapopulation, with implications for longer-term population dynamics. Domoic acid poisoning can kill foraging animals in the seasonal span immediately before and into the breeding season. Female mortality would further undermine rookeries' success through such cycles.

ECOLOGICAL DYNAMICS, REPRODUCTION, AND ARCHAEOFAUNAL AGE/SEX RATIOS

Today, the total *Callorhinus* stock is decreasing for reasons that remain a matter of debate among experts (Lander 1981; National Marine Fisheries Service 2007; Towell et al. 2006; Trites and Larkin 1989). With hindsight, many see the killing of upward of 330,000 female

Pribilof fur seals between 1956 and 1968 in a fisheries management strategy as a contributing factor in the decline, but recovery of the population after the take of females ceased has not been as expected (Gentry 1981). High juvenile mortality in specific cohorts also plays a role in constricting replacement of the breeding population, as does declining pregnancy rates among surviving females, and both may be tied to ecosystemic changes throughout the North Pacific (Trites and York 1993). Some of these changes may be due to human overfishing and environmental destruction, while others may be products of longer-term climatic change.

At the same time, northern fur seals are moving into new breeding colonies. The *Callorhinus* population on Bogoslof Island in the Aleutians, on the margin of the Bering Sea, is the fastest-growing northern colony, growing from 0 in 1910 to 12,000 in the mid-1970s, gaining immigrants from the dwindling stocks on the eastern Pribilofs (National Marine Fisheries Service 2007). On South Farallon Island, at the edge of the continental shelf break due west of San Francisco, *Callorhinus* subadult males and breeding age females have been doubling in numbers since the 1990s (Martin 2006). Located in a highly productive zone for an epipelagic forager, the Farallons once had upward of 100,000 northern fur seals, all taken by sealers between 1807 and 1812, when the colony was extinguished (Pyle et al. 2001).

Such cycles of abandonment and colonization may come as less of a surprise to zooarchaeologists accustomed to longer time cycles and to species appearing in unanticipated places (Lyman and Cannon 2004) than they do to marine mammal biologists. Perhaps the species' uncanny degree of breeding-site fidelity has led biologists to discount its flexibility in foraging and breeding site choices over time. Gentry (1998), however, stressed the dynamic nature of the North Pacific basin through the Pleistocene and the fact that *Callorhinus* has weathered major changes in the terrestrial-oceanic interface, a point to be discussed later in this chapter.

Moreover, this is a relatively long-lived species, and there are hints that females and other age/sex classes may know of more than one breeding or foraging ground to which they can resort if one becomes untenable (DeLong and Antonelis 1991). Pribilof pups up to 2 years of age go on a kind of "walkabout" along the continental shelf margins and even closer inshore (York 1991). If *Callorhinus* pups can recall their mothers' vocalizations several years after they are separated from them (Insley 2001), why should they not recall rewarding feeding grounds? Some juvenile males entering their reproductive years display the pattern of scouting beaches in their foraging ranges, as was the case with the South Farallon colonization. When they coincide with females landing to pup, such subadult male pioneers gain several more years of reproductive success, while still smaller than fully mature territorial males. San Miguel Island was colonized by animals from both Siberian and Alaskan colonies (Peterson et al. 1968) coming ashore on an island that presented an acceptable range of conditions for breeding: isolation from human disturbance, adequate beach space, proximity of the continental shelf break, and strong onshore winds and/or heavy fog cover aiding thermoregulation. The more recent recolonization of South Farallon by females and subadult males from SMI repeats such a scouting pattern.

In this connection, I return to the issue of female-and-YOY-dominant harvest patterns typical of California south of Humboldt County. Rather than infer that human foresight and ability to manage a high-value resource stopped around the 40th parallel, we might ask whether there might be underlying, nonanthropogenic regional processes that conditioned the availability of these age/sex classes to aboriginal groups seeking to exploit them. The multiple ENSO-related effects outlined earlier would render California *Callorhinus* colonies more fragile than those in the far North Pacific, which is less affected by the vagaries of the California Current. This in turn might predispose them to behaviors not observed in present-day populations.

If coastal central California colonies were truly "satellites" of the Farallon rookeries, then in less-favored sites for breeding locales, southern subpopulation juvenile males might have gravitated to the denser Farallon rookeries for their haul-outs, although this would be a different behavior than generally reflected in Pribilof juvenile males. In fact, although arctic juvenile males display high natal-site fidelity, they have deserted their birthplaces in the eastern Aleutians for Bogoslof Island rookeries. This at least suggests a propensity toward relocation among juvenile males of the species. Such behavior would leave *Callorhinus* territorial males, reproductive age females, and YOY on the mainland rookeries as the main prey, rendering such colonies extremely vulnerable to rapid depletion and population collapse. Cooper and Etnier (2006) modeled predator-prey relations using a cumulative distribution function (CDF), predicting northern fur seal population declines with differing harvesting levels. The CDF model specified the number of years over which a hypothetical population of 1000 animals would decline to near extinction, set at two individuals. Harvest rates of 10% of females per year are modeled to drive the population to extinction within 100 years, and 20 to 30% harvesting rates resulted in extinction within 50 years. We plan to undertake more realistic modeling, based on metapopulation estimates derived from aDNA, when those become available to us.

An alternative explanation for the lack of younger juveniles in California sites would involve a slightly different life-history pattern. North Pacific juvenile males continue offshore foraging until around 3 years of age before beginning their "apprenticeship" in male-male competition during the breeding season. Southern population males would have faced life history trade-offs that differ from those of their northern cousins; given the extreme lows in marine productivity during El Niño years, California juvenile males might have spent an extra 1 or 2 years at sea, coming to land around the age of spermatogenesis. The trade-off would have been between that of a few years' sparring experience with peers and a more rapid gain in body mass, ultimately a key factor in male-male competition. While it might seem far-fetched at first glance, juvenile year-round foraging would, in fact, simply extend the behavior entrained in the first 2 years of life for a few years more. In any case, the undeniable fact of the lack of juvenile male remains in archaeofaunas south of Humboldt Bay requires some explanation.

How can we test hypotheses drawn from these speculations? In the face of negative archaeological evidence, we need to attend closely to the onshore-offshore behaviors of juveniles from the SMI colony and, over time, the South Farallon colony. These data may help us at least assess whether either proposal has merit.

Finally, the ancient metapopulation size estimates from aDNA diversity need to be considered seriously when they are complete, especially in light of paleoclimatic information. New aDNA findings suggest the *Callorhinus* population of 1000 to 2000 years ago was about the same size as historically reported levels (Pinsky et al. 2010). The aDNA sample derives just from the end of the Neoglacial span during which Crockford and Frederick (2007) propose sea ice blocked North Pacific rookery sites such as the Pribilof Islands. If this proves to be the case with further targeted research, then we may best understand northern fur seal presence in coastal sites in the greater Monterey Bay, northern California, Oregon, and perhaps even the Olympic Peninsula, Vancouver Island, and other parts of British Columbia as part of the long-term adaptation of a species to climate change during the Holocene. If northern fur seals' most southern colony locales more negatively affected net reproductive success than did those to the far north, the disappearance of *Callorhinus* from the California coast may have been climatically forced, as animals moved north toward less risky breeding grounds when these became open. Local disappearances may have come about with some human help, especially given the age- and sex-specific offtake patterns typical of California sites, but the impacts of aboriginal people may be less relative to these larger-scale processes.

CONCLUSION

To conclude, it is useful to sketch some ways that collaboration between paleoecologists and zooarchaeologists, on the one hand, and marine mammal biologists studying northern fur seals, on the other, might shed more light on the species' past and hence its potentials in the present.

1. Study of historic Farallon fur seal midden remains could elucidate anatomical and genetic diversity and maternal attendance patterns in central California. Wake (personal communication 2008) reported he was able to excavate only two *Callorhinus* elements from post-sealing contexts, so collections made earlier (Riddell 1955; Pyle et al. 2001) may be critical to this research.
2. Morphological studies of weight-bearing bones of the skeleton may shed light on the degree of terrestriality of members of the southern subpopulations.
3. Stable isotopic study of an age series of bones in the fashion of Newsome S. D., M. A. Etnier, D. Gifford-Gonzalez et al. (2007) from Ozette and other large samples may reveal diversity in foraging patterns and life histories across subarctic populations.
4. Data collected on the modern SMI population over its dynamic history may provide clues to the fate of California *Callorhinus* in the past, including rates of recruitment and life history parameters of juvenile males.

Northern fur seals are a surprisingly durable and resourceful species that deserves our best efforts both to understand all aspects of its ecological history and to conserve the species today. Archaeological records can contribute a great deal to understanding the historical ecology of *Callorrhinus* in a preindustrial world. There, humans may not have had more than a low-level impact, but they sampled enough animals into the fossil record to help chart the species' behavior and population dynamics over time. The interface of complex contemporary behavioral ecological research and detailed paleoecological analysis promises to be a productive zone, where a certain amount of turbulence produces rich rewards.

ACKNOWLEDGMENTS

The author's research has been funded by NSF Archaeology BCS-0320168, Earth Sciences EAR-000895, California State Department of Parks & Recreation contracts C0468040 and CSP 105-04, and the University of California, Santa Cruz, Academic Senate Committee on Research. She is grateful to the following persons for facilitating her access to research materials: Gary Breschini and Trudy Haversat, Archaeological Research Consulting; Randy Milliken, Bill Hildebrandt, Far Western Anthropological Research; Mark Hylkema, Santa Cruz District Archaeologist, CA Dept of Parks and Recreation; Kenneth Coale, Director, and Joan Parker, Librarian, Moss Landing Marine Labs. Radiocarbon date for CA-SCR-3 donated by Albion Environmental, Inc. For colleagueship, collaboration, and critical feedback, the author thanks Mike Etnier, Paul Koch, Rob Burton, Bill Hildebrandt, Seth Newsome, and Herbie Lee, Applied Math and Statistics, UC Santa Cruz. Paul Koch kindly read and commented on a draft of this paper. Thanks are due to Bob DeLong for his constructive and informative criticisms. Finally, the author thanks her senior zooarchaeological analysts, Cristie Boone, Ben Curry, Charlotte Sunseri, and UC Santa Cruz undergraduate assistants, 2003–2008, Natalie Bagley, Kira Bonomo, Jen Bower, Stella Doro, Danny Gilmour, Carrie Howard, Kambiz Kamrani, Jenni Kraft, Josh Noyer, Patrick O'Meara, Amanda Rankin, and Albert Valdivia. All errors of fact and judgment are the author's own.

REFERENCES CITED

Arntz, W., W. G. Pearcy, and F. Trillmich
1991 Biological Consequences of the 1982–83 El Niño in the eastern Pacific. In *Pinnipeds and El Niño. Responses to Environmental Stress*, edited by F. Trillmich and K. A. Ono, pp. 22–42. Springer-Verlag, New York.

Baker, C. M.
1978 The Size Effect: An Explanation of Variability in Surface Artifact Assemblage Content. *American Antiquity* 43:288–293.

Baker, J. D.
2007 Post-Weaning Migration of Northern Fur Seal *Callorhinus ursinus* Pups from the Pribilof Islands, Alaska. *Marine Ecology Progress Series* 341:243–253.

Baker, J. D., and M. J. Donohue
2000 Ontogeny of Swimming and Diving in Northern Fur Seal (*Callorhinus ursinus*) pups. *Canadian Journal of Zoology* 78:100–109.

Breschini, G. S., and T. Haversat
1989 *Archaeological Investigations at CA-SCR-44, Northeast of Watsonville*. Coyote Press.
2000 *Archaeological Data Recovery at CA-SCR-44 at the Site of the Lakeview Middle School, Watsonville, Santa Cruz County, California*. Coyote Press.

Burton, R. K., D. Gifford-Gonzalez, J. J. Snodgrass, and P. L. Koch
2002 Isotopic Tracking of Prehistoric Pinniped Foraging and Distribution along the central California Coast: Preliminary Results. *International Journal of Osteoarchaeology* 12:4–11.

Burton, R. K., and P. L. Koch
1999 Isotopic Tracking of Foraging and Long-Distance Migration in Northeastern Pacific Pinnipeds. *Oecologia (Berlin)* 119(4):578–585.

Burton, R. K., J. J. Snodgrass, D. Gifford-Gonzalez, T. Guilderson, T. Brown, and P. L. Koch
2001 Holocene Changes in the Ecology of Northern Fur Seals: Insights from Stable Isotopes and Archaeofauna. *Oecologia* 128(1):107–115.

Cooper C. K., and M. A. Etnier
2005 Poster: Mathematical Modeling of Human and Marine Mammal Interaction in the Prehistoric Monterey Bay. 69th Annual Meetings, Society for American Archaeology Annual Conference, Salt Lake City.

Costa, D. P.
2008 A Conceptual Model of the Variation in Parental Attendance in Response to Environmental Fluctuations: Foraging Energetics of Lactating Sea Lions and Fur Seals. *Aquatic Conservation: Marine and Freshwater Ecosystems* 17:S44–S52.

Costa, D. P., M. J. Weise, and J. P. Y. Arnould
2006 Potential Influences of Whaling on the Status and Trends of Pinniped Populations. In *Worldwide Pinniped Population Status and Trends in Whales, Whaling and Ocean Ecosystems*, edited by J. A. Estes, T. M. Williams, D. Doak and D. DeMaster, pp. 342–357. University of California Press, Berkeley.

Crockford, S., G. Frederick, and R. Wigen
2002 The Cape Flattery Fur Seal: An Extinct Species of *Callorhinus* in the Eastern North Pacific? *Canadian Journal of Archaeology* 26(3):152–174.

Crockford, S. J., and S. G. Frederick
2007 Pacific Sea Ice Expansion in the Bering Sea during the Neoglacial: Evidence from Archaeozoology. *The Holocene* 17(6):699–706.

DeLong, R. L., and G. A. Antonelis
1991 Impact of the 1982–1983 El Niño on the Northern Fur Seal Population at San Miguel Island, California. In *Pinnipeds and El Niño. Responses to Environmental Stress*, edited by F. Trillmich and K. A. Ono, pp. 75–83. Springer-Verlag, New York.

Dietz, S. A., W. R. Hildebrandt, and T. L. Jones
1988 *Archaeological Investigations at Elkhorn Slough : CA-MNT-229: A Middle Period Site on the Central California Coast*. Papers in Northern California Anthropology No. 3. Coyote Press, Salinas, CA.

Dietz, S. A., and T. L. Jackson
1981 *Final Report of Archaeological Excavations at Nineteen Archaeological Sites for the Stage 1 Pacific Grove-Monterey Consolidation Project of the Regional Sewerage System*. Submitted to Engineering-Science Inc. On file at California Regional Information Center, Sonoma State University, Rohnert Park, CA.

Etnier, M. A.
2002 *The Effects of Human Hunting on Northern Fur Seal (*Callorhinus ursinus*) Migration and Breeding Distributions in the Late Holocene*. Ph.D. dissertation, Anthropology, University of Washington, Seattle.
2007 Defining and Identifying Sustainable Harvests of Resources: Archaeological Examples of Pinniped Harvests in the Eastern North Pacific. *Journal for Nature Conservation* 15: 196–207.

Gentry, R. L.
1981 Northern Fur Seal *Callorhinus ursinus* (Linnaeus, 1758). In *Handbook of Marine Mammals, Volume 1: The Walrus, Sea Lions, Fur Seals and Sea Otter*, edited by S. H. Ridgeway and R. Harrison, pp. 143–160. Academic Press, New York.
1991 El Niño Effects on Adult Northern Fur Seals at the Pribilof Islands. In *Pinnipeds and El Niño. Responses to Environmental Stress*, edited by F. Trillmich and K. A. Ono, pp. 84–93. Springer-Verlag, New York.
1998 *Behavior and Ecology of the Northern Fur Seal*. Princeton University Press, Princeton.

Gifford-Gonzalez, D., and C. K. Sunseri
2008 An Earlier Extirpation of Fur Seals in the Monterey Bay Region: Recent Findings and Social Implications. *2007 Proceedings of the Society for California Archaeology*.

Hildebrandt, W. R., and P. J. Mikkelson
1993 *Archaeological Test Excavations at Fourteen Sites along Highways 101 and 152, Santa Clara and San Benito Counties.* Sonoma State University.
Hobson, K. A., J. L. Sease, R. L. Merrick, and J. F. Piatt
1997 Investigating Trophic Relationships of Pinnipeds in Alaska and Washington Using Stable Isotope Ratios of Nitrogen and Carbon. *Marine Mammal Science* 13:114–132.
Hylkema, M. G.
1991 *Prehistoric Native American Adaptations along the Central California Coast of San Mateo and Santa Cruz Counties.* M.A. thesis, San Jose State University, San Jose, CA.
Hylkema, M. G., W. H. Hildebrandt, J. Farquhar, D. Gifford-Gonzalez, and K. Gobalet (editors)
2006 *Archaeological Investigations at CA-SMA-18: A Study of Prehistoric Adaptations at Año Nuevo State Reserve.* Far Western Anthropological Research Group, Davis, Davis CA.
Insley, S. J.
2001 Mother–Offspring Vocal Recognition in Northern Fur Seals is Mutual but Asymmetrical. *Animal Behaviour* 61:129–137.
Jackson, J. B. C., M. X. Kirby, W. H. Berger, K. A. Bjorndal, L. W. Botsford, B. J. Bourque, R. H. Bradbury, R. Cooke, J. Erlandson, J. A. Estes, T. P. Hughes, S. Kidwell, C. B. Lange, H. S. Lenihan, J. M. Pandolfi, C. H. Peterson, R. S. Steneck, M. J. Tegner, and R. R. Warner
2001 Historical Overfishing and The Recent Collapse of Coastal Ecosystems. *Science (Washington D. C.)* 293(5530):629–638.
Jones, T. L.
2002 Chapter 6. Archaeology and Prehistory. In *Changes in a California Estuary: A Profile of Elkhorn Slough,* edited by J. Caffrey, M. Brown, B. Tyler and M. Silberstein, pp. 55–91. Elkhorn Slough Foundation, Moss Landing, CA.
Jones, T. L., N. E. Stevens, D. A. Jones, R. T. Fitzgerald, and M. G. Hylkema
2007 The Central Coast: A Multilatitude Milieu. In *California Prehistory: Colonization, Culture, and Complexity,* edited by T. L. Jones and K. A. Klar, pp. 125–146. AltaMira Press, Walnut Creek CA.
Jones, T. L., T. Van Bueren, S. Grantham, J. Huddleson, and T. W. Fung
1996 *Archaeological Test Excavations for the State Highway 1 Widening Project near Castroville, Monterey County, California.* Cultural Studies Office, CALTRANS Environmental Program.
Kurle, C. M., and G. A. J. Worthy
2001 Stable Isotope Assessment of Temporal and Geographic Differences in Feeding Ecology of Northern Fur Seals (*Callorhinus ursinus*) and Their Prey. *Oecologia* 126:254–265.
Lander, R. H.
1981 A Life Table and Biomass Estimate for Alaskan Fur Seals. *Fisheries Research* 1(1981/1982): 55–70.
Liu, Z., J. Kutzbach, and L. Wu
2000 Modeling Climate Shift of El Niño Variability in the Holocene. *Geophysical Research Letters* 27(15):2265–2268.
Lyman, R. L.
2003 Pinniped Behavior, Foraging Theory, and the Depression of Metapopulations and Nondepression of a Local Population in the Southern Northwest Coast of North America. *Journal of Anthropological Archaeology* 22:376–388.
Lyman, R. L., and K. P. Cannon (editors)
2004 *Zooarchaeology and Conservation Biology.* University of Utah Press, Salt Lake City.
Martin, G.
2006 Rare Fur Seals Reclaim Place on Farallon Islands. Animals Fled 1834 Slaughter; Now They're Back and Breeding. *San Francisco Chronical* Monday, September 11, 2006:A. San Francisco.
Milliken, R., J. Nelson, W. R. Hildebrandt, and P. Mikkelsen
1999 *The Moss Landing Hill Site. A Technical Report on Archaeological Studies at CA-MNT-234.* Far Western Anthropological Research Group, Inc.
Moss, M. L., D. Y. Yang, S. D. Newsome, C. F. Speller, I. McKechnie, A. D. McMillan, R. J. Losey, and P. L. Koch
2006 Historical Ecology and Biogeography of North Pacific Pinnipeds: Isotopes and Ancient DNA from Three Archaeological Assemblages. *Journal of Island & Coastal Archaeology* 1:165–190.
National Marine Fisheries Service
2007 Conservation Plan for the Eastern Pacific Stock of Northern Fur Seal *(Callorhinus ursinus).* Protected Resources Division, Alaska Region, U.S. Department of Commerce, Washington DC.
Newsome, S. D., M. A. Etnier, D. Aurioles-Gamboa, and P. L. Koch
2006 Using Carbon and Nitrogen Isotopes to Investigate Reproductive Strategies in Northeast Pacific Otariids. *Marine Mammal Science* 22(3): 556–572.
Newsome, S. D., M. A. Etnier, D. Gifford-Gonzalez, D. L. Phillips, M. van Tuinen, E. A. Hadly, T. P. Guilderson, and P. L. Koch
2007 The Shifting Baseline of Fur Seal Ecology in the Northeast Pacific Ocean. *Proceedings of the*

National Academy of Sciences: Biological/Ecological Sciences 104:9709–9714.
Newsome, S. D., M. A. Etnier, C. M. Kurle, J. R. Waldbauer, C. P. Chamberlain, and P. L. Koch
2007 Historic Declines in Primary Productivity in Western Gulf of Alaska and Eastern Bering Sea: Isotopic Analysis of Northern Fur Sea Teeth. *Marine Ecology Progress Series* 332:211–224.
Newsome, S. D., D. L. Phillips, B. J. Culleton, T. P. Guilderson, and P. L. Koch
2004 Dietary Reconstruction of an Early to Middle Holocene Human Population from the Central California Coast: Insights from Advanced Stable. *Journal of Archaeological Science* 31:1101–1115.
Peterson, R. S., B. J. LeBoeuf, and R. L. DeLong
1968 Fur Seals from the Bering Sea Breeding in California. *Nature* 219:899–901.
Pinsky, Malin L., Seth, D. Newsome, B. R. Dickerson, Y. Fang, Marcel Van Tuinen, Douglas J. Kennett, Rolf R. Ream, and Elizabeth A. Hadly
2010 Dispersal provided resilience to range collapse in a marine mammal: insights from the past to inform conservation biology. *Molecular Ecology* 19(12):2418–2429.
Polovina, J. J., E. Howell, D. R. Koayashi, and M. P. Seki
2001 The Transition Zone Chlorophyll Front, a Dynamic Global Feature Defining Forage Habitat for Marine Resources. *Progress in Oceanography* 49:469–483.
Potenzone, K.
2007 *Analysis of the Faunal Assemblage from CA-SCR-9*. Senior thesis, Anthropology and Earth Science, University of California, Santa Cruz.
Pyle, P., D. J. Long, J. Schonewald, R. E. Jones, and J. Roletto
2001 Historical and Recent Colonization of the South Farallon Islands, California, by Northern Fur Seals (*Callorhinus ursinus*). *Marine Mammal Science* 17(2):297–402.
Ragen, T., G. A. Antonelis, and M. Kiyota
1995 Early Migration of Northern Fur Seal Pups from St. Paul Island, Alaska. *Journal of Mammalogy* 76(4):1137–1148.
Ream, R. R.
2002 *Molecular Ecology of North Pacific Otariids: Genetic Assessment of Northern Fur seal and Steller Sea Lion Distributions*, Ph.D. dissertation, University of Washington, Seattle.
Riddell, R. A.
1955 Archaeological Excavations on the Farallon Islands, California. *Report of the University of California Archaeological Survey* 32:1–18.
Robson, B. W., M. E. Goebel, J. D. Baker, R. R. Ream, T. R. Loughlin, R. C. Francis, G. A. Antonelis, and D. P. Costa
2004 Separation of Foraging Habitat among Breeding Sites of a Colonial Marine Predator, the Northern Fur Seal (*Callorhinus ursinus*). *Canadian Journal of Zoology* 82:20–29.
Savelle, J. M., T. M. Friesen, and R. L. Lyman
1996 Derivation and Application of an Otariid Utility Index. *Journal of Archaeological Science* 23:705–712.
Service, N. W.
2008 Distribution of Precipitation Totals during El Niño Events by Climate Division November–December: 1914 1918 1940 1941 1957 1963 1965 1972 1982 1986 1987 1991 1994, edited by C. P. C. I. Team, vol. 2008. Climate Prediction Center, National Oceanic and Atmospheric Administration, Camp Springs MD.
Sunseri, C. K.
2009 Spatial Economies of Precontact Exchange in the Greater Monterey Bay Area, California, Ph.D. dissertation, Anthropology, University of California, Santa Cruz.
Towell, R. G., R. R. Ream, and A. E. York
2006 Declines in Northern Fur Seal (*Callorhinus ursinus*) Pup Production on the Pribilof Islands. *Marine Mammal Science* 22(2):486–491.
Trainer, V. L., N. G. Adams, B. D. Bill, C. M. Stehr, and J. C. Wekell
2000 Domoic Acid Production near California Coastal Upwelling Zones, June 1998. *Limnology and Oceanography* 45(6):1818–1833.
Trillmich, F.
1990 The Behavioral Ecology of Maternal Effort in Fur Seals and Sea Lions. *Behaviour* 114(1–4):3–20.
Trites, A. W.
1990 Thermal Budgets and Climate Spaces: The Impact of Weather on the Survival of Galapagos (*Arctocephalus galapagoensis* Heller) and Northern Fur Seal Pups (*Callorhinus ursinus* L.). *Functional Ecology* 4:753–768.
Trites, A. W., and P. A. Larkin
1989 The Decline and Fall of the Pribilof Fur Seal (*Callorhinus ursinus*): A Simulation sSudy. *Canadian Journal of Aquatic Sciences* 46: 1437–1445.
Trites, A. W., and A. E. York
1993 Unexpected Changes in Reproductive Rates and Mean Age of First Birth during the Decline of the Pribilof Northern Fur Seal (*Callorhinus ursinus*). *Canadian Journal of Fisheries and Aquatic Science* 50:858–864.
Vladimirov, V. A.
1987 Age-Specific Reproductive Behavior in Northern Fur Seals on the Commander Islands. In *Status, Biology, and Ecology of Fur Seals*, edited by J. P. Croxall and R. L. Gentry, pp. 113–120, vol.

51, NOAA Technical Report National Marine Fisheries Service.
Washington Department of Fish and Wildlife Management Program
1997 Forage Fish Information: Pacific herring, *Clupea harengus pallasi* In: Koenings, JP (ed). Washington Department of Fish and Wildlife, Olympia, WA.
York, A. E.
1983 Average Age of First Reproduction of the Northern Fur Seal (*Callorhinus ursinus*). *Canadian Journal of Aquatic Sciences* 40:121–127.
1991 Sea Surface Temperatures and Their Relationship to the Survival of Juvenile Male Northern Fur Seals from the Pribilof Islands. In *Pinnipeds and El Niño. Responses to Environmental Stress*, edited by F. Trillmich and K. A. Ono, pp. 94–111. Springer-Verlag, New York.
Zeppelin, T. K., and R. R. Ream
2006 Foraging Habitats Based on the Diet of Female Northern Fur Seals (*Callorhinus ursinus*) on the Pribilof Islands, Alaska. *Journal of Zoology* 270:565–576

11

Toward a Prehistory of the Southern Sea Otter (*Enhydra lutris nereis*)

Terry L. Jones, Brendan J. Culleton, Shawn Larson, Sarah Mellinger, and Judith F. Porcasi

It's clear that within the sea otter's stabilized foraging range there can be virtually no human harvest of abalones (*Haliotis* spp.) (except for a few taken intertidally).

D.J. MILLER, 1974, CALIFORNIA DEPARTMENT OF FISH AND GAME MARINE RESOURCES LEAFLET NO. 7

THE SOUTHERN SEA OTTER (*Enhydra lutris nereis*) is one of the mostly widely recognized and highly cherished marine mammals on the coast of California. In seaside communities up and down the state, images of sea otters are ubiquitous on T-shirts, coffee mugs, and bumper stickers. Tourists flock in droves to watch otters from cliffs and jetties, and to peer at them underwater at the Monterey Aquarium. Not surprisingly, scientific research on sea otters has been commensurate with this interest, and much is known about their basic biology, behavior, and ecology.

The prehistory of sea otters, however, is much less well understood and has been the subject of longstanding speculation by fishermen, biologists, and California Fish and Game representatives. Sea otters are a keystone predator in kelp forests along the central California coast, and they are voracious consumers of shellfish. Since their return from the brink of extinction early in the 20th century, their impact on shellfish populations has been obvious, and the most robust populations of abalone are found only in areas where otters have not reestablished their populations. This pattern prompted the statement by the California Department of Fish and Game quoted above.

Nonetheless, casual observations of the archaeological record and ethnohistoric accounts of Native people indicate that sea otters, abalone, and humans coexisted before the arrival of Euro-Americans in California in 1769 (Davis 1981; Walker 1982). Along the central coast (between San Francisco Bay and Point Conception) where sea otters have reestablished their populations, archaeological research was not sufficiently advanced in the 1970s (with some key exceptions [e.g., Greenwood's 1972 study at Diablo Canyon]) to allow for anything other than informed speculation about the relationships between abalones, sea otters, and humans. In the last 2 to 3 decades, however, hundreds of archaeological sites have been excavated in this region, including many that contain the remains of abalone and sea otters (for a history of this work see Jones et al. 2007). A regional culture history extending back to 8000 cal BC is now well established based on over 1000 radiocarbon dates. Here we rely on this dataset to develop a preliminary outline of sea otter prehistory for the central California coast that emphasizes the ecological relationships between key predators and their prey. First, we summarize some of the important traits of this animal with respect to its suitability as a prey item. We then summarize the spatial and temporal distribution of archaeological otter remains with particular regard to diachronic variability. Finally, we've supplemented this basic historical and biological information with the results of two specialized studies on otter remains: DNA analysis that reveals the sex of the otters that were exploited by Native Californians, and isotope analysis that provides insights into their diet.

THE SOUTHERN SEA OTTER (*ENHYDRA LUTRIS NEREIS*) AS PREY • *Biology, Ecology, History*

The sea otter is the smallest marine mammal on the planet, and its unique characteristics and history of near extinction and reemergence make it particularly interesting. Sea otters are members of the Mustelidae family, which bear and nurse live young. Three geographically segregated subspecies are generally recognized within *Enhydra lutris*: *E. lutris lutris*, *E. lutris gracial*, and, *E. lutris nereis* (Kenyon 1981; Nickerson 1984; Riedman and Estes 1990). *E. lutris lutris* (the Asian sea otter) is found in the western Pacific, and *E. lutris gracial* (also *E. lutris kenyoni*), the Alaskan sea otter, is found in the Aleutian Islands and along the Alaskan mainland coast. The focus of this paper, *Enhydra lutris nereis* (the California or southern sea otter), inhabits the central coast of California.

BIOLOGY

According to Riedman and Estes (1990), adult male sea otters weigh 34% more than their female counterparts, and males are 8% longer than females. An average sea otter pup weighs about 5 pounds and is 22 to 24 inches long; fully mature male sea otters weigh 60 to 85 pounds and are about 58 inches in length, whereas females only weigh 35 to 60 pounds and are 55 inches long (Kenyon 1981). Their forelimbs have retractable claws and sensitive pads that allow the animal to accurately find and consume prey. The sea otter swims slowly compared to other marine mammals and travels through the water at 1.5 nautical miles per hour (Kenyon 1981).

The sea otter's coat is unlike that of any other marine mammal, as it is unusually dense. The denseness creates a layer of trapped air that provides warmth for the animal, which is essential because sea otters lack the fatty blubber that most other marine mammals depend on (Kenyon 1981; Riedman and Estes 1991). This layer of trapped air also provides the sea otter with extra buoyancy. The pelage of an adult can range in color from light brown to nearly black, and pups are born with a "woolly" coat that is replaced within the first few months (Kenyon 1981). It was the otter's pelt that was the object of commercial exploitation in the 18th and 19th centuries. It was also a main focus of indigenous hunting before that; ethnohistoric accounts of both the Costanoan (Levy 1978:493) and Chumash (Greenwood 1978:523) describe use of otter skins for blankets and robes, and as important trade items. Their use for these purposes is in fact much more heavily attested to than their exploitation as a food source, but it is unlikely that California Indians did not also take advantage of the animal's meat.

REPRODUCTION

Sea otters can breed throughout the year; however, Riedman and Estes (1991:59) suggest that "the general yearly reproductive pattern in California . . . consists of a winter pupping season and a summer-fall breeding season." Males and females often form pair bonds, but this is not necessary for breeding and is often not observed in California (Riedman and Estes 1991; Woodhouse et al. 1977). The mating male and female spend one to three days participating in daily activities together, and the female often retreats ending any subsequent interactions (Kenyon 1981; Riedman and Estes 1991; Woodhouse et al. 1977). It should be noted that the reproduction patterns of Alaskan sea otters are better understood because most information gathered on otters in California is based on a limited number of occurrences.

Californian male sea otters likely reach sexual maturity around 5 years, but territoriality occurs around 8 years; males pursue females in estrus and copulation takes place in the water (Riedman and Estes 1991). Males appear to be able to reproduce until death (Riedman and Estes 1991). A male's territorial boundaries change seasonally depending on food resources, security of coastline (e.g., storm protection and available kelp), and the number of reproductive females (Riedman and Estes 1991). Although males have been observed interacting with pups, they do not participate in rearing the young (Kenyon 1981; Riedman and Estes 1991).

Californian female sea otters reach sexual maturity at approximately 4 to 5 years, with estrus lasting about 3 to 4 days, but information on the frequency of estrus is limited (Riedman and Estes 1991). Sea otters have single-pup births (mostly in the water, but land births have been observed), and California sea otters typically give birth to one offspring per year (Kenyon 1981; Riedman and Estes 1991). Gestation length is still under investigation, as Kenyon (1981) asserts gestation lasts for 8 months, while Riedman and Estes (1991:66) cite several sources that indicate gestation requires only 4 to 6 months. Sea otter pups are raised solely by the female and are dependent on their mother for survival for several months (Kenyon 1981). After giving birth, sea otters temporarily separate themselves from other females; after a few days, they accompany other females with dependent pups (Riedman and Estes 1991).

BEHAVIOR

Sea otters often spend time in groups commonly referred to as rafts. When in rafts, sea otters mostly rest, groom, and tend to pups. Foraging, mating, and parturition all occur away from rafts, and once these activities are completed, otters return to their group. Female rafts are smaller than male groupings; female rafts have been observed to typically contain two to a dozen members, whereas male rafts are typically larger, with hundreds of otters having been observed grouped together (Riedman and Estes 1991). Otters are relatively social mammals and engage in a variety of close interactions.

Male and female sea otters spend most of their time separated and segregated into single-sex areas; however, Riedman and Estes (1991:53) suggest that in "California, differences in the degree of exposure to rough sea conditions and availability of food resources characterizing established male and female areas are less pronounced than they are in Alaska." Juvenile females may occupy areas near or in male rafts, and juvenile males maintain independence and isolation farther offshore for most of their activities. Territorial adult male otters remain near the periphery of female rafts, and contact is sought for mating. Within the sea otter's habitat range, females and dependent pups occupy the center area and males occupy the fringe to take advantage of "unoccupied habitat" to expand their territory; however, throughout the season territorial males travel from the periphery to the center in search of a mate (Riedman and Estes 1991:53).

All sea otters exhibit haul-out behavior in which the sea otter takes refuge on exposed rocks or beaches. Common haul-out locations are composed of "low-relief, algal covered rocks" and, less frequently, "sand and cobble beaches" (Riedman and Estes 1991:24). Otters usually haul out in small groups of up to six, and during estrus females frequently haul out (Riedman and Estes 1991). Southern sea otters haul out less often than their northern counterparts, a result of near-shore human contact (Riedman and Estes 1991; Woodhouse et al. 1977).

ECOLOGY

HABITAT

California sea otters predominantly occupy areas with rocky bottoms, less frequently inhabiting soft sandy bottom areas (Riedman and Estes 1991; Woodhouse et al. 1977), including estuaries. It seems that sea otters prefer areas with dense kelp, but it is not a habitat requirement. California sea otters play a complex role in their environment, influencing many other species, and thus are known as a keystone species (Riedman and Estes 1991). Sea otter predation can stabilize nearshore ecological communities by limiting invertebrate species (e.g., sea urchins, *Strongylocentrorus* spp.) that, void of limitation, would destroy marine plants (e.g., kelp), which provide habit for benthic fish species (Riedman and Estes 1991; Simenstad et al. 1978; Woodhouse et al. 1977). Sea otters do not migrate, but rather maintain residence within a 5-to-10-mile territory along the coastline (Kenyon 1981).

PREDATION AND PREDATORS

While most carnivorous animals have sharp teeth, sea otter teeth are flat and blunt, and are ideal for crushing the shells of their invertebrate prey (Kenyon 1981). Sea otters typically specialize on about three species (sea urchins, abalones, and rock crabs); as these species become sparse, otters prey on turban snails, kelp crabs, mussels, and other intertidal species (Riedman and Estes 1991; Woodhouse et al. 1977). A sea otter's diet may also include fish, with pursuit of fish usually limited to males who possess the required strength to catch and kill this prey (Kenyon 1981). The sea otter's dexterity and collection techniques allow them to adapt and exploit a changing environment. An average adult sea otter must eat 23 to 33% of its body weight daily (Riedman and Estes 1991).

California sea otters dive to depths of 25 m while foraging for food, and, on average, each dive lasts for around a minute, but they can dive to greater depths and stay underwater longer when they are being hunted. Males and females maintain different foraging strategies; subadult males forage with deeper and lengthier dives farther from shore, while females with pups foraged closer to shore with quicker and shallower dives (Riedman and Estes 1991).

Food stealing among sea otters is commonplace. Adult females (mostly females with dependent pups) are the most frequent targets. When a female enters a foraging area near a territorial male, the male often steals her catch; the female easily complies (Riedman and Estes 1991). Pair-bonded males frequently steal food from females, and the female is known to offer prey to the pair-bonded male (Riedman and Estes 1991). Worthy of note is a male tactic termed "hostage behavior" in which a male seizes an otter pup while the female is diving, and essentially the pup is exchanged for the female's collected prey (Riedman and Estes 1991). Males are not the only perpetrators, as females also occasionally steal from one another. Female thieves use a different strategy. Males steal any prey that is caught, but females are more selective and target one species at a time (Riedman and Estes 1991).

Killer whales and great white sharks prey upon sea otters, and otter pups sometimes fall victim to bald eagles, but all occur relatively infrequently. Humans are the only known population-limiting species (Riedman and Estes 1991; Walker 1982; Woodhouse et al. 1977).

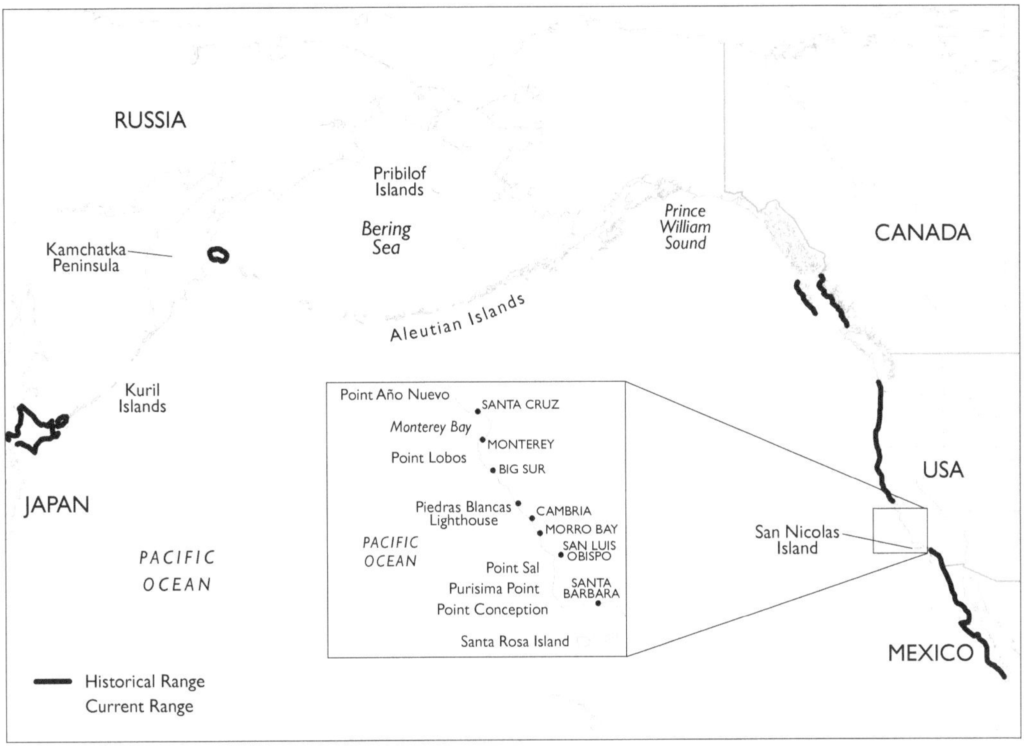

FIGURE 11.1. Distribution of sea otters in the Pacific (current range was part of the historic range).

HISTORY

The history of market-based overexploitation of the sea otter, which nearly rendered the species extinct by the end of the 19th century, was exceptionally well documented by Ogden (1941). Before and during the fur trade, sea otters maintained an expansive, contiguous territory throughout the North Pacific that extended from the northern Islands of Japan to the southern shores of Baja California (Figure 11.1). The more restricted distribution of sea otters seen today is a direct result of the intensive hunting of the animals for pelts during the 19th century by the Spanish, Mexicans, and most significantly, the Russians, who established Fort Ross on the coast of Sonoma County in northern California as a base of operations for marine mammal hunting. The numbers of marine animals caught during the fur trade in the Pacific Ocean are staggering. The Russian-American Company reported their colonies exported 72,894 sea otters and over 1.2 million fur seals from 1797 to 1821 and another 25,899 sea otters and 372,894 fur seals from 1842 to 1862 (Tikhmenev 1978 [1888]). Gibson (1992) reports even more astronomical numbers, stating that from 1804 to 1837, American vessels alone imported 158,070 sea otter pelts to the Asian market at Canton. Of course, the California fur trade was only a portion of the overall enterprise. The sea otter fur trade was most productive between 1801 and 1819 at the height of the Russian-American partnership. During those two decades, nearly 80% of California sea otters were taken. Nonetheless, numbers of sea otters caught in California are staggering in their own right. According to Ogden (1941), fur traders caught over 45,000 sea otters along California's coast, estuaries, and islands in just over six decades. This figure serves as conservative estimate since it does not include contraband and native catches.

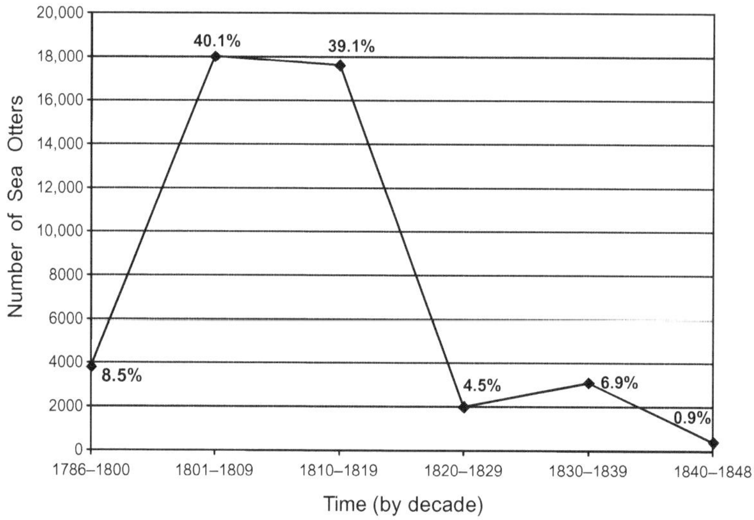

FIGURE 11.2. Total sea otters caught in California during the fur trade from 1786 to 1848 with percentages of the total harvested population (43,035) from Ogden (1941).

Today, sea otters have been extirpated south of Prince William Sound to Baja California with the exception of the central California coast, where they maintain residence from Point Año Nuevo above Santa Cruz in the north to Point Conception near Santa Barbara in the south. Riedman and Estes (1991) argue the southern range limit seems to be near the Santa Maria River in San Luis Obispo County. Sea otters also live near San Nicolas Island off the coast halfway between Santa Barbara and Los Angeles, where they are part of a conservation plan to reestablish the population.

The California Department of Fish and Game and the U.S. Fish and Wildlife Service estimate that the pre–fur trade population of otters in California was between 16,000 and 20,000 animals. Estimates for the entire population in the Pacific generally range between 150,000 and 300,000. According to Ogden (1941), historic fur trade records indicate sea otter populations were greatest in central California from San Francisco Bay to the Santa Barbara Channel Islands. Records of the fur trade exploitation show a precipitous decline in numbers of animals taken over the course of the 19th century, reflecting extreme overexploitation (Figure 11.2). By 1914, only 30 to 50 otters were thought to remain in California waters. In 1938, approximately 50 otters were discovered off the coast of Monterey, and these have been credited with the reemergence of California's sea otter. Riedman and Estes (1991) calculate that from 1914 to 1984 the California sea otter population and range increased 5% annually. Currently, scientists conduct a census of sea otters twice a year, and California's sea otter population for 2007 was 3,026, which is the highest number recorded since the census began in 1982 (Figure 11.3).

PRE–FUR TRADE HUMAN-OTTER INTERACTION

ETHNOHISTORY

Historic records and ethnographic accounts indicate that sea otters were hunted with various techniques and technology. According to Ogden, people along the coast of California were known to hunt sea otters by the following technique:

> When the parent otter left its young on the surface of the water, which it did only when it dived for food, the Indian hunter would slip up and tie a cord to the foot of the baby. Fastened to the cord, close to the body of the ani-

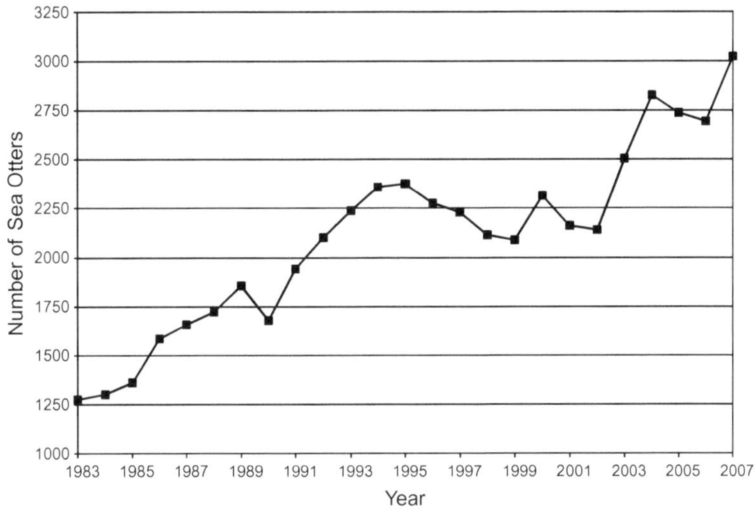

FIGURE 11.3. California sea otter census from 1983 to 2007. Source: Modified from USGS and Western Ecological Research Center 2007.

> mal, would be placed a couple of fishhooks. Retiring in his canoe to a short distance, the Indian would pull his cord and thus hurt the small otter so that he would cry. The mother would rush to the rescue and could be easily approached, either because she was occupied in freeing her offspring or because she herself would become caught in the line and hooks. (Ogden 1941:14)

Woodhouse et al., citing La Perouse (1799), refer to another hunting technique employed by Native Californians:

> The Indians whose boats at Monterey are only made of reeds, catch them on land with snares or by knocking them down with large sticks when they find them at a distance from the shore; for this purpose, they keep themselves concealed among the rocks, for this animal is frightened at the least noise and immediately plunges into the water. (1977:56–57)

Ogden (1941:14) also argues that hunting in California was often done without chasing the sea otter with watercraft; she states, "In California and certain places in the North Pacific, nets were spread out on the kelp beds, snares, and clubs were used." Ogden (1941:142) even refers to an account in which three or four Native Americans caught 30 sea otters by lassoing them while the otters were hauled out on shore. In the latter part of the fur trade, guns were used to hunt sea otters. Ogden (1941:145) reported a typical hunt included three men (two rowers and one shooter) in each of three canoes. The canoes would create a triangle formation, and, in form, the team would follow a fleeing otter and shoot at it each time it came up for air.

Sea otter pelts were clearly important exchange items in Native California. Driver and Massey (1957:376) documented 55 materials that were traded in California and ranked them in order of importance. Hide or furs, tied with salt, were the second-most-important commodity. Jones (1996) also argued that sea otter pelts were a valued trade commodity and noted a tentative correlation between the frequency of otter bones and obsidian in archaeological sites.

ARCHAEOLOGY AND PREHISTORY OF SEA OTTER EXPLOITATION

Although lacking a substantive dataset, some important observations were made by archaeologists in the 1970s and 80s about possible relations among humans, sea otters, and abalone during the prehistoric era. Davis (1981) noted that midden sites suggest California Indians harvested large abalones at the same time that healthy populations of sea otters existed. She made this point specifically in reference to the view of California Fish and Game biologists that the presence of one species precluded the existence of large populations of the other. Walker (1982) discussed many of the same issues but suggested that human exploitation of otters would have changed the ecological structure of nearshore communities in that prior to the arrival of humans, otter populations would have been limited by available food (a "bottom-up" situation), but that after the arrival of humans their populations were controlled by human predation. Walker (1982) discussed some ethnographic and archaeological findings, but concluded, like Davis (1981), that significant archaeological information was not yet available.

Around the same time, findings from Alaska demonstrated the importance of archaeological data for defining the long-term ecological relationships between species. A midden site on the Aleutian Islands showed a stratigraphic sequence in which the basal layer produced abundant remains of sea otters and an overlying layer exhibited a dearth of otter remains and higher frequencies of sea urchins and remains of other shellfish (Simenstad et al. 1978). This transition suggested that hunting of otters represented in the basal layer led to near elimination of the species from the waters surrounding the island which, in turn, encouraged the expansion of sea urchin populations. The uppermost layers investigated by Simensted et al. (1978) showed a return of otters, a renewal of their hunting, and a decrease in urchin remains. Overall, these findings clarified the keystone role of otters in kelp forest habitats: when otters are present, they reduce populations of sea urchins that compete for space on the bottom with kelp. If sea otters are overharvested, their removal from these systems can be followed by marked increases in sea urchins and reduction in kelp forests. The relationships recognized in this study establish the blueprint for investigating diachronic variability in otter exploitation elsewhere.

More recently, Hildebrandt and Jones (1992) used an optimal-foraging model to argue that sea otters were highly elusive prey which would have been more lowly ranked than larger, less mobile marine taxa (e.g., seal and sea lions in rookeries). They argued that Native Californians would have turned to sea otters (and harbor seals) only after populations of the more vulnerable, highly ranked terrestrially breeding seals and sea lions (e.g., elephant seals, northern fur seals, Guadalupe fur seals, California sea lions and Steller sea lions) decreased as a result of overexploitation. They used faunal data from over 40 sites in California and Oregon to demonstrate this trend. For the central coast, data from 20 sites generally showed early (before 1000 BC) exploitation of terrestrial breeders (otariids and elephant seals) and later (after AD 1000) exploitation of sea otters. (Hildebrandt and Jones 1992) defined terrestrial breeders as pinnipeds who use terrestrial settings to accomplish breeding. The category included otariids (fur seals and sea lions) and elephant seals. Because these species generally migrate substantial distances and occupy breeding sites only seasonally, the term "migratory breeders" was also used to refer to this group. In central coast archaeological collections, however, remains of terrestrial breeders are almost entirely limited to otariids; only two elephant seal bones have been recovered from the region. Otariid is considered synonymous with terrestrial breeders in this chapter. The oldest evidence for sea otter exploitation in this study came from CA-MNT-229 at Elkhorn Slough, where a component dating ca. 4000 cal BC yielded 17 sea otter bones (Hildebrandt and Jones 1992).

Since then, the antiquity of sea otter exploitation has been more firmly established along the central and southern coasts based largely on findings from the Channel Islands, where Erlandson et al (2005) document sea otter exploitation as early as ca. 7000 cal BC. On the mainland, the Duncans Landing Rockshelter in Sonoma County (CA-SON-348/H; Kennedy et al. 2005; Schwaderer 1992; Simons and Wake 2000) and the Diablo Canyon site in San Luis Obispo County (CA-SLO-2; Jones, Porcasi, Gaeta et al. 2008) have produced sea otter remains from contexts dating from 7000 to 5000 cal BC. Both the island and mainland records are still limited by small samples sizes, and it is likely that the antiquity of otter exploitation will eventually be pushed back coeval with the earliest exploitation of marine shellfish and fish on the islands (ca. 10,000 cal BC) and mainland (8000 cal BC). For now, the record of sea otter exploitation begins at 7000 cal B.C. Given the antiquity of unequivocal evidence for watercraft use on the Channel Islands (ca. 11,000 to 10,000 cal BC), it is reasonable to assume that boats of some type were available for the pursuit of marine animals on the mainland at this time.

Archaeological sites that have produced otter remains are fairly common on the central coast; no fewer than 46 sites have produced at least a single otter bone, and 19 have produced more than ten (Table 11.1; Figure 11.4). A total of 1194 otter bones has been reported, although only one site, CA-SMA-115 (north of Año Nuevo), produced an assemblage dominated by otter bones (57.3%). At most other sites, deer or rabbit bones tend to be most abundant. The mean representation of otters for the central coast as a whole is 8.99%. The largest collection of otter bones (NISP = 421; 14.1%) comes from CA-SLO-2 at Diablo Canyon, which also provides the longest sequence of otter exploitation (ca. 5000 cal BC to cal AD 1800; Jones, Porcasi, Gaeta et al. 2008). Putting aside small samples (mammal and bird NISP < 100), three other sites produced greater than 14% sea otter remains: CA-MNT-101, CA-MNT-391, and CA-SLO-832/1420. Overall, sites with the highest frequency of otter remains are found on exposed headlands, including the Monterey Peninsula, the Pecho Coast, and the San Simeon area. Some sites at the Elkhorn Slough estuary have also produced substantial samples of otter bones. It should be noted that middens in the San Francisco Bay area have also produced substantial quantities of otter remains (see, for example, Broughton 1999; Simons 1992), but the focus of the current study is the open coast where otters are still found today. Otter bones have also been recovered as far inland as the upper Carmel Valley (Breschini and Haversat 1992), but, not surprisingly, their frequency at inland sites is generally very low since there would have been no reason for bones to accompany pelts during forays to inland settlements or along exchange routes.

Of course, diachronic comparisons are ultimately more meaningful (Table 11.2; Figure 11.5). Fortunately, the corpus of available data on marine mammal frequencies is now substantial enough that such comparisons can be made with reasonable confidence. Overall, the record from the region shows clearly that sea otters were the preferred marine mammal along central coastal California throughout the Holocene contradicting the earlier assessment by Hildebrandt and Jones (1992) based on a more limited sample. While central coast middens generally exhibit a certain degree of mixing due to the impacts of small burrowing animals, diachronic trends can still be recognized with a reasonable degree of certainty. During the Milling Stone Period (5000–3500 cal BC), there was little interest in marine mammals in general; sites at Elkhorn Slough, Morro Bay, and the open coast have produced low frequencies of otter bones (5.4% of all mammals and birds) and very few remains of otariids and elephant seals (terrestrial breeders). Sea otters outnumber remains of these larger marine taxa during this period by a ratio of nearly 3:1. This situation changed significantly after 3500 cal BC with the onset of the Hunting Culture, when there was a marked increase in marine mammal hunting in general and otariids in particular. Otters were still the preferred marine prey overall (increasing to 11.1% of regional NISP), but the larger otariids show a fivefold increase to 9.9%. This trend continued through the Middle Period (600 cal BC to cal AD 1250), when the greatest number of otter (NISP = 704) and otariid bones (NISP = 560) occur. It is important to acknowledge that significant spatial variability is also apparent during this period; sites near Año Nuevo (CA-SMA-218), Elkhorn Slough (CA-MNT-234), and the Monterey Peninsula (CA-MNT-115) produced assemblages

TABLE 11.1

*Archaeological Sites with Sea Otter (*Enhydra lutris*) Bones on the Central California Coast*

TRINOMIAL	LOCATION	TOTAL NISP[a]	SEA OTTER NISP	%	REFERENCE
CA-SMA-115	Año Nuevo	96	55	57.29	Hylkema 1991
CA-MNT-391	Monterey Peninsula	241	54	22.40	Cartier 1993
CA-MNT-170	Monterey Peninsula	36	8	22.22	Dietz 1991
CA-SLO-71	San Simeon	51	11	21.60	Joslin 2006
CA-MNT-1942	Big Sur	69	13	18.84	Wolgemuth et al. 2002
CA-SLO-832/1420	Pismo Beach	124	22	17.74	D. Jones et al. 2004
CA-MNT-101	Monterey Peninsula	384	64	16.70	Dietz 1987
CA-SCR-7	Santa Cruz coast	36	6	16.66	D. Jones and Hildebrandt 1990
CA-SLO-585	Pecho Coast	49	8	16.32	T. Jones et al. 2009
CA-MNT-116	Monterey Peninsula	28	4	14.30	Dietz and Jackson 1981
CA-SLO-2	Pecho Coast	3049	431	14.13	T. Jones et al. 2008
CA-SLO-9	Pecho Coast	139	18	12.94	Codding and Jones 2007
CA-MNT-228	Elkhorn Slough	283	35	12.40	Jones et al. 1996
CA-SLO-179	San Simeon	631	70	11.10	Jones and Ferneau 2002
CA-SCR-132	Santa Cruz Coast	28	3	10.71	Hylkema 1991
CA-SLO-165	Morro Bay	307	33	10.75	Mikkelson et al. 2000
CA-MNT-234	Elkhorn Slough	365	37	10.13	Milliken et al. 1999
CA-SMA-18	Año Nuevo	683	67	9.80	Hildebrandt et al. 2006
CA-MNT-831	Monterey Peninsula	290	28	9.65	Breschini and Haversat 2006
CA-SLO-274	San Simeon (?)	94	9	9.60	Hildebrandt et al. 2002
CA-SMA-97	Año Nuevo	84	8	9.52	Hylkema 1991
CA-MNT-1765	Elkhorn Slough	11	1	9.10	Fitzgerald et al. 1995
CA-MNT-63	Big Sur	154	14	9.10	Jones 2003
CA-MNT-229	Elkhorn Slough	1013	89	8.80	Dietz et al. 1988; Jones and Jones 1992
CA-SMA-118	Año Nuevo	103	9	8.73	Hylkema 1991
CA-SCR-93	Santa Cruz Coast	71	6	8.45	Bourdeau 1986

TABLE 11.1 *(continued)*

TRINOMIAL	LOCATION	TOTAL NISP[a]	SEA OTTER NISP	%	REFERENCE
CA-MNT-1570	Elkhorn Slough	147	12	8.16	Jones et al. 1996
CA-MNT-115	Monterey Peninsula	74	6	8.11	Dietz and Jackson 1981
CA-SLO-273/H	San Simeon	14	1	7.14	Hildebrandt et al. 2002
CA-SLO-2357	Pismo Beach/ Halcyon Bay	65	4	6.20	Jones and Mikkelsen 2006
CA-MNT-112	Monterey Peninsula	114	6	5.30	Dietz and Jackson 1981
CA-MNT-73	Big Sur	49	2	4.10	Jones 2003
CA-SLO-267	San Simeon	308	12	3.90	Jones and Ferneau 2002
CA-MNT-238	Big Sur	862	24	2.78	Mikkelson et al. 2005
CA-SMA-218	Año Nuevo	197	5	2.53	Hylkema 1991
CA-MNT-1277/H	Big Sur	124	3	2.41	Jones 2003
CA-SLO-175	San Simeon	49	1	2.04	Jones and Waugh 1989
CA-MNT-108	Monterey Peninsula	305	4	1.31	Breschini and Haversat 1989
CA-MNT-1233	Big Sur	238	2	0.84	Jones 2003
CA-MNT-1227	Big Sur	126	1	0.79	Jones 2003
CA-MNT-1223	Big Sur	119	1	0.84	Jones 2003
CA-MNT-1232/H	Big Sur	151	1	0.66	Jones 2003
CA-SLO-215	Morro Bay	165	1	0.60	T. Jones et al. 2004
CA-MNT-1486/H	Monterey Peninsula (inland)	485	2	0.41	Breschini and Haversat 1995
CA-SCR-9	Santa Cruz Coast	619	2	0.32	Hylkema 1991
CA-MNT-1485/H	Monterey Peninsula (inland)	646	1	0.15	Breschini and Haversat 1995
Totals		13276	1194	8.99[b]	

[a] All nonfish animal bones except intrusive and ground-burrowing species (e.g., pocket gophers) identified to the genus level or better.
[b] Mean.

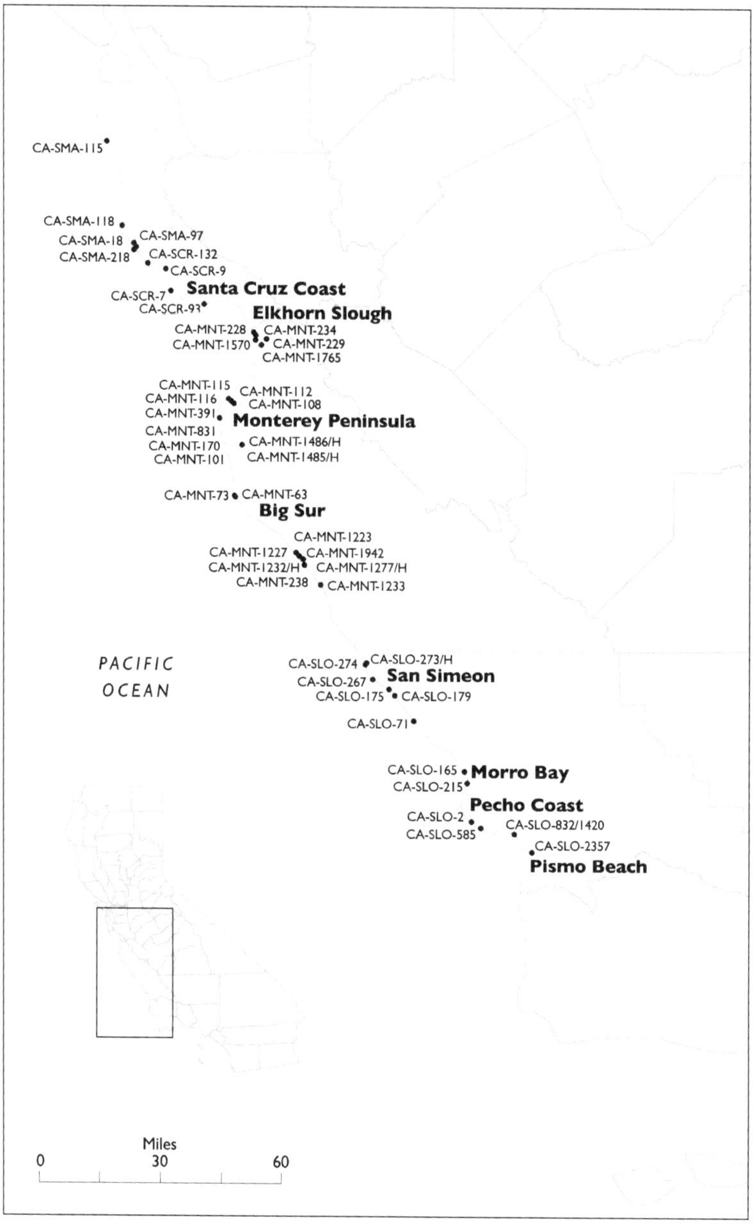

FIGURE 11.4. Archaeological sites with sea otter bones on the central California coast.

with heavy proportions (25–72%) of otariids and few otter bones,while other sites at Elkhorn Slough (CA-MNT-228) and the Pecho Coast (CA-SLO-2) show high frequencies of otters (14–17%) and few remains of the larger otariids. The bones at the majority of these sites represent northern fur seals (*Callorhinus ursinus*), and it is fairly apparent that this species had a greater onshore presence (e.g., breeding colonies) during the Middle Period near Año Nuevo, Elkhorn Slough, and the Monterey Peninsula than it does today (see discussion by Gifford-Gonzalez, Chapter 10).

With the end of the Hunting Culture and the onset of the Late Period ca. cal AD 1250, trends in otter and otariid remains diverge dramatically; otter bones continue to increase, while the remains of large otariids and elephant

TABLE 11.2

Sea Otter and Other Marine Mammal Remains through Time on the Central California Coast

COMPONENT	TOTAL MAMMAL AND BIRD NISP	SEA OTTER NISP	%	OTARIIDS NISP	%
MILLINGSTONE (5000–3500 BC)					
CA-MNT-228 Area B	3	0	0.00	0	0.00
CA-MNT-229	155	17	10.96	7	4.51
CA-MNT-1232/H	31	0	0.00	2	6.45
CA-SLO-2 Component II	419	17	4.06	5	0.12
CA-SLO-165	26	1	3.84	0	0.00
CA-SLO-215	165	1	0.60	0	0.00
CA-SLO-585	45	8	17.80	1	2.22
CA-SLO-832/1420	69	5	7.25	0	0.00
CA-SLO-2357	65	4	6.20	4	6.20
Total	978	53	5.41	19	1.94
HUNTING (EARLY PERIOD: 3500–600 BC)					
CA-MNT-73	49	2	4.08	1	2.04
CA-MNT-108	305	4	1.31	16	5.25
CA-MNT-116	28	4	14.30	5	17.86
CA-MNT-170	36	8	22.22	15	41.66
CA-MNT-234	124	22	17.74	28	22.58
CA-MNT-391	241	54	22.40	32	13.27
CA-MNT-831	290	28	9.65	29	10.00
CA-SCR-7	36	6	16.66	3	0.83
CA-SLO-165	263	26	9.88	13	4.94
CA-SLO-274	94	9	9.60	5	5.32
CA-SLO-273/H	14	1	7.14	0	0.00
Total	1480	164	11.08	147	9.93
HUNTING (MIDDLE: 600 CAL BC TO CAL AD 1250)					
CA-MNT-63	154	14	9.10	21	13.63
CA-MNT-101	384	64	16.70	39	10.15
CA-MNT-115	74	6	8.11	49	66.21
CA-MNT-229	669	90	13.45	20	2.98
CA-MNT-234	97	8	8.24	25	25.77
CA-MNT-238	804	23	2.98	38	4.72

(continued)

TABLE 11.2 *(continued)*

COMPONENT	TOTAL MAMMAL AND BIRD NISP	SEA OTTER NISP	%	OTARIIDS NISP	%
CA-MNT-1233	238	2	0.84	2	0.84
CA-SCR-9	619	2	0.32	11	1.77
CA-SCR-93	71	6	8.45	1	1.41
CA-SCR-132	28	3	10.71	3	10.71
CA-SLO-2 Component III	1638	278	16.97	20	1.22
CA-SLO-9	139	18	12.24	2	1.43
CA-SLO-165	2	0	0.00	1	50.00
CA-SLO-175	49	1	2.04	0	0.00
CA-SLO-179	631	70	11.10	48	7.60
CA-SLO-267	308	12	3.90	19	6.16
CA-SMA-18	683	67	9.80	112	16.39
CA-SMA-218	197	5	2.53	142	72.08
Total	7030	704	10.01	560	7.96
LATE					
CA-MNT-112	114	6	5.30	7	6.14
CA-MNT-234	25	2	8.00	6	24.00
CA-MNT-1223	119	1	0.84	0	0.00
CA-MNT-1227	126	1	0.79	0	0.00
CA-MNT-1277/H	124	3	2.41	1	0.80
CA-MNT-1765	11	1	9.10	0	0.00
CA-MNT-1942	69	13	18.84	0	0.00
CA-SLO-2 Component IV	688	136	19.77	17	2.47
CA-SLO-165	4	0	0.00	1	25.00
CA-SLO-71	51	11	21.60	4	7.84
CA-SMA-97	84	8	9.52	12	14.28
CA-SMA-115	96	55	57.29	6	6.25
CA-SMA-118	103	9	8.73	6	5.82
Total	1614	246	15.24	60	3.71

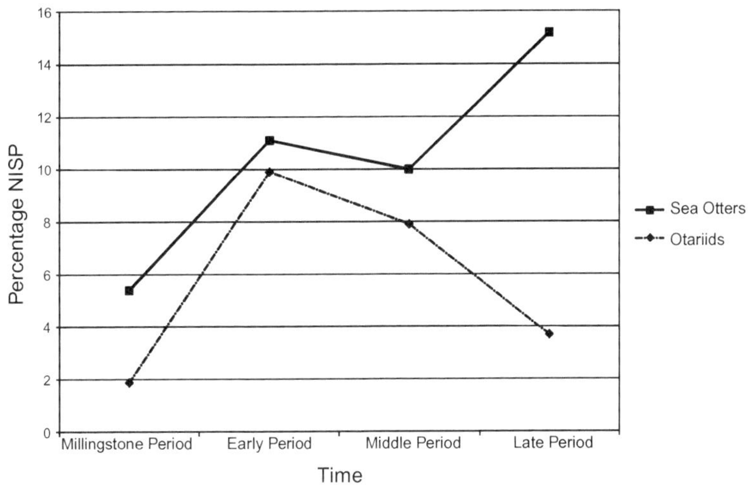

FIGURE 11.5. Sea otter and terrestrial breeder remains (northern fur seal, Guadalupe fur seal, California sea lion, Steller sea lion, and elephant seals) over time along the central California coast based on percentage of identified bird and mammal remains (NISP).

seals decrease. Regional variability is apparent, as some sites on the Pecho Coast (CA-SLO-2) and near Año Nuevo (CA-SMA-115) produced substantial quantities of otter remains (19–57% NISP) while others at Big Sur (CA-MNT-1223 and -1227) yielded almost none. The overall regional increase runs counter to a trend described by Jones (1996) based on preliminary findings from Big Sur where a decrease in otter bones during the Late Period (ca. AD 1250–1769 give approximate dates) was thought to reflect decreased interregional trade. A subsequent excavation in Big Sur by Wolgemuth et al. (2002) produced previously underrepresented otter bones, offsetting the earlier pattern.

Both the combined regional record and the large collection from CA-SLO-2 show the same relative pattern: otters were the most heavily exploited marine mammal over time and their exploitation increased incrementally through the Holocene. That these two datasets are mutually concordant suggests that the overall pattern is empirically legitimate and provides some justification for using data from CA-SLO-2 to represent the region as a whole. Several aspects of the patterning in these data also suggest they are the products of incremental human population growth over the course of the Holocene. At CA-SLO-2, the increase in sea otter bones correlates inversely with the disappearance of the flightless sea duck (*Chendytes lawi*; Figure 11.6), which was caused by overexploitation throughout the duck's range over an 8000-year period (Jones, Porcasi, Erlandson et al. 2008). Increasing exploitation of otters likewise seems to reflect increased use of watercraft and greater presence of humans in offshore habitats over time. Importantly, the CA-SLO-2 data set also includes values for exploitation of both abalone and sea urchins over the course of the Holocene (Figure 11.7) that show no evidence of the kind of species replacement that Simenstad et al. (1978) documented in the Aleutians. Rather, sea urchins remained insignificant over time, while abalone exploitation increased concomitantly with otter exploitation. This suggests that some reduction in the nearshore populations of otters as a result of human harvest led to increased availability of abalones for Native people, with local kelp forests appearing to remain fairly stable over the long term.

One other recent archaeological finding has contributed to knowledge of sea otter hunting of a different type. Langenwalter II et al. (2001) recovered a sea otter femur embedded with a chert projectile point tip at a site in Long Beach,

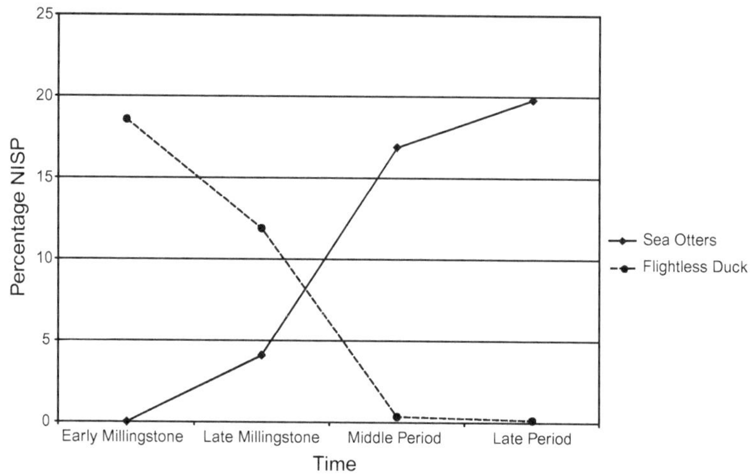

FIGURE 11.6. Remains of the sea otter and the flightless duck (*Chendytes lawi*) from CA-SLO-2 over time (percentage of NISP).

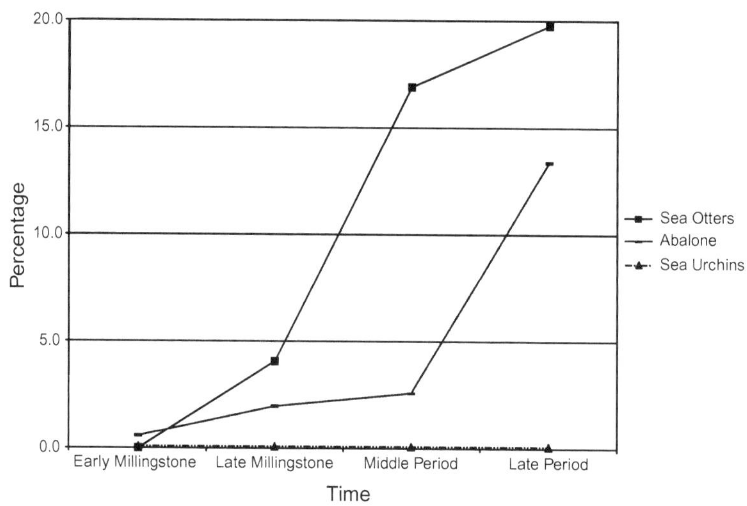

FIGURE 11.7. Sea otter, abalone, and sea urchin remains through time from CA-SLO-2 (percentage NISP for sea otters; percentage of shell weight for abalone and urchins).

California, that had been occupied from ca. AD 1420 to contact. This evidence suggests that prehistoric people hunted sea otters with stone-tipped projectiles. Langenwalter II et al. (2001) assert the otter was killed while swimming and not on land due to the projectile point's angle of entry. Erlandson and Braje (2007) have also speculated that small Channel Island Barbed Points (aka Arena points) dated to ca. 8000 to 6000 BC may have been used to hunt sea otters on the Channel Islands.

PREHISTORIC GENETICS

A sample of 41 prehistoric bones from CA-SLO-2 (Table 11.3), representing approximately 10% of the 431 sea otter NISP from this site, was analyzed genetically at the Seattle Aquarium to determine the sex of the animals. The method used focused on variation at four microsatellite loci (Mvi 57 and Mvi 87 [ÕConnell et al. 1996] and Mvis 72 and 75 [Fleming et al. 1999]). Caution was used to prevent multiple sampling from the same individual and to prevent sam-

TABLE 11.3

Results of DNA and Isotopic Analysis of CA-SLO-2 Sea Otter Bones

SAMPLE	UNIT	LEVEL (CM)	μG N	δ^{15}N (AIR)	μG C	δ^{13}C (PDB)	C:N	SEX	COMPONENT
S2	N10/W17&18	50–60						Male	Late Period
S3	S1/W1	20–30						Female	Late Period
S4	N4/W1	30–40	190.5	14.01	498.2	–10.23	3.05	Female	Late Period
S5	N4/W2	20–30	79.1	13.63	216.3	–10.45	3.19	Female	Late Period
S6	N11/W14	50–60	189.6	15.31	510.1	–10.12	3.14	Female	Late Period
EL1	N1/W1	40–50						Female	Late Period
EL2	N1/W1	40–50						Female	Late Period
EL3	N1/W1	40–50	149.3	13.32	411.4	–10.92	3.21	Male	Late Period
EL4	N1/W1	60–70	163.9	14.03	444.4	–9.86	3.16	Female	Late Period
EL5	N1/W1	60–70						Female	Late Period
EL6	N10/W1	50–60						Female	Late Period
EL8	N10/W17	0–10						Female	Late Period
EL11	N10/W4	50–60						Female	Late Period
EL12	N10/W4	50–60						Female	Late Period
EL14	N4/W1	10–20						Female	Late Period
EL15	N4/W1	10–20	201.6	17.11	522.0	–10.50	3.02	Female	Late Period
EL16	N4/W1	50–60						Female	Late Period
EL17	N4/W1	50–60						Female	Late Period
EL18	N4/W1	60–70	135.6	12.91	371.3	–10.62	3.19	Female	Late Period

(*continued*)

TABLE 11.3 *(continued)*

SAMPLE	UNIT	LEVEL (CM)	μG N	$\delta^{15}N$ (AIR)	μG C	$\delta^{13}C$ (PDB)	C:N	SEX	COMPONENT
EL19	N4/W1	60–70						Female	Late Period
EL23	N4/W2	20–30	173.3	13.36	473.4	–10.70	3.19	Female	Late Period
EL24	N4/W2	40–50	147.6	13.60	412.4	–11.60	3.26	Female	Late Period
EL25	N4/W2	40–50	131.3	13.75	375.3	–11.78	3.34	Female	Late Period
EL26	N4/W2	40–50	151.9	13.28	413.4	–10.12	3.18	Male	Late Period
EL27	N4/W2	40–50	187.9	12.63	485.3	–10.15	3.01	Female	Late Period
EL 28	N4/W2	40–50	216.2	12.94	559.6	–10.18	3.02	na	Late Period
EL29	N4/W2	50–60	182.8	14.03	473.4	–9.92	3.02	Female	Late Period
EL30	N4/W2	60–70	159.6	14.19	436.4	–10.79	3.19	Female	Late Period
EL10	N10/W17	80–90						Female	Middle Period
EL20	N4/W1	70–80						Female	Middle Period
EL21	N4/W1	70–80	262.6	14.45	675.5	–9.46	3.00	Female	Middle Period
EL22	N4/W1	70–80	166.5	16.29	454.4	–11.63	3.18	Female	Middle Period
EL9	N10/W17	80–90	102.1	13.14	280.4	–9.69	3.20	Female	Middle Period
EL13	N10/W4	70–80	140.7	13.53	378.3	–10.55	3.14	Female	Middle Period
EL31	N1/W1	180–190	167.3	14.30	447.4	–9.34	3.12	Male	Middle Period

EL33	N10/W4	190–200	110.7	17.26	306.7	–12.00	3.23	Female	Middle Period
EL40	S4/W9	230–240	185.3	15.12	511.1	–10.96	3.22	Female	Millingstone
EL41	S4/W9	230–240	175.9	15.21	471.4	–12.62	3.13	Female	Millingstone
EL44	N1/W4	240–250	170.8	18.43	473.4	–14.53	3.23	Female	Millingstone
EL45	S1/W12	250–260	22.4	15.02	87.5	–15.26	4.56*	Female	Millingstone
EL46	N4/W1	250–260						Female	Millingstone
MOD1			186.3	12.96	463.3	–12.25	2.90	Male	Modern
M-1788			227.9	14.47	572.1	–11.85	2.93	Male	Modern
M-1793			193.9	13.66	486.1	–10.82	2.92	Male	Modern
M-1797			178.4	15.06	445.2	–10.39	2.91	Male	Modern
M-1784			198.0	16.43	498.3	–11.58	2.94	Female	Modern
M-1899			175.3	15.48	439.1	–12.12	2.92	Female	Modern
M-2211			206.8	14.72	517.7	–12.48	2.92	Female	Modern

ple contamination. To minimize the chances of obtaining more than one sample per individual, three precautions were taken: (1) samples were taken from a wide array of locations on each bone; (2) a narrow range of skeletal elements (femur, humerus, mandible, maxilla) was utilized; and (3) after amplification, samples were compared for identical genotypes and if found one was removed. Control of potential contamination of the ancestral bone samples followed aspects of protocols described previously (Hagelberg and Clegg 1991; Hoss and Paabo 1993; Rosenbaum et al. 1997). All materials and equipment that could potentially come into contact with the samples (cotton gauze, tips, tubes, etc.) were treated with UV light for 10 minutes. Each bone sample was cleaned repeatedly with ethanol and 10% bleach and rinsed with RNA- and DNA-free water prior to sampling. A variable-speed Dremel rotary tool was used, with a new UV-treated drill bit for each sample, to collect bone dust. Samples were collected in a sterile 1.5 mL microcentrifuge tube and stored at ambient temperature until extraction. Bone samples were decalcified in 1 mL of 0.5 m EDTA for at least 24 hours at 37°C. Several changes of EDTA supernatant were made to remove pigmented humic acids absorbed from the sediments. Once relatively clear EDTA supernatant was obtained, the EDTA was removed and the resulting bone pellet was rinsed with sterile water, and the DNA was extracted using the DNeasy tissue extraction kit (Qiagen, Valencia, CA). Blank controls during DNA extraction, PCR, and fragment visualization were employed to determine potential contamination. Bone sexing followed the methods described in Hattori et al. (2003) and utilized pcr primers ZFX/ZFY, digested with NlaIII and visualized on a 1.5 % agarose gel.

A determination could not be made for one specimen. Sex determinations for the remaining 40 specimens showed that 36 (90%) were female and 4 were male. Five specimens representing the Early to Middle Holocene (Millingstone) occupation (5000–3000 cal BC) were all female. The Late Holocene samples (Middle and Late periods; ca. 1500 cal BC to cal AD 1800) included 31 (89%) females and 4 males. While the available sample is small, it suggests that prehistoric exploitation favored females over males at least at CA-SLO-2.

STABLE ISOTOPE ANALYSES

Collagen stable isotopes reflect the average isotopic composition of the protein portion of an individual's diet in the years before death. The duration over which the values are averaged depends upon the rate of collagen turnover, which varies by species, life history stage, diet quality, and skeletal element (Tieszen et al. 1983). The heavier stable isotopes of carbon and nitrogen (i.e., ^{13}C and ^{15}N) are preferentially taken up during protein synthesis, causing a widely recognized but variable trophic enrichment between diet and consumer tissues (DeNiro and Epstein 1978, 1981; Hedges and Reynard 2007; Minagawa and Wada 1984; Sealy et al. 1987). Collagen isotope values represent a proportional mixture of the isotope signatures of the various prey species consumed by an individual. In a simple system, for example, an individual eating 50% prey A and 50% prey B will have a corrected collagen value halfway between the two food sources. Organisms with relatively monotonous diets (e.g., specialized feeders) show less intraspecific variation in collagen isotope values than animals capable of foraging in multiple isotopically distinct habitats (e.g., a mixed marine-terrestrial diet) or at a variety of trophic levels (e.g., a diet including both herbivores and carnivores). Some pinniped species, as dedicated piscivores, are examples of the former, whereas sea otters are consistent with the latter, given the variety of invertebrates from different functional classes they may consume (e.g., filter feeders, grazers, scavengers, etc.).

Among contemporary central coast otter populations in the Monterey Bay National Marine Sanctuary (MBNMS), recent feeding studies observe six diet specializations defined by prey size and foraging habitats: two large prey specializations in abalone and crab or *Cancer* crabs; three medium prey specializations in varied kelp forest prey, in urchins and/or mussels on rocky hard substrates, and in clams, worms, and other invertebrates on soft-bottom substrates; and a small prey specialization in turban snails (*Tegula* sp.), and some kelp crabs and sea stars (Oftedal et al. 2007:160). Foraging specialization has the effect of dividing finite prey resources, thus reducing intraspecific competition and increasing foraging efficiency. In the case of the modern MBNMS, the diversity of diets is thought to be an adaptation to increasing food limitation, which is currently manifest in the relatively poor body condition of central coast otters compared to populations to the north and south (Oftedal 2007:174–175). Although specialists may increase foraging efficiency, not all observed diet specializations may be viable, with the specialization in turban snails and other small prey appearing to be nutritionally inadequate (Oftedal et al. 2007). Therefore, it may not be expected to persist if resource competition is reduced, either as higher-quality prey becomes more abundant and the suboptimal diet is abandoned, or otter populations decrease due to mortality related to the inadequacy of the small prey diet. Thus, on longer time scales dietary specialization will change in response to changing ecological parameters.

The prey species that compose each of these diets have distinct stable isotopic signatures, so an otter population with a variety of diet specializations will exhibit a greater spread of collagen isotope values than a population of generalists or a population with a single specialization (Bearhop et al. 2004; Oftedal et al. 2007:187–189). Therefore, diachronic change in the scatter of prehistoric otter isotope values can be used as a proxy for foraging specialization. The ultimate driver of specialization may be difficult to identify: for example, food limitation due to declines in basal productivity cannot be distinguished from food limitation as a population reaches carrying capacity. However, relevant archaeological data on human foraging patterns, technological adaptations, and demography can be used to evaluate multiple working hypotheses generated by patterns of isotope variability.

ANALYTICAL METHODS

Analyses of stable C and N isotopes were conducted on bone collagen extracted from 25 of the sea otter bones from CA-SLO-2 that were subjected to DNA analysis, as well as 7 modern samples collected from the central California coast. General procedures follow those of Newsome et al. (2004). Samples ranging from ~250 to 750 mg dry weight were removed with single-use Dremel cut-off wheels, which were changed between each sample. To reduce potential cross-contamination, cutting was done under a hood onto aluminum foil that was disposed of along with the dust after each sample was taken, and the work area was wiped down with 70% ethanol. Samples were physically cleaned of adhering sediment, and the outer surfaces of the bone were scraped away with an X-acto blade. Bone was demineralized over 2 to 5 days in 0.5N HCl at ~5°C in scintillation vials. Lipids were removed by soaking demineralized bone in a methanol-chloroform solution (2:1) and rinsed in multiple baths of deionized water while being sonicated. The resulting collagen was lyophilized and weighed, and % collagen yield was determined. Of an original 43 specimens from CA-SLO-2 that were subject to extraction, 25 produced usable collagen, ranging from ~2.5 to 10% yield by weight. Yields in these ranges reflect good preservation for isotopic analyses (Ambrose 1990; DeNiro and Weiner 1988; van Klinken 1999). Many of the unusable specimens were slightly charred bones that disintegrated during demineralization or lipid extraction. Collagen yield from the modern specimens ranged between ~17 to 24% by weight.

Subsamples of 1.0 ± 0.2mg of lyophilized, extracted collagen were packed into tin capsules and submitted to the UC Davis Stable Isotope Facility (Dept. of Plant Sciences) where C and

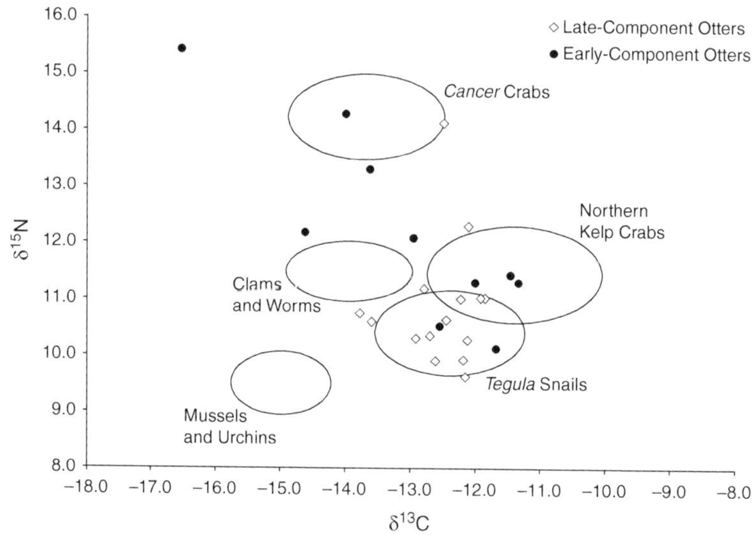

FIGURE 11.8. Stable isotope results for Early component (closed circles) and Late component (open diamonds) sea otter collagen from CA-SLO-2. The range of values for seven modern central coast otters is depicted with a gray ellipse. Typical prey isotope values for the MBNMS (open ellipses) are redrawn from Oftedal et al. (2007:202, figure 6.6). Collagen values are corrected for fractionation to place them in diet space ($\delta^{15}N_{diet} = \delta^{15}N_{collagen} - 3‰$; $\delta^{13}C_{diet} = \delta^{13}C_{collagen} - 2‰$), and modern otters and prey $\delta^{13}C$ values are increased 1‰ to correct for modern ^{13}C depletion due to fossil fuel burning (i.e., Suess effect).

N stable isotope concentrations were analyzed using a PDZ Europa ANCA-GSL elemental analyzer interfaced to a PDZ Europa 20-20 isotope ratio mass spectrometer. Sample isotope ratios are reported as $\delta^{13}C$ or $\delta^{15}N$ values, where $\delta R_{sample} = [(R_{sample}/R_{standard}) - 1] \times 1000$, and R_{sample} and $R_{standard}$ are the $^{13}C/^{12}C$ and $^{15}N/^{14}N$ ratios of measured samples and standards, respectively. Results are reported in ‰ notation (per mil or parts per thousand) with respect to the Pee Dee Belemnite (PDB) scale for $\delta^{13}C$ and AIR (Ambient Inhalable Reservoir; atmospheric N_2) scale for $\delta^{15}N$.

RESULTS

Stable isotope results for sea otter bone collagen are presented in Table 11.3 with measured abundances of C and N (in µg) for each sample. The C:N ratio of collagen in terms of elemental abundance [i.e., (µgC × µmol/12µg)/(µgN × µmol/14µg)] gives an indication of collagen preservation, with values between 2.9 and 3.6 being consistent with modern, unaltered collagen (DeNiro 1985). C:N ratios reported in Table 11.3 indicate good collagen preservation in these samples, with the exception of the high value of 4.56 for EL45 that is attributable to a very small sample size (87.5 µgC, 22.4 µgN). The modern otter samples have lower average C:N ratios than archaeological specimens, but all are within the recommended range.

Isotope results for modern and archaeological otters ($n=31$, excluding EL45) are plotted in Figure 11.8 along with isotope ranges for typical otter prey groups recently measured for the Monterey Bay National Marine Sanctuary (Oftedal et al. 2007). All data are presented in "diet space," i.e., measured collagen values are corrected for trophic fractionation to reflect the isotopic content of the consumer's diet by reducing $\delta^{13}C$ by 2‰ and $\delta^{15}N$ by 3‰ (Kelly 2000; Oftedal et al. 2007). Additionally, $\delta^{13}C$ values for modern prey groups and otter specimens are increased 1‰ to correct for isotopic depletion of the atmospheric carbon reservoir due to historic fossil fuel burning (i.e., the Se-

TABLE 11.4
Variance in Stable Isotopes by Component at CA-SLO-2

COMPONENT	N	VAR. $\delta^{15}N$	VAR. $\delta^{13}C$	TOTAL VARIANCE
Modern	7	1.31	0.60	1.91
Late	15	1.22	0.32	1.54
Early	9	3.02	2.90	5.92
Early (excl. EL44)	8	1.88	1.52	3.40

uss effect). Accounting for these offsets allows for the direct comparison of prehistoric and modern otter diet with a variety of potential prey species.

The number of potential food sources precludes unique solutions for individual diets, because multiple diet combinations could lead to the same mixture (Newsome et al. 2004; Phillips and Koch 2002; Phillips et al. 2005). For example, assuming a linear mixing model, a diet composed of ~50% clams and worms and ~50% kelp crabs would appear similar to a diet of ~25% *Cancer* crabs and ~75% *Tegula* snails. Dietary specialization could therefore be underestimated when considering individuals that fall in the center of the diet space plot. By contrast, any values close to one of the more distinct diet sources must contain a large proportion of that source in the diet.

With the exception of one outlier (EL44), all of the modern and prehistoric otters fall within the diet space defined by the prey species (Figure 11.8), which suggests this is a reasonable first approximation of the prehistoric foraging regime. Comparing the scatter between the otter subsamples, there is clearly a greater spread in the Early component compared to the Late component or the modern samples (their range indicated by the grey ellipse), consistent with decreased dietary specialization within otter populations through time. Reduced scatter can also be expressed by comparing the variance in each stable isotope by component (Table 11.4). Again, modern and Late components show lower variance than the Early component, even when the sample falling outside of the prey-defined diet space is excluded. The shift in variability is accompanied by a decline in mean $\delta^{15}N$ values (Early=14.91‰ vs. Late=13.87‰) indicating that the trend towards diet homogeneity in the Late component occurred through the removal of higher trophic-level diet specializations among otters in the CA-SLO-2 assemblage.

DISCUSSION

Assuming that intraspecific variation in stable isotope values is a proxy for the variety of diet specializations within a population, the data indicate that individual otter diets became less differentiated in the vicinity of CA-SLO-2 over the course of the Holocene. That is, there are fewer distinct foraging strategies evident in the Late component versus the Early component. For otters, the effect of adopting different foraging strategies is to reduce intraspecific competition by dividing available resources, and thereby increasing foraging efficiency (Oftedal et al. 2007). In the modern context, the degree of observed specialization in central coast otter populations is interpreted as a response to food limitation. The behavioral response of foraging specialization could be predicted to occur as a population approaches the effective carrying capacity of its habitat, which could be caused by several processes working singly or in combination: increased otter population; decreased prey abundance due to external environmental change; or increased interspecific competition for resources. Conversely, a population exhibiting fewer foraging specializations would be predicted to be below its effective carrying capacity because of suppressed numbers, increased abundance of high-quality prey, or less direct competition from other predators for resources.

SUMMARY AND CONCLUSIONS

The archaeological data on otter hunting by Native Californians summarized in this chapter provides evidence for growing predatory pressure on otter populations through the Holocene, while at the same time, the historic record indicates that a robust population of otters (ca. 20,000 animals as opposed to only 3000 today) was present along California shores at the end of the prehistoric era. It is fairly apparent that growth in human populations over the Holocene led to increased exploitation of otters for food and their pelts as trade goods. While the available sample is small and results must be considered preliminary, DNA analysis of sea otter bones shows that Native exploitation focused heavily on females rather than males. Females spend more time on land, dive to shallower depths, and occupy territories closer to shore than males, so a focus on females would be consistent with attempts to maximize hunting yields relative to pursuit costs. A modest increase in males during the Late Holocene suggests a slight decrease in foraging efficiency over time as lower-ranked males were eventually targeted in addition to females although the sample available for the Early to Middle Holocene is very small. Since otters often occupy single-sex areas, it is possible that the overabundance of females at CA-SLO-2 reflects the site's proximity to a female-dominated territory. Nonetheless, heavy focus on females at any location would be antithetical to stewardship or conservation of otter populations. In this seemingly paradoxical case, nonconservative hunting over a 9000-year period culminated at the end of prehistory in an extremely large population of sea otters along the California coast. Findings from Diablo Canyon also show no evidence for major alterations in the structure or distribution of kelp forests as a consequence of otter exploitation; there is no evidence for any significant increase in sea urchin remains akin to that associated with otter overexploitation in the Aleutians (Simensted et al. 1978), and no evidence for any collapse of kelp forest habitats.

Preliminary results from isotope studies suggest, however, that increasingly intensive otter exploitation over the course of the Holocene was not without its effects. Molluscan remains from CA-SLO-2 show increased abalone harvesting concomitant with increased exploitation of otters, suggesting a greater human presence in nearshore habitats. Both direct competition between humans and otters for the same food sources (e.g., abalone and sea urchins), and the increased human presence in the otter's offshore habitat (as a result of increased hunting) would have reduced the foraging options for otters. From the otters' perspective, habitat occupied by human hunter-gatherers would have become less accessible. Isotopically, this could lead to increased variability because greater diet specialization would be required. However, if certain isotopically distinct foraging specializations are excluded by human competition (e.g., nearshore foraging), the range of values within a component could be seen to decrease. The observed pattern of decreased diet specialization from the Early to Late components at CA-SLO-2 argues that human predation of otters both suppressed local otter populations and decreased the available foraging habitats through competitive exclusion.

Overall, the combined archaeological, historic, genetic, and stable isotopic records show unequivocally that humans, otters, and abalone coexisted for 9000 years along the central California coast and that robust populations of one species did not preclude the existence of the other. Paradoxically, otters were harvested regularly along the entire central coast in a nonconservation-oriented manner, yet this increasingly intensive harvest was sustainable, as a robust population of otters was available for commercial hunters at the beginning of the 19th century and there is no archaeological evidence for the collapse of local kelp forests. This seems to speak to a level of productivity in California's precontact nearshore environments that is almost inconceivable in comparison with that of recent times. During the prehistoric era, conservation seems to have been an epiphenomenon that resulted from this richness and human populations that were still below the carrying capacity of these remarkably productive habitats.

ACKNOWLEDGEMENTS

We are deeply indebted to Todd Braje and Torben Rick for inviting us to participate in the symposium in Vancouver. Analyses of the otter remains from Diablo Canyon were funded by California Sea Grant R/CZ-187. Figure 11.1 was composed by Rusty van Rossmann. We also thank Francis Villablanca from the Cal Poly Department of Biology for allowing us to sample modern sea otter specimens and Mike Harvis from California Fish and Game for supplying references.

REFERENCES CITED

Ambrose, S.H.

1990 Preparation and Characterization of Bone and Tooth Collagen for Isotopic Analysis. *Journal of Archaeological Science* 17:431–451.

Bearhop, S., C. E. Adams, S. Waldron, R. A. Fuller, and H. Macleod

2004 Determining Trophic Niche Width: A Novel Approach Using Stable Isotope Analysis. *Journal of Animal Ecology* 73:1007–1012

Bourdeau, L.

1986 *Report on Archaeological Investigations at Sunflower House: CA-SCR-93 and CA-SCR-93/H with Recommendations for Cultural Resource Management, Project Parcel APN# 05*114-01, Santa Cruz County, California.* Pacific Museum Consultants, Boulder Creek, California. Copies available from Northwest Archaeological Information Center, Department of Anthropology, Sonoma State University, Rohnert Park, California.

Breschini, G. S., and T. Haversat

1989 Archaeological Investigations at CA-MNT-108 at Fisherman's Wharf, Monterey, Monterey County, California. *Archives of California Prehistory* 29. Coyote Press, Salinas.

1992 *Baseline Archaeological Studies at Rancho San Carlos, Carmel Valley, Monterey County, California.* Archaeological Consulting, Salinas, California. Copies available from Northwest Archaeological Information Center, Department of Anthropology, Sonoma State University, Rohnert Park, California.

2006 *An Offramp on the Kelp Highway: Archaeological Investigations at CA-MNT-831, Pacific Grove, Monterey County, California.* Archaeological Consulting, Salinas. Copies available from Northwest Archaeological Information Center, Department of Anthropology, Sonoma State University, Rohnert Park, California.

Broughton, J. M.

1999 Resource Depression and Intensification during the Late Holocene, San Francisco Bay: Evidence from the Emeryville Shellmound Vertebrate Fauna. *University of California Anthropological Records* 32.

Cartier, R. R.

1993 *The Saunders Site: MNT-391, a Littoral Site of the Early Period.* Scotts Valley Historical Society Monograph 1. Scotts Valley, California.

Codding, B. F., and T. L. Jones

2007 History and Behavioral Ecology during the Middle-Late Transition on the Central California Coast: Findings from the Coon Creek Site (CA-SLO-9), San Luis Obispo County. *Journal of California and Great Basin Anthropology* 27: 23–49.

Davis, B. S.

1981 Indian, Sea Otter, and Shellfish Interrelationships. *Society for California Occasional Papers* 3:33–41.

DeNiro, M. J.

1985 Postmortem Preservation and Alteration of *in Vivo* Bone Collagen Isotope Ratios in Relation to Palaeodietary Reconstruction. *Nature* 317:806–809.

DeNiro, M. J., and S. Epstein

1978 Influence of Diet on the Distribution of Carbon Isotopes in Animals. *Geochimica et Cosmochimica Acta* 42:495–506.

1981 Influence of Diet on the Distribution of Nitrogen Isotopes in Animals. *Geochimica et Cosmochimica Acta* 45:341–351.

DeNiro, M. J., and S. Weiner

1988 Chemical, Enzymatic and Spectroscopic Characterization of "Collagen" and Other Organic Fractions from Prehistoric Bones. *Geochimica et Cosmochimica Acta* 52:2197–2206.

Dietz, S. A.

1987 *Archaeological Test Excavations at CA-MNT-101, CA-MNT-298, CA-MNT-929, and El Castillo at the Presidio and City of Monterey, California.* Archaeological Consulting and Research Services, Santa Cruz. Copies available from

Northwest Archaeological Information Center, Department of Anthropology, Sonoma State University, Rohnert Park, California.

1991 *Final Report of Archaeological Investigations at Pescadero Point, Data Recovery Excavations and Monitoring of CA-MNT-170.* Archaeological Consulting and Research Services, Santa Cruz. Copies available from Northwest Archaeological Information Center, Department of Anthropology, Sonoma State University, Rohnert Park, California.

Dietz, S. A., and T. L. Jackson

1981 *Report of Archaeological Excavations at Nineteen Archaeological Sites for the Stage 1 Pacific Grove-Monterey Consolidation Project of the Regional Sewerage System.* Archaeological Consulting and Research Services, Santa Cruz. Copies available from Northwest Archaeological Information Center, Department of Anthropology, Sonoma State University, Rohnert Park, California.

Dietz, S. A., W. R. Hildebrandt, and T. L. Jones

1988 *Archaeological Investigations at Elkhorn Slough: CA-MNT-229, a Middle Period Site on the Central California Coast.* Papers in Northern California Anthropology 3. Northern California Anthropological Research Group, Berkeley.

Driver, H. E., and W. C. Massey

1957 Comparative Studies in North American Indians. *Transactions of the American Philosophical Society* 47(2):165–456.

Erlandson, J. M., and T. J. Braje

2007 Early Maritime Technology on California's San Miguel Island: Arena Points from CA-SMI-575-NE. *Current Research in the Pleistocene* 24:85–86.

Erlandson, J. M., T. C. Rick, J. A. Estes, M. H. Graham, T. J. Braje, and R. L. Vellanoweth

2005 Sea Otters, Shellfish, and Humans: 10,000 Years of Ecological Interactions on San Miguel Island, California. In *Proceedings of the Sixth California Islands Symposium,* edited by D.K. Garcelon & C.A. Schwemm, pp. 58–69. Institute for Wildlife Studies and National Park Service, Arcata, California.

Fitzgerald, R. T., J. L. Edwards, J. M. Farqhuar, and K. Loefller

1995 *Archaeological Test Excavation at CA-MNT-1765, for the Moro Cojo Standard Subdivision Project (SH 93001), Monterey County, California.* Copies available from Northwest Archaeological Information Center, Department of Anthropology, Sonoma State University, Rohnert Park, California.

Fleming, M. A., E. A. Ostrander, and J.A. Cook

1999 Microsatellite Markers for American Mink (*Mustela vison*) and Ermine (*Mustela erminea*). *Molecular Ecology* 8:1352–1354.

Gibson, J. R.

1992 *Otter Skins, Boston Ships, and China Goods: the Maritime Fur Trade of the Northwest Coast, 1785–1841.* McGill –Queen's University Press, Montreal.

Greenwood, R. S.

1972 *9000 Years of Prehistory at Diablo Canyon, San Luis Obispo County, California.* San Luis Obispo County Archaeological Society Occasional Papers No. 7. San Luis Obispo.

1978 Obispeño and Purisimeño Chumash. In *California,* edited by R. F. Heizer, pp. 520–523. Handbook of North American Indians, vol. 8, W. G. Sturtevant, general editor, Smithsonian Institution, Washington D. C

Hagelberg, E., and J.B. Clegg

1991 Isolation and Characterization of DNA from Archaeological Bone. *Proceedings of the Royal Society of London* 244 45–50.

Hattori K, A. M. Burdin, M. Onuma, M. Suzuki, and N. Ohtaishi,

2003 Sex Determination in the Sea Otter (*Enhydra lutris*) from Tissue and Dental Pulp Using PCR amplification. *Canadian Journal of Zoology* 81:52–56.

Hedges, R. E. M., L. M. Reynard

2007 Nitrogen isotopes and the trophic level of humans in archaeology. *Journal of Archaeological Science* 34:1240–1251.

Hildebrandt, W. R., J. Farquhar, and M. Hylkema

2006 *Archaeological Investigations at CA-SMA-218: A Study of Prehistoric Adaptations at Año Nuevo State Reserve.* Far western Anthropological Research Group. Copies available from Northwest Archaeological Information Center, Department of Anthropology, Sonoma State University, Rohnert Park, California.

Hildebrandt, W. R., and T. L. Jones

1992 Evolution of Marine Mammal Hunting: A View from the California and Oregon Coasts. *Journal of Anthropological Archaeology* 11:360–401.

Hildebrandt, W. R., D. A. Jones, and K. Carpenter

2002 *Prehistoric Coastal Occupations at Arroyo de los Chinos.* Far Western Anthropological Research Group, Davis California. Prepared for California Department of Transportation, 50 Higuera Street, San Luis Obispo. Copies available from Far Western Anthropological Research Group.

Hoss, M, and S. Paabo
1993 DNA Extraction from Pleistocene Bones by a Silica-based Purification Method. *Nucleic Acids Research* 21:3913–3914.
Hylkema, M. G.
1991 *Prehistoric Native American Adaptations along the Central California Coast of San Mateo and Santa Cruz Counties.* Unpublished master's thesis, Department of Anthropology, San Jose State University.
Jones, D. A., and W. R. Hildebrandt
1990 *Archaeological Excavation at Sand Hill Bluff: Portions of Prehistoric Site CA-SCR-7, anta Cruz County, California.* Far Western Anthropological Research Group, Davis California. Copies available from Northwest Archaeological Information Center, Department of Anthropology, Sonoma State University, Rohnert Park, California.
Jones, D. A., and P. A. Mikkelsen,
2006 *Extended Phase I: Archaeological Investigations at CA-SLO-1243, -2356, and -2357, and Phase II Investigations at CA-SLO-2357.* Far Western Anthropological Research Group, Davis California. Prepared for San Luis Obispo County Department of Public Works, County Government Center, Rm. 207, San Luis Obispo. Copies available from Far Western Anthropological Research Group.
Jones, D. A., P. A. Mikkelsen, and W.R. Hildebrandt
2004 Prehistoric Occupations on Ancient Halcyon Bay/Estuary: Excavation Results from CA-SLO-832, and 1420. In *Emerging from the Ice Age: Early Holocene Occupations on the California Central Coast,* edited by Ethan Betrando and Valerie A. Levulett, pp. 71–80. San Luis Obispo County Archaeological Society Occasional Papers No. 17. San Luis Obispo.
Jones, T. L.
1996 Mortars, Pestles, and Division of Labor in Prehistoric California: A View from Big Sur. *American Antiquity* 61:243–264.
2003 Prehistoric Human Ecology of the Big Sur Coast, California. *Contributions of the University of California Archaeological Research Facility* No. 61. Berkeley.
Jones, T. L., and J. A. Ferneau
2002 *Prehistory at San Simeon Reef: Archaeological Data Recovery at CA-SLO-179 and -267, San Luis Obispo County, California.* San Luis Obispo County Archaeological Society Occasional Papers No. 16. San Luis Obispo.
Jones, T. L., and D. A. Jones
1992 Elkhorn Slough Revisited: Reassessing the Chronology of CA-MNT-229. *Journal of California and Great Basin Anthropology* 10:163–186.
Jones, T. L., and G. Waugh
1995 *Central California Coastal Prehistory: A View from Little Pico Creek.* Cotsen Institute of Archaeology, University of California, Los Angeles.
Jones, T. L., J. F. Porcasi, J. Gaeta, and B. F. Codding
2008 The Diablo Canyon Fauna: A Course-Grained Record of Trans-Holocene Foraging from the Central California Mainland Coast. *American Antiquity* 73:289–316.
Jones, T. L., J. F. Porcasi, K. W. Gobalet, and L. T. Laurie
2004 CA-SLO-215, A Late Milling Stone Site at Morro Bay, San Luis Obispo County, California. In *Out of the Ice Age: Papers in Honor of Roberta Greenwood,* edited by E. Bertrando and V. Levulett, pp. 57–70. San Luis Obispo County Archaeological Society Occasional Papers No. 17. San Luis Obispo.
Jones, Terry L., Sebastian C. Garza, Judith F. Porcasi, Jerome W. Gaeta
2009 Another Trans-Holocene Sequence from Diablo Canyon: New Faunal and Radiocarbon Findings from CA-SLO-585, San Luis Obispo County, California. *Journal of California and Great Basin Anthropology* 29:19–31.
Jones, T. L., N. E. Stevens, D. A. Jones, R. T. Fitzgerald, and M. G. Hylkema
2007 The Central Coast: A Mid-Latitude Milieu. In *California Prehistory: Colonization, Culture, and Complexity,* edited by T. L. Jones and K. L. Klar, pp. 125–146. Alta Mira Press, New York.
Jones, T. L., T.M. Van Bueren, S. Grantham, J. Huddleston, and T. Fung
1996 *Archaeological Test Excavations for the Castroville Bypass Project, Monterey County, California.* Office of Cultural Studies, California Department of Transportation. Copies available from Northwest Archaeological Information Center, Department of Anthropology, Sonoma State University, Rohnert Park, California.
Jones, T. L., J. F. Porcasi, J. M. Erlandson, H. Dallas, Jr., T. A. Wake, R. Schwaderer
2008 The Protracted Holocene Extinction of California's Flightless Sea Duck *(Chendytes lawi)* and its Implications for the Pleistocene Overkill Hypothesis. *Proceedings of the National Academy of Science* 105:4105–4108.
Joslin, T. L.
2006 *Late Prehistoric Coastal Adaptations along the San Simeon Reef, San Luis Obispo County, California.* Unpublished master's thesis, Department of Anthropology, University of California, Santa Barbara.

Kelly, J. R.
2000 Stable Isotopes of Carbon and Nitrogen in the Study of Avian and Mammalian Trophic Ecology. *Canadian Journal of Zoology* 78: 1–27.
Kennedy, M. A., A. D. Russell, and T. Guilderson
2005 A Radiocarbon Chronology of Hunter-Gatherer Occupation from Bodega Bay, California, U.S.A. *Radiocarbon* 47:1–29.
Kenyon, K. W.
1981 Sea Otter—*Enhydra lutris*. In *Handbook of Marine Mammals*, edited by Sam H. Ridgway and Richard J. Harrison F.R.S., pp. 209–223. Academic Press, San Francisco.
Langenwalter II, P. E., M. A. Boxt, L. M. Boxt, and T. T. Miller
2001 A Sea Otter (*Enhydra lutris*) Femur with Embedded Projectile Point Fragment from a Late Prehistoric Camp Site in Long Beach, California. *Pacific Coast Archaeological Society Quarterly* 37(1):51–59.
Levy, R.
1978 Costanoan. In *California*, edited by R. F. Heizer, pp. 485–495. Handbook of North American Indians, vol. 8, W. G. Sturtevant, general editor, Smithsonian Institution, Washington D. C.
Mikkelsen, P., W. R. Hildebrandt, and D. A. Jones
2000 *Prehistoric Adaptations on the Shores of Morro Bay Estuary: Excavations at Site CA-SLO-165, Morro Bay, California*. San Luis Obispo County Archaeological Society Occasional Papers No. 14.
Mikkelsen, P., W. R. Hildebrandt, D. A. Jones, J. Rosenthal, and R. Gibson
2005 *Thirty Years After: 1974 Excavations at Kirk Creek, CA-MNT-238, on the Big Sur Coast*. Far Western Anthropological Research Group, Davis. Submitted to California Department of Transportation, District 5, San Luis Obispo. Report on file, Northwest Archaeological Information Center, Department of Anthropology, Sonoma State University, Rohnert Park, California.
Milliken, R. T., J. Nelson, W. R. Hildebrandt, and P. Mikkelsen
1999 *The Moss Landing Hill Site: A Technical Report on Archaeological Studies at CA-MNT-234*. Far Western Anthropological Research Group, Davis. Copies available from Northwest Archaeological Information, Department of Anthropology, Sonoma State University, Rohnert Park, California.
Miller, D. J.
1974 California Department of Fish and Game Marine Resources Leaflet No. 7. Sacramento.
Minagawa, M., and E. Wada
1984 Stepwise Enrichment of ^{15}N Along Food Chains: Further Evidence and the Relation between $\delta^{15}N$ and Animal Age. *Geochimica et Cosmochimica Acta* 48:1135–1140.
Newsome, S. D., D. L. Phillips, B. J. Culleton, T. P. Guilderson, and P. L. Koch
2004 Dietary Reconstruction of an Early to Middle Holocene Human Population from the Central California Coast: Insights from Advanced Stable Isotope Mixing Models. *Journal of Archaeological Science* 31:1101–1115.
Nickerson, R.
1984 *Sea Otters: A Natural History and Guide*. Chronicle Books, San Francisco, California.
O'Connell, M., J. M. Wright, and A. Farid
1996 Development of PCR Primers for Nine Polymorphic American Mink, *Musela vison*, Microsatellite Loci. *Molecular Ecology* 5:311–312.
Oftedal, O.T, K. Ralls, M.T. Tinker and A.Green
2007 *Nutritional Constraints on the Southern Sea Otter in the Monterey Bay National Marine Sanctuary, and a Comparison to Sea Otter Populations at San Nicolas Island, California, and Glacier Bay, Alaska*. Joint Final Report to Monterey Bay National Marine Sanctuary and the Marine Mammal Commission.
Ogden, A.
1941 *The California Sea Otter Trade: 1784–1884*. University of California Press, Los Angeles, California.
Phillips, D. L., and P. L. Koch
2002 Incorporating Concentration Dependence in Stable Isotope Mixing Models. *Oecologia* 130:114–125.
Phillips, D. L., S. D. Newsome, and J. W. Gregg
2005 Combining Sources in Stable Isotope Mixing Models: Alternative Methods. *Oecologia* 144:520–527.
Riedman, M. L. and J. A. Estes
1990 The Sea Otter (*Enhydra lutris*): Behavior, Ecology, and Natural History. U.S. Department of the Interior, Fish and Wildlife Service, Biological Report 90(14) September 1990, Washington D.C.
Rosenbaum H. C., M. G.Egan, P. J. Clapham, R. L. Brownell, and R. DeSalle
1997 An Effective Method for Isolating DNA from Historical Specimens of Baleen. *Molecular Ecology* 6 677–681.
Schwaderer, R.
1992 Archaeological Test Excavation at the Duncans Point Cave, CA-SON-348/H. In *Essays on the Prehistory of Maritime California*, edited by

T. L. Jones, pp. 55–71. Center for Archaeological Research at Davis no. 10. University of California, Davis.

Sealy, J. C., N. J. van der Merwe, J. A. Lee-Thorp, J. L. Lanham
1987 Nitrogen isotopic ecology in southern Africa: implications for environmental and dietary tracing. *Geochimica et Cosmochimica Acta* 51: 2707–2717.

Simensted, C. A., J. A. Estes, and K. W. Kenyon
1978 Aleuts, Sea Otters, and Alternative Stable State Communities. *Science* 200:403–411.

Simons, D. D.
1992 Prehistoric Mammal Exploitation in the San Francisco Bay Area. In *Essays on the Prehistory of Maritime California,* edited by T. L. Jones, pp. 73–103. Center for Archaeological Research at Davis no. 10. University of California, Davis.

Simons, D. D., and T. A. Wake
2000 Trans-Holocene Subsistence Strategies and Topographic Change on the Northern California Coast: The Fauna from Duncan's Point Cave. *Journal of California and Great Basin Anthropology* 22:295–320.

Tieszen, L. L., T. W.Boutton, K. G. Tesdahl, and N. A. Slade
1983 Fractionation and Turnover of Stable Carbon Isotopes in Animal Tissues: Implications for the $\delta^{13}C$ Analysis of Diet. *Oecologia* 57: 32–37.

Tikhmenev, P. A.
1978 [1888] A History of the Russian-American Company. Translated and edited by Richard A. Pierce and Alton S. Donnelly. University of Washington Press, Seattle.

USGS and Western Ecological Research Center
2007 Sea Otter Research at WERC. Electronic document, http://www.werc.usgs.gov/otters/, accessed January 1, 2008.

van Klinken, G.J.
1999 Bone Collagen Quality Indicators for Palaeodietary and Radiocarbon Measurements. *Journal of Archaeological Science* 26:687–695.

Walker, P. L.
1982 California Indians, Sea Otters, and Shellfish: The Prehistoric Record. In *Social Science Perspectives on Managing Conflict between Marine Mammals and Fisheries,* edited by Biliana Cicin-Sain, Phyllis M. Grifman, and John B. Richards. Proceeding from a Conference on Management of Sea Otters and Shellfish Fisheries in California, pp. 17–46. Arroyo Grande, California.

Wohlgemuth, E., W. R. Hildebrandt, and K. Ballantyne
2002 *Data Recovery Excavations for Unanticipated Discovery at CA-MNT-1942, Big Creek Bridge (BR. NO. 44–56), Monterey County, California Highway 1, P.M. 28.1.* Far Western Anthropological Research Group, Davis. Copies available from the Northwestern Information Center, Sonoma State University, Rohnert Park.

Woodhouse, C. D., R. K. Cowen, and L. R. Wilcoxon
1977 *A Summary of Knowledge of the Sea Otter Enhydra Lutris, L., in California and an Appraisal of the Completeness of Biological Understanding of the Species.* Santa Barbara Museum of Natural History. Santa Barbara, California.

12

Resilience and Reorganization

ARCHAEOLOGY AND HISTORICAL ECOLOGY OF CALIFORNIA CHANNEL ISLAND MARINE MAMMALS

Todd J. Braje, Torben C. Rick, Robert L. DeLong, and Jon M. Erlandson

DEBATES ABOUT THE NATURE and intensity of marine mammal hunting by ancient peoples and the degree of anthropogenic forcing on northeastern Pacific pinniped populations have been hotly contested. Several models have been proposed to explain patterns observed in marine mammal faunal assemblages from mainland and island localities along North America's Pacific Coast, with little agreement on the initial antiquity and chronology of marine mammal exploitation, the focus of prey species, the distribution of rookery locations, and the effects humans had on sea mammal abundance, behavior, and distributions (e.g., Colten and Arnold 1998; Erlandson et al. 1998, 2004; Etnier 2002; Hildebrandt and Jones 1992; Jones and Hildebrandt 1995; Lyman 1995; Rick et al. 2008). Such studies suffer from a number of historic, cultural, natural, and methodological problems. While researchers have recognized many of these shortcomings, others have been largely overlooked or the severity of their influence on modern interpretations has been underestimated.

Due to the good stratigraphic integrity and preservation of faunal remains, many archaeological models proposed for North Pacific marine mammal hunting have been based, at least partly, on data from California's Channel Islands and the larger Santa Barbara Channel area (e.g., Hildebrandt and Jones 1992). Within this area many mainland coast sites have been damaged or destroyed by historic development and compromised by bioturbation, but island sites have yielded long stratigraphic sequences with well-preserved sea mammal remains that can be more easily identified to genus or species. Even under these conditions, however, incomplete comparative collections and small sample sizes make detailed zooarchaeological analysis difficult (see Monks 2005:171). Recent DNA studies on some archaeological marine mammal bones suggest that a small number have been misidentified by zooarchaeologists, posing further questions about the reliability of data used to generate models of ancient marine mammal exploitation (Moss et al. 2006). Despite these and other problems, zooarchaeological collections remain the primary means of understanding the deep history of human use of marine mammals in the North Pacific and around the world.

Perhaps the most difficult obstacles to overcome in interpreting the ancient record of sea mammal exploitation are those that have affected northeastern Pacific and global sea mammal populations at large. The most obvious may be the decimation of worldwide marine mammal populations during the historic fur trade and commercial hunting period from about the mid-17th to early 20th centuries (see Ellis 2003). With the emergence of the first truly global industrialized economies, many pinnipeds, cetaceans, and other marine mammals were relentlessly hunted to the brink of extinction (or beyond) worldwide—for their blubber, furs, ivory, and other products or purposes. This decimation fundamentally altered the structure and distribution of marine mammal populations and their recovery over the past century has sometimes occurred under very different ecological conditions from the more "natural" conditions in which they evolved. Modern communities of pinnipeds and marine mammals are a product, therefore, of severe historic overhunting and their subsequent protection and recovery in a demographic vacuum, often in food webs and ecosystems altered by overfishing, pollution, climate change, and other anthropogenic disruptions (Jackson et al. 2001). Without evidence for historical continuity, we cannot be confident that their modern distributions or abundances reflect those of the ancient past (see Etnier 2002; Newsome et al. 2007), and in some cases their feeding, breeding, and migration patterns may have been changed by historic ecological disruptions (e.g., DeMaster et al. 2001; Fritz et al. 1995). While the relationship between overfishing and sea mammal behavior is poorly understood, some species may have undergone dramatic behavioral changes as a result of anthropogenic food shortages. Natural environmental fluctuations (e.g., El Niño) probably have had similar effects on sea mammal behavior and distributions (Trillmich and Ono 1991; DeLong and Melin 2002). Archaeological and paleontological evidence suggest that sea surface temperature (SST) fluctuations have influenced ancient distributions of marine mammals. Cold-water-adapted species such as the sea cow and walrus, for example, have been identified in California paleontological assemblages dating to the Pleistocene, when colder than average waters allowed increased southern limits on their ranges (Clementz 2002; Harington 1984). Changes in kelp forest ecosystems from the Late Pleistocene to the Holocene (Kinlan et al. 2005; Steneck et al. 2002) also may have influenced pinniped and sea otter distributions and abundance, but such changes are not well understood.

Today California's Channel Islands shelter more than 200,000 pinnipeds of six different species (DeLong and Melin 2002) and more than a dozen large and small cetacean species also swim through or are seasonally resident in island waters. Under federal and state protection, the behavior, demographics, and ecology of many of these marine mammal populations have been intensively studied for several decades. During those same decades, archaeologists studying island shell middens spanning the past 10,000 years have recovered extensive faunal and technological records relevant to understanding the deep history of human exploitation of sea mammals. However, only rarely are these ecological and archaeological data sets compared to examine the long, intertwined history of humans, pinnipeds, and other marine mammals in the Southern California Bight. In this paper, with a variety of methodological and interpretive limitations in mind, we synthesize the available historical and archaeological evidence for marine mammal hunting on the California Channel Islands. In the process we use archaeological, historical, and ecological data to examine the resilience and reorganization of marine mammal populations in the area over the past 12,000 years. We also summarize key aspects of the natural history of those North Pacific marine mammals that reside in or visit southern California waters today and were the target of ancient Native hunters and scavengers in the past. We believe a nuanced interpretation of the archaeological record, which incorporates knowledge of the mating, breeding, and weaning behavior of individual species, offers the most robust model for reconstructing ancient human-marine mammal dynamics (see Etnier 2002).

We begin with a brief environmental and cultural overview of the Channel Islands, and then describe current models of ancient marine mammal exploitation on the islands. After a brief summary of modern marine mammal behavior, natural history, and distribution in the Santa Barbara Bight, we synthesize the archaeological record of island marine mammal hunting. Finally, we use a case study approach and modern, historic, and archaeological data from Anacapa, Santa Cruz, Santa Rosa, and San Miguel islands to model the long-term history of human-marine mammal dynamics along the southern California coast.

ENVIRONMENTAL AND CULTURAL CONTEXT

Located off the California coast between Point Conception to the north and San Diego to the south, the Channel Islands are divided into a northern chain (Anacapa, Santa Cruz, Santa Rosa, and San Miguel) and a more dispersed southern group (San Clemente, Santa Catalina, San Nicolas, and Santa Barbara; Figure 12.1). The islands and the broader Santa Barbara Channel region are characterized by a Mediterranean climate with warm, dry summers and cool, wet winters. While the islands have undergone considerable physiographic and biotic changes due to environmental fluctuations, sea level rise, and natural and cultural transformations, the islands were never connected to the mainland during the Quaternary, making for a unique flora and fauna. Today, the islands range in size from about 2.6 to 249 km^2 and are located between about 20 and 98 km from the mainland coast (Schoenherr et al. 1999:7).

Due to their relative isolation, the Channel Islands lack many of the terrestrial flora and fauna common on the mainland (Schoenherr et al. 1999:7–17). Historically, the largest endemic land mammals were the diminutive island fox (*Urocyon littoralis*), found on all the islands except Anacapa and Santa Barbara, and the island spotted skunk (*Spilogale gracilis*), found historically on Santa Rosa, Santa Cruz, and possibly San Miguel. Vegetation communities on the islands also are relatively impoverished compared to the mainland and consist primarily of introduced grasses and native coastal sage scrub, some oak and pine woodland, and chaparral. The Channel Islands exist in a coastal marine upwelling environment characterized by cool, nutrient-rich waters. The nearshore waters have dense kelp forests that provide basal productivity for a variety of aquatic resources including diverse populations of shellfish, fishes, sea birds, and one pinniped, the harbor seal. The remaining pinnipeds and cetaceans feed in offshore waters on coastal pelagic forage fish assemblages, which flourish in the nutrient-rich California Current waters. The location of the Channel Islands at the intersection of warm southerly and cool northerly ocean currents promotes a diverse mix of cold and warm water fauna. Some of the richest and most productive marine habitats in the world are located here, including kelp forests, rocky coasts and reefs, sandy beaches, and deep-water pelagic and benthic communities.

Historically, two different indigenous groups occupied the Channel Islands. The Chumash, who spoke an island dialect of the Chumashan language family, controlled the northern islands (Jones and Klar 2005:472), while the

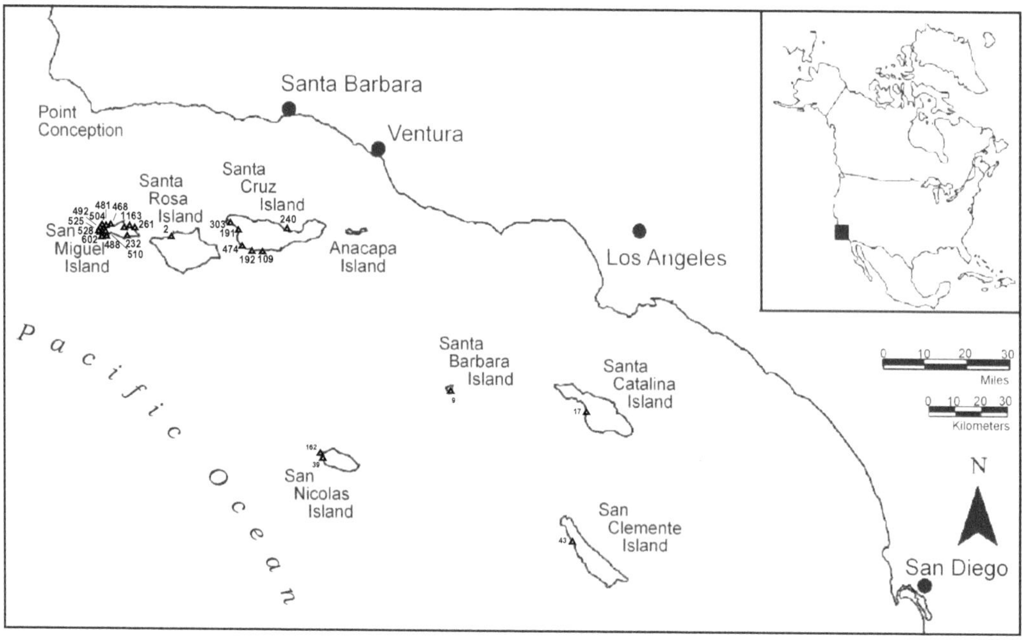

FIGURE 12.1. Map of the Santa Barbara Channel, the Channel Islands, and excavated archaeological sites with abundant sea mammal remains.

Uto-Aztecan-speaking Gabrielino, or Tongva, lived on the southern islands. Ethnographic accounts and archaeological data suggest that distinctive cultural traditions existed among these groups, but they shared a strong maritime orientation and technology. Over a century of archaeological research has demonstrated a long history of coastal and maritime lifeways on the islands, which focused on shellfish harvesting, fishing, bird and marine mammal hunting, and inter-island and mainland trading networks (Rick et al. 2005).

Evidence of marine mammal exploitation for subsistence, tool manufacture, and ornament production has been documented on each of the Channel Islands. Native Channel Islanders hunted or scavenged pinnipeds and sea otters, and occasionally dolphins and other small cetaceans (Glassow 2005; Porcasi and Fujita 2000). Although whale bones are found in many Channel Island archaeological sites, the Chumash and Tongva did not appear to actively engage in whaling, with most specimens probably scavenged from beached individuals (Heizer 1978). While considerable spatial and temporal variation exists, the general archaeological pattern on the islands seems to be a shift from a heavy emphasis on intertidal shellfish harvesting during the Early (ca. 11,000 to 7500 cal BP) and Middle (ca. 7500 to 3500 cal BP) Holocene to an increasing dependence on nearshore fishing and pinniped hunting during the Late Holocene (after ~3500 cal BP; Braje 2010; Erlandson et al. 2009; Rick et al. 2005). Intensive sea mammal hunting seems to occur at about 1500 cal BP, especially on San Miguel Island, when populations were growing and new boating and hunting technologies provided greater access to offshore sea mammal populations (Braje 2010; Colten 2002; Kennett 2005; Rick 2007; Rick et al. 2005; Walker et al. 2002).

CURRENT MODELS OF MARINE MAMMAL EXPLOITATION ON THE CHANNEL ISLANDS

A number of archaeologists have discussed ancient marine mammal hunting on the Pacific coast of North America and the Channel Islands. Perhaps the most widely debated has been

Hildebrandt and Jones' (1992) "tragedy of the commons" model (see also Jones and Hildebrandt 1995; Jones et al. 2004; Lyman 1995; Porcasi et al. 2000). Drawing partly on Channel Islands data, they employed optimal foraging theory to suggest that Native peoples hunted larger, more accessible marine mammals from rookery locations first (migratory/terrestrial breeders—fur seals and California sea lions) and turned to smaller, more difficult to capture resident/aquatic breeders (harbor seals and sea otters) after these populations were overexploited. Hildebrandt and Jones (1992) argued that Native hunters intensified their predation of marine mammals by developing more stable, ocean-going watercraft. However, offshore rookeries and breeding populations persisted due to the increased associated cost with pursuing and transporting marine prey.

Other researchers have questioned aspects of this model, pointing out that migratory breeders appear to be the most abundant species in Late Holocene assemblages from Santa Cruz, Santa Rosa, and San Miguel islands (Colten 2002; Colten and Arnold 1998; Rick 2007), even after Jones and Hildebrandt's (1995:85) AD 500 impact transition date. Colten and Arnold (1998) suggested that natural climate change and reduced marine productivity were the driving forces behind intensified sea mammal predation and no causal link can be made between technological innovations and pinniped predation. With a few exceptions, there is little zooarchaeological evidence to suggest that people were hunting sea mammals, either migratory or resident in the Channel Islands, intensively prior to about 1500 cal BP (Braje 2010; Erlandson et al. 2004; Kennett 2005; Rick 2007).

Kennett (2005), relying on behavioral ecological principles and optimal foraging theory, suggested that increased marine mammal hunting about 1500 cal BP was related to the introduction of the bow and arrow and the development of the plank canoe, or *tomol*, which made pinniped hunting a more energy-efficient strategy to feed growing island populations. These technologies may have been necessary for effective pinniped exploitation because modern behavioral observations suggest that human hunters (and possibly their domesticated dogs) on the islands would quickly force pinniped populations to offshore islets or rocks (Allen et al. 1971; Gerrodette and Gilmartin 1990; Riedman 1990; Wilkinson and Bester 1988), where access would be more difficult. Today, the consistent strength of northwesterly winds and swells on the islands makes landing on such offshore islets quite difficult even with modern technologies (e.g., inflatable boats and outboard motors), suggesting that accessing such landforms with earlier technologies would have been even more challenging.

Archaeologists have recently identified technological evidence from San Miguel, Santa Rosa, and Santa Cruz islands that Northern Channel islanders may have hunted marine mammals during the earliest well-documented occupations of the islands. Braje (2010), Erlandson and Braje (2007), Erlandson et al. (2005), Glassow et al. (2008), and Rick (2008) have recovered Channel Islands Barbed (a.k.a. Punta Arena) points from terminal Pleistocene or Early Holocene sites. These finely made, stemmed and barbed projectile points may be evidence of Paleocoastal sea mammal hunting between ca. 12,000 and 8500 years ago. Although faunal evidence for this early exploitation is currently lacking—possibly lost to sea level rise, coastal erosion, and the "Schlepp Effect"—early island hunters may have focused on pinnipeds and/or sea otters, disrupting these populations at a relatively early date.

Until recently, modeling human impacts on pinnipeds and sea otters was hindered by a dearth of archaeological data from the Channel Islands, especially on San Miguel Island, where the major breeding colonies occur today. Ongoing research at sites on or near the Point Bennett rookery are helping to better understand the nature of ancient human hunting of pinnipeds (see Walker et al. 2002). Studies in Late Holocene sites with fairly dense marine mammal assemblages elsewhere on San Miguel Island are also providing useful data that can help derive new models of ancient human hunting of pinnipeds (Braje 2010; Rick 2007). Finally, syntheses of unpublished and published data have also looked at ancient human hunting of sea otters to better understand changes in nearshore kelp forest ecosystems throughout the last 10,000 years and the role of Native peoples in the long-term structure and functioning of such habitats (Erlandson et al. 2005). Below we evaluate these data to build a comprehensive framework for understanding the trans-Holocene record of human exploitation of marine mammals on the Channel Islands.

MARINE MAMMAL BEHAVIOR, NATURAL HISTORY, AND MODERN DISTRIBUTIONS

Studies of pinniped and marine mammal ecology have been of great interest to a variety of scientists, especially since the advent of worldwide conservation efforts over the last 50 years. Restoration protocols have been so successful that seal and sea lion encounters with boaters, fishers, and sightseers have become a major point of concern (see Gales et al. 2003; Lunneryd et al. 2003; Yodzis 2001). While increasing populations of seals and sea lions have been documented, especially on the Channel Islands and in southern California (see Carretta et al. 2006; DeLong and Melin 2002), pinnipeds tend to recover slowly from overexploitation due to small litter size, large body mass, and long life span (K-selectionist strategy; Delany 1982:33). Nonetheless, the northern elephant seal has made a remarkable demographic recovery during the past 100 years, expanding from a small relict population on Isla de Guadalupe, Baja California, Mexico, in AD 1892 to a current population of approximately 150,000 animals (see Busch 1985; Ellis 2003:189–194). Other pinnipeds, the California sea otter, and several cetaceans have also made dramatic recoveries since the early- to mid-1900s.

Six species of pinnipeds are found today in southern California, including the California sea lion (*Zalophus californianus*), the northern elephant seal (*Mirounga angustirostris*), the northern fur seal (*Callorhinus ursinus*), the harbor seal (*Phoca vitulina*), and the relatively rare Steller sea lion (*Eumetopias jubatus*) and Guadalupe fur seal (*Arctocephalus townsendi*). Pinniped communities tend to cluster in large rookeries on San Miguel's west coast, but large populations of California sea lions, harbor seals, and elephant seals can be found clustered on beaches and rocky outcrops on the north, south, and east coasts. Equivalent-sized populations of sea lions, elephant seals, and harbor seals are also found at San Nicolas Island. Elephant seals breed on Santa Rosa, Santa Barbara, and San Clemente islands, and harbor seals breed on all the Channel Islands. Sea otters were once abundant throughout southern California, and a small population has been successfully reintroduced to San Nicolas Island. They have not been allowed to establish populations at the other islands, however, in an effort to reduce impacts by this voracious shellfish predator on sea urchin and abalone fisheries.

A number of cetacean species reside or visit Channel Island waters throughout the year. The largest of these include blue (*Balaenoptera musculus*), humpback (*Megaptera novaeangliae*), gray (*Eschrichtius robustus*), and killer whales (*Orcinus orca*), all of which make migratory or occasional visits to the Southern California Bight and Channel Islands (Carretta et al. 2006; Schoenherr et al. 1999:136–139). A variety of dolphins are found in island waters, with three species especially abundant: the short-beaked common dolphin (*Delphinus delphis*), Risso's dolphin (*Grampus griseus*), and the northern right-whale dolphin (*Lissidelphus borealis*; Carretta et al. 2006; Schoenherr et al. 1999:136–137). Except for a few Middle Holocene sites (Glassow 2005; Porcasi and Fujita 2000), however, cetacean remains are generally found in low frequencies in Channel Island middens—except where large whale bones were used for technological (i.e., house supports) or ceremonial (e.g., burial-related) purposes.

McLaren and Smith (1985) identified three primary drivers of marine mammal distribution and density: food availability, breeding space, and predation. These factors have resulted in the evolution of unique natural histories and feeding, breeding, and fight-flight behaviors in all of these species. In Table 12.1, we summarize the unique behavioral aspects of many Channel Island marine mammal species, providing data that may help to explain the archaeological record of Channel Island marine mammal exploitation.

Native peoples affected the demography of Pacific Coast marine mammals for millennia, but dramatic reductions of sea mammal populations in the North Pacific occurred following European contact. Spanish navigator Juan Rodríguez Cabrillo is credited with the first historic contact by Europeans of southern California and the Channel Islands. Cabrillo and three Spanish ships made landfall on Catalina and San Miguel islands in AD 1542/1543, ultimately wintering on San Miguel Island for 3 months (Wagner 1929). The next recorded contact occurred in 1602 when Spanish soldier and merchant Sebastían Vizcaíno was commissioned to locate safe harbors and map the California coastline. Beginning in AD 1769, Spanish colonization of Alta California began in earnest with the establishment of a series of missions, presidios, and pueblos on the mainland (Lightfoot 2004). With the decimation of Native peoples from Old World diseases and the incorporation of the survivors into colonial enterprises, wildlife populations in Alta and Baja California were temporarily freed from human predation pressure and may have undergone dramatic recoveries (Preston 2002).

The Spanish government showed little interest in establishing settlements along the dry mainland coast, but the luxuriant fur of the North Pacific sea otter attracted great attention (Ellis 2003:139–151). By the 18th century, Spanish traders were part of a frantic race to harvest as many sea otter skins as possible, a commodity fetching exorbitant prices in overseas Chinese markets. Fierce competition developed between the Spanish moving up from the south and the Russians, who brought skilled Native Alaskan hunters, moving down from the north. American, British, and other colonial and commercial interests soon joined the fray, and annual harvest averages rapidly declined due to overharvesting (Busch 1985; Ellis 2003:145; Ogden 1941). Cetaceans, pinnipeds, and other sea mammals suffered a similar fate, slaughtered by the thousands for skins, oil, fat, and medicinal purposes. Ironically, the rise of the petroleum industry early in the 20th century contributed to the restoration of North Pacific sea mammal populations, as the market for "train oil" was greatly reduced. Ultimately, however, federal protection has provided the key to the dramatic recovery of many marine mammal species along the California coast. Despite the transplantation of California sea otters to San Nicolas Island, most Channel Island waters continue to be largely devoid of otters.

ARCHAEOLOGICAL RECORDS FROM THE CHANNEL ISLANDS

Pinniped, sea otter, and cetacean remains have been documented in numerous sites from the Northern and Southern Channel Islands, but the importance of marine mammals varied both spatially and temporally (e.g., Braje 2010; Cassidy et al. 2004; Colten 1993, 2002; Colten and Arnold 1998; Erlandson et al. 2009; Lyman 1995; Meighan 1959; Porcasi et al. 2000, 2002; Raab et al. 1995; Rick 2007; Rick et al. 2005; Walker and Craig 1978; Walker et al. 2002; Walker and Snethkamp 1984). Excluding sites that contain only minor concentrations of sea mammal remains, archaeologists have sampled and documented significant densities (≥1 kg/m^3 and/or ≥100 $NISP/m^3$) of marine mammal bone at 24 sites from the Northern and Southern Channel Islands (see Figure 12.1 and Table 12.2): six (25.0%) from Santa Cruz, one (4.2%) from Santa Rosa, 13 (54.2%) from San Miguel, two (8.3%) from San Nicolas, one (4.2%) from Santa Catalina, and one (4.2%) from San Clemente. While these data provide a general distribution of ancient marine mammal

TABLE 12.1

Major Marine Mammals Identified in Channel Islands Archaeological Sites

TAXA	ADULT BEHAVIOR	BREEDING HABITS	NEWBORN BEHAVIOR	HISTORIC/MODERN DISTRIBUTION	ARCHAEOLOGY
PINNIPEDS					
Arctocephalus townsendi (Guadalupe fur seal)	Visit Channel Islands 12 months per year	Breeds on rocks at Guadalupe Island	Mothers leave helpless pups for up to 13 days at a time to feed, pups nurse for 8–9 months	Rare on islands today, 1 pup born on SMI in 1997	Common in sites after 1500 cal BP
Callorhinus ursinus (Northern fur seal)	Visit Channel Islands 4–5 months per year, skittish	Breeds on sand and rock beaches	Pups helpless for several weeks, weaning after 4 months	Common and breeds on SMI today	Common in sites after 1500 cal BP
Eumetopias jubatus (Steller sea lions)	Frequent long-range travel at least 8 months per year, skittish	Breeds on beaches and haul-outs	Pups able to swim at birth	Common on San Miguel in early 1900s, rare today	Rare in sites of any age
Mirounga angustirostris (Elephant seal)	Biannual long-range travel at least 8 months per year, relatively tolerant of humans	Breeds on beaches and haul-outs	Pups weaned at 1 month, helpless, remain on land for <4 months	Common and breeds on SMI, SRI, SBI, SNI today	Rare in sites of any age
Phoca Virtulina (Harbor seal)	Year round resident of Channel Islands, skittish	Breeds in water	Pups able to swim	Common and breeds on San Miguel today	Relatively rare in sites of any age
Zalophus californianus (California sea lion)	Females resident; males migrate for <9 months per year, skittish	Breeds on beaches	Pups helpless, left for extended periods for <11 months while mother feeds	Common and breeds on SMI, SNI, SBI, SCI today	Common in sites after 1500 cal BP
CARNIVORES					
Enhydra lutris (Sea otter)	Year-round residents, skittish, primarily in water	Breeds in water	Pups able to swim and highly mobile at birth	Reintroducted to San Nicolas, restricted populations elsewhere	Common in sites after 9000 cal BP

CETACEANS					
Delphinus delphis (Common dolphin)	Social, travels in groups, feed on coastal pelagic fish	Breeds in water	Calves able to swim and highly mobile at birth	Resident in Channel Island waters	Rare in most sites of any age
Grampus griseus (Risso's dolphin)	Travels in groups in nearshore and deep water	Breeds in water	Calves able to swim and highly mobile at birth	Common visitors to Channel Island waters	Rare in most sites of any age
Lisodelphis borealis (Northern right whale dolphin)	Travels in groups in nearshore and deep water	Breeds in water	Calves able to swim and highly mobile at birth	Common visitors to Channel Island waters	Rare in most sites of any age
Orcinus orca (Killer whale)	Travels in pods in nearshore or deep water	Breeds in water	Calves able to swim and highly mobile at birth	Rarely visits the Channel Islands	Rare in most sites of any age, likely scavenged
Tursiops truncate (Bottlenose dolphin)	Travels in groups in nearshore and deep water	Breeds in water	Calves able to swim and highly mobile at birth	Common visitors to Channel Island waters	Rare in most sites of any age
Eschrichtius robustus (Gray whale)	Migrate past in near-shore waters	Breeds in water	Calves able to swim and highly mobile at birth	Migrant through Channel Island waters in winter and spring; occasionally reside during summer	Rare in sites, likely scavenged
Balaenoptera musculus (Blue whale)	Island offshore waters	Breeds in water	Calves able to swim and highly mobile at birth	Resident May-October	Rare in sites, likely scavenged
Megaptera movaeangliae (Humpback whale)	Island offshore waters	Breeds in water	Calves able to swim and highly mobile at birth	Resident April-October	Rare in sites, likely scavenged

NOTE: Data compiled from Belcher and Lee (2002); Carretta et al. (2006); Estes (1980); Loughlin et al. (1987); Schoenherr et al. (1999); and Stewart and Huber (1993).

exploitation across the Channel Islands, significant gaps and methodological biases exist in the record. Scientific archaeological research has been discontinuous across the islands, faunal assemblages from some sites have not been completely analyzed or reported on, and early archaeological projects tended to focus on artifacts and features and not faunal materials.

Despite these shortcomings, a synthesis of the current data suggests some interesting spatial and temporal patterns on the Channel Islands. The most extensive and intensive record of marine mammal hunting—both zooarchaeological and technological—on the Channel Islands is found on San Miguel Island (see below). Most of the earliest sites contain very small amounts of marine mammal bone, but several have produced technological evidence that hunting was more important than indicated by faunal evidence (see Braje 2010; Erlandson et al. 2005, 2008). Daisy Cave (CA-SMI-261), on San Miguel Island, currently contains the earliest evidence of marine mammal hunting (i.e., sea otters ~9000 years ago), although dolphin hunting occurs nearly as early on San Clemente Island at the Eel Point site (CA-SCLI-43).

Table 12.2 provides a summary of sites on the Channel Islands with significant faunal evidence for marine mammal predation. Some of these sites lack tight chronological controls and consistent reporting of data, so a comparison of relative abundances through time and between sites is often not possible. However, the overall pattern seems to be an increase in marine mammal predation during the Late Holocene (after ca. 3500 cal BP), with some evidence of Early or Middle Holocene predation on Santa Cruz, San Miguel, Santa Catalina, and San Clemente islands.

HUMAN-MARINE MAMMAL DYNAMICS • *A Historical Ecology for the Northern Channel Islands*

Historical ecology is a rapidly expanding field that involves the synthesis of ecological, historical, archaeological, and paleontologial data to document the structure of past and present ecosystems and identify sources of anthropogenic and natural change. Having reviewed the general ecology, history, and archaeology of marine mammal populations on the Channel Islands, we now examine spatial and temporal variation in their distribution on the Northern Channel Islands. We use the Northern Channel Islands as a case study to identify instances of continuity and disruption of marine mammals as a means of better understanding the long-term history of local marine mammal populations.

The Northern Channel Islands have been of interest to antiquarians and archaeologists for over a century, with a history of archaeological research using modern scientific methods dating back at least 40 years (Baldwin 1996). In the mid-20th century, seminal work by Phil Orr (1968) on Santa Rosa, Charles Rozaire (1978) on San Miguel, and other scholars contributed to our understanding of the chronology, cultural history, material culture, and ecological change on the islands. By now, all the islands have been extensively (but not completely) surveyed by archaeologists, but the history of subsurface excavations has varied. San Miguel Island has the most extensive coverage of archaeological site sampling and research. Phillip Walker, Mike Glassow, and their students from the University of California at Santa Barbara have conducted investigations for several decades on San Miguel and Santa Cruz islands. Jon Erlandson, Douglas Kennett, and their students at the University of Oregon have worked on a variety of San Miguel Island and Santa Rosa Island sites for over two decades, and Torben Rick, at the Smithsonian, and his students recently completed excavations on Anacapa Island. Jeanne Arnold, working with her students at the University of California-Los Angeles and other colleagues, has a long history of research on Late Holocene sites of western and southern Santa Cruz Island, where Glassow and his students have also done important work on trans-Holocene sequences (e.g., Glassow et al. 2008; Sharp 2000). Eastern Santa Cruz Island and Santa Rosa Island have seen significantly less

TABLE 12.2

Archaeological Sites on the Channel Islands with Significant Evidence of Sea Mammal Predation

SITE	AGE (CAL BP)	FAUNAL EVIDENCE OF SEA MAMMAL HUNTING	SOURCE
SANTA CRUZ ISLAND			
CA-SCRI-109	6300–5300	Abundant northern right whale dolphin, common dolphin, Pacific white-sided dolphin, and Risso's dolphin, lesser amounts of pinniped	Glassow 2005; Glassow et al. 2008
CA-SCRI-191	2600–0	Abundant otariids and sea otter, lesser amounts of harbor seal	Colten 1995, 2002; Colten and Arnold 1998
CA-SCRI-192	650–0	Abundant otarriids, Guadalupe fur seal, and sea otter, lesser amounts of harbor seal	Colten 1995, 2002; Colten and Arnold 1998
CA-SCRI-240	5300–500	Abundant fur seal, Guadalupe fur seal, and sea lion, lesser amounts of Northern fur seal, harbor seal, and sea otter	Colten 1995, 2002; Colten and Arnold 1998
CA-SCRI-330	650–0	Abundant otarriids and sea otter, lesser amounts of fur seal and harbor seal	Colten 1995, 2002; Colten and Arnold 1998
CA-SCRI-474	2600–800	Abundant otariids and sea otter, lesser amounts of fur seal and harbor seal	Colten 1995, 2002; Colten and Arnold 1998
SANTA ROSA ISLAND			
CA-SRI-2	260–0	Abundant sea otter and otariid, lesser amounts of Guadalupe fur seal, northern fur seal, and cetacean	Rick 2004
SAN MIGUEL ISLAND			
CA-SMI-1	7000–3000	Abundant otariids and fur seal, lesser amounts of sea otter and harbor seal	Colten 2002; Colten and Arnold 1998
CA-SMI-163	700–400	Small amounts of Guadalupe fur seal, northern fur seal, otariid, sea otter, and harbor seal	Rick 2004, 2007
CA-SMI-232	1300–1050	Abundant Guadalupe fur seal, lesser amounts of California sea lion, cetacean, harbor seal, and sea otter	Braje 2010
CA-SMI-261	9800–600	Abundant otariids, lesser amounts of sea otter, fur seal, and harbor seal	Colton 2002
CA-SMI-468	900–500	Small amounts of otariid, northern fur seal, California sea lion, phocid, and elephant seal	Rick 2007
CA-SMI-481	1300–900	Abundant Guadalupe fur seal, lesser amounts of California sea lion, northern fur seal, sea otter, and harbor seal	Rick 2007

(continued)

TABLE 12.2 *(continued)*

SITE	AGE (CAL BP)	FAUNAL EVIDENCE OF SEA MAMMAL HUNTING	SOURCE
CA-SMI-488	<800	Small amounts of sea otter, sea lion, and harbor seal	Colten 2002; Hildebrandt and Jones 1992
CA-SMI-492	5600–1250	Small amounts of sea lion, sea otter, Guadalupe fur seal, fur seal, and harbor seal	Colten 2002
CA-SMI-504	3050–2850	Small amounts of sea lion and sea otter	Colten 2002
CA-SMI-510	1300–1100	Small amounts of fur seal, sea otter, and harbor seal	Colten 2002
CA-SMI-525	2700–100	Small amounts of fur seal, sea lion, sea otter, elephant seal, and harbor seal	Hildebrandt and Jones 1992
CA-SMI-528	1600–1100	Abundant Guadalupe fur seal, lesser amounts of sea otter, northern fur seal, California sea lion, Steller sea lion, harbor seal, and elephant seal	Walker et al. 2002
CA-SMI-602	500–0	Abundant Guadalupe fur seal, lesser amounts of sea otter, northern fur seal, elephant seal, California sea lion, harbor seal, and Steller sea lion	Walker et al. 2002
SANTA BARBARA ISLAND			
CA-SBI-9	2680–2120	Abundant otariids, seals, and sea lions, lesser amounts of cetacean and sea otter	Rick 2001
SAN NICOLAS ISLAND			
CA-SNI-39	1500–270	California sea lion most common along with lesser amounts of sea otter, whale, dolphin, and porpoise	Fagan et al. 2006
CA-SNI-162	2000–750	California sea lion most common along with lesser amounts of sea otter	Fagan et al. 2006
SANTA CATALINA ISLAND			
CA-SCAI-17	7900–500	Abundant dolphin, California sea lion, and Guadalupe fur seal	Raab et al. 1995
SAN CLEMENTE ISLAND			
CA-SCLI-43	9000–7000	Abundant dolphin, lesser amounts of sea otter and pinnipeds	Porcasi et al. 2000
CA-SCLI-43	7000–3500	Abundant dolphin and pinnpeds, with lesser amounts of large cetacean and sea otter	Porcasi et al. 2000
CA-SCLI-43	3500–500	Abundant sea otter and pinniped, lesser amounts of dolphin and large cetacean	Porcasi et al. 2000

archaeological excavation, but Jennifer Perry (Ponoma College) and Torben Rick are working to fill these respective gaps.

One problem in comparing existing datasets on marine mammal exploitation from the Channel Islands is the lack of systematic and consistent reporting of faunal data. Colten (1993, 2002) and Colten and Arnold (1998) report only densities of combined bird and mammal bone from Late Holocene sites on western Santa Cruz (CA-SCRI-191, 192, 330, and 474) by coarse-grained cultural time periods (Middle, Transitional, Late, and Historic periods), for example, without reporting specific MNI, NISP, or bone weight values. Temporal changes of sea mammal bone density within individual sites cannot be precisely calculated and only approximations can be made from the published data. Other sites such as CA-SMI-1 were excavated by early archaeologists without tight stratigraphic controls (Rozaire 1965; see Erlandson 1991 for a description of difficulties in interpretation), and others remain unpublished or incompletely analyzed. Another significant problem has been the heavy reliance of Channel Island archaeologists on column or bulk sampling of sites (e.g., Glassow 1980; Kennett 2005), with excavation volumes too small to adequately represent the remains of marine mammals or the technologies used to procure them, which are often widely scattered in shell middens (Erlandson et al. 2008).

Such methodological problems limit the ability of archaeologists to effectively evaluate some important models related to human hunting of marine mammal populations along the Pacific coast of North America. One of the more influential examples is the debate over the "tragedy of the commons" model proposed by Hildebrandt and Jones (1992). Since this model was first published, a number of zooarchaeologists, paleoecologists, and other scientists have published responses critiquing or supporting this model with faunal, isotopic, and/or paleoecological evidence (e.g., Erlandson et al. 1998; Etnier 2002; Gifford-Gonzalez et al. 2005; Jones and Hildebrandt 1995; Jones et al. 2004; Lyman 1995; Newsome et al. 2007; Porcasi et al. 2000). Arguments have tended to center on the types and timing of prey species—migratory or resident breeders—and the identification and accessibility of rookery locations. In the process, many researchers have reported data aggregated in a manner suitable for an evaluation of the "tragedy of the commons" model (i.e., as resident vs. migratory taxa or by broad swaths of space and time) but not for other research issues that require species-specific data for individual components or sites.

Despite these problems, a substantial set of zooarchaeological data exists from the Northern Channel Islands for evaluating spatial and temporal patterns of marine mammal hunting. We scanned published and unpublished data from scores of Channel Island archaeological sites to compile data from 15 Northern Channel Island sites that contain at least one dated shell midden component with abundant densities of sea mammal bone, as described by the project investigators (Figure 12.2). Ten of these (58.8%) are on San Miguel Island, one (5.9%) on Santa Rosa, and six (35.3%) on Santa Cruz.

These data suggest that the density (g/m^3 or NISP/m^3) of sea mammal bone in archaeological sites generally decreases from west to east, with the highest densities on San Miguel Island (range=43.0 to 0.9 kg/m^3; 15,337 to 107 NISP/m^3), with four sites containing more than 10.0 kg/m^3 and lower densities in Santa Rosa and Santa Cruz island sites (Table 12.3; Figure 12.2). While more quantitative data are available from San Miguel Island, investigations at four sites on Santa Cruz Island and one site on Santa Rosa Island support this pattern. Colten (1995, 2002), Colten and Arnold (1998), and Glassow (2005) investigated a series of sites with relatively abundant marine mammal bone on western and southern Santa Cruz, for example, and found that none of the sites produced marine mammal bone densities greater than 2.9 kg/m^3. Investigators at each of these sites noted the conspicuously high densities of sea mammal bone (Colten 1995, 2002; Colten and Arnold 1998; Glassow 2005; Glassow et al.

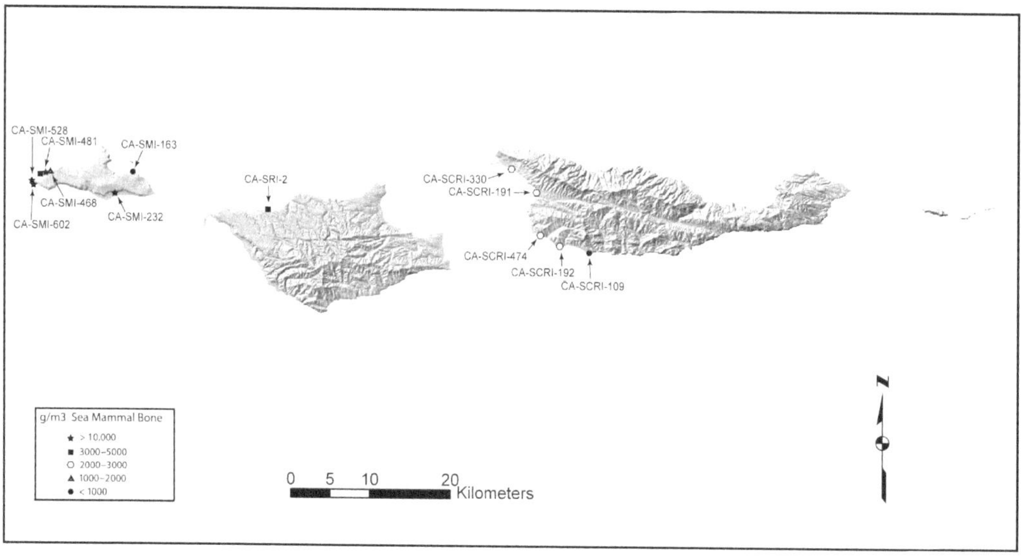

FIGURE 12.2. Location map of excavated archaeological sites on the Northern Channel Islands with intensive sea mammal predation.

2008) compared to other sites nearby. Compared to four San Miguel Island sites (CA-SMI-232, CA-SMI-481, CA-SMI-528, and CA-SMI-602), however, these densities are quite low.

The highest densities of sea mammal remains seem to cluster near western San Miguel Island (CA-SMI-481: 37.3 kg/m^3, CA-SMI-528: 43.1 kg/m^3, and CA-SMI-602 10.9 kg/m^3) and the modern Point Bennett rookery area, along with one site on the south-central coast of San Miguel Island (CA-SMI-232: 32.7 kg/m^3). Today, Point Bennett is one of the largest and most diverse pinniped rookeries in the world, occupied by over 150,000 animals yearly (DeLong and Melin 2002). Six different pinniped species regularly visit the island, with breeding populations dominated by northern elephant seals, California sea lions, northern fur seals, and harbor seals, as well as occasional visits by Steller's sea lion and Guadalupe fur seals. While there is no evidence to suggest that this area functioned as a prehistoric rookery (see Braje 2010; Braje and DeLong 2009; Rick 2007; Walker et al. 2002), the high density of sea mammal bone suggests a long-term continuity in the distribution of pinnipeds with the densest populations centered near western San Miguel and declining abundances trending east.

Northern Channel Islands data also indicate that the late Middle Period, from about 1600 to 1100 cal BP, was a critical time for the increased exploitation of sea mammals. Three sites (CA-SMI-232, CA-SMI-481, and CA-SMI-528) dating to this interval, all on San Miguel Island, contain dramatically higher densities of sea mammal bone than any other Northern Channel Island sites. Transitional Period (800–650 cal BP) and Late Period (after 650 cal BP) sites show dramatic reductions in sea mammal bone density and a number of researchers have documented a shift toward an economy based on intensive fishing (Arnold 1991, 1992, 2000, 2001; Arnold and Graesch 2001; Kennett 2005; Rick 2007; Walker et al. 2002). These data suggest that pinniped hunting was a particularly important strategy for San Miguel Islanders during the late Middle Period, likely because of the greater accessibility of marine mammals in San Miguel waters.

Across the Northern Channel Islands, marine mammal hunting seems to have focused primarily on Guadalupe fur seals (Rick et al. 2009). While researchers at a number of sites did not identify pinniped remains beyond the family level (otariid or fur seal, usually due to the fragmentary nature of the bone), when specific identifications could be made the dominant species was usually Guadalupe fur seal. Two exceptions have been documented, dolphins in the Middle Holocene component at the Punta Arena site (CA-SCRI-109) on Santa Cruz Island and northern fur seal at CA-SMI-468 on San Miguel Island. The Punta Arena site is located adjacent to a submarine canyon, which probably allowed for greater access to dolphin populations (Glassow 2005), and CA-SMI-468 contained a relatively small sea mammal bone assemblage, a high percentage of which was undifferentiated suggesting that this pattern may be a result of sampling error (see Rick 2007).

This pattern of heavy Guadalupe fur seal exploitation may be best explained by an understanding of the modern behavior of North Pacific sea mammals (see Table 12.1). Today, Guadalupe fur seals visit the Channel Islands, albeit in small numbers, 11 months out of the year. Other species of pinnipeds such as northern fur seals, Steller sea lions, elephant seals, and California sea lions make frequent long-range travels for approximately 8 months out of the year. Sea otters, harbor seals, Guadalupe fur seals, and various cetaceans, then, were probably the only species available for nearly year-round harvest. While some large cetaceans are only seasonal visitors to the Channel Islands today (killer, sperm, and baleen whales), all of the cetaceans were probably difficult to capture due to the high mobility and elusiveness of both calves and adults. Harbor seals and sea otters were probably more difficult to harvest than Guadalupe fur seals because they breed in water, pups are able to swim at birth, and mothers do not need to haul out to feed their young. The conclusion that Guadalupe fur seals were the most abundant prehistoric pinniped species on the Channel Islands, then, may not be accurate. Rather, their abundance in ancient Channel Islands marine mammal assemblages may be a reflection of their year-round availability due to their breeding and feeding behaviors and not just their prehistoric abundance.

Species level identifications of marine mammal remains from Channel Island archaeological sites are also striking for what is not found. Today, northern elephant seals are ubiquitous on Channel Island beaches, with breeding populations on San Miguel, Santa Rosa, Santa Barbara, and San Nicolas islands. These creatures are somewhat tolerant of humans, breed on beaches, and leave their weaned young on the rookery beach unprotected for several months. Despite these behavioral characteristics that would have made them easy prey for Native hunters, elephant seal remains are rare in island archaeological sites of any age (Rick et al. 2010). It may be that the first human visitors to the Channel Islands targeted elephant seals during the terminal Pleistocene as easy-to-capture meat and oil sources. Direct human predation on the population, even by small numbers of people, and particularly heavy human settlement adjacent to prime beaches used by elephant seals, may have driven elephant seal populations from Santa Barbara Channel waters to remote island refugia such as Isla de Guadalupe where they were safe from human hunters. Evidence of this early hunting may have been lost to rising sea levels and the inundation of low-lying Pleistocene island landscapes where hunting processing camps were established.

California sea lions are also common on the Channel Islands today, with breeding colonies on San Miguel, San Nicolas, Santa Barbara, and San Clemente islands, where they breed on beaches and haul-outs, and leave their young helpless for extended periods. While California sea lions are fairly common in archaeological sites postdating 1500 cal BP, sex identifications suggest that most all of these remains from the Northern Channel Islands are from adult males (see Braje 2010; Rick 2007). Unlike adult male California sea lions, which make long-distance migrations for up to 9 months of the year, females and dependent young are resident and should have been more available to island hunters. Evidence that early humans hunted

TABLE 12.3

Archaeological Sites on the Northern Channel Islands with Intensive Sea Mammal Predation

SITE	AGE (CAL BP)	SEA MAMMAL BONE G/M^3	SEA MAMMAL NISP/M^3	DOMINATE MARINE MAMMAL	SOURCE
SANTA CRUZ ISLAND					
CA-SCRI-109	6300–5300	N/A	301	Dolphin	Glassow 2005
CA-SCRI-191[a]	2600–0	2247[b]	2729[b]	Otariid or sea otter	Colten 1995; Colten and Arnold 1998
CA-SCRI-192[a]	650–0	2809[c]	4443[c]	Otariid	Colten 1995; Colten and Arnold 1998
CA-SCRI-240	5300–500	N/A	N/A	Fur seal	Colten 1995, 2002
CA-SCRI-330[a]	650–0	2851	3275	Otariid or sea otter	Colten 1995; Colten and Arnold 1998
CA-SCRI-474[a]	2600–800	2525	2951	Otariid	Colten 1995; Colten and Arnold 1998
SANTA ROSA ISLAND					
CA-SRI-2	260–0	4322	3020	Sea otter	Rick 2004
SAN MIGUEL ISLAND					
CA-SMI-1	7000–3700	N/A	N/A	Fur seal	Colten and Arnold 1998
CA-SMI-1	<3700	N/A	N/A	Fur seal	Colten and Arnold 1998
CA-SMI-163	700–400	922	107	Guadalupe fur seal	Rick 2004, 2007
CA-SMI-232	1300–1050	32,700	10,000	Guadalupe fur seal	Braje 2010
CA-SMI-468	900–500	1641	260	Northern fur seal	Rick 2004, 2007

CA-SMI-481	6200–5800	3855	N/A	N/A	Vellanoweth et al. 2006
CA-SMI-481	1300–900	37,297	15,337	Guadalupe fur seal	Rick 2004, 2007
CA-SMI-525	2700–100	N/A	N/A	Guadalupe fur seal	Walker and Snethkamp 1984
CA-SMI-528	1600–1100	43,063	N/A	Guadalupe fur seal	Walker et al. 2002
CA-SMI-602	500–0	10,854	N/A	Guadalupe fur seal	Walker et al. 2002

[a] Reported as density of bird and mammal bone combined, although the majority is probably sea mammal bone (Colten and Arnold 1998).
[b] Average of Colten and Arnold's (1998) Middle, Transitional, and Late Periods.
[c] Average of Colten and Arnold's (1998) Late and Historic Periods.

California sea lion females is still found in their profound fright-flight reaction to human presence. That this is a response conditioned by exposure to human predation is evident in the lack of fear of humans in the Galapagos Island populations, which evolved in the absence of humans. Variation is evident in this pattern, however, as the majority of the California sea lion remains from Eel Point on San Clemente Island have been identified as females (Porcasi et al. 2000). Again, evidence of heavy hunting of female California sea lions on the Northern Channel Islands may be found in terminal Pleistocene age deposits that are now submerged or destroyed.

CONCLUSIONS

Synthesis of the archaeological record of marine mammal hunting from California's Channel Islands reveals a long history of human exploitation that began more than 12,000 years ago. Faunal evidence for an emphasis on marine mammal hunting is largely absent in terminal Pleistocene and Early Holocene components, although the presence of barbed and serrated projectile points in several early sites suggests that pinniped or sea otter hunting may have been more important than faunal remains suggest. There is limited faunal evidence for a possible intensification of sea mammal hunting at some Middle Holocene sites, but a dramatic intensification of pinniped hunting appears to occur between about 1500 and 1200 cal BP, especially at San Miguel Island. This was a turbulent time in Channel Islands prehistory, when human populations appear to have been expanding, people were coalescing in large coastal villages, important technological changes were taking place, trade and interaction were intensifying, and interpersonal violence and other health problems increased significantly (Kennett 2005; Rick et al. 2005). By the late Middle Period (~1200 cal BP) if not earlier, it seems likely that pinniped populations were largely confined to offshore islets, especially given the long history of maritime hunters (and domesticated dogs) on the islands. Sea mammal hunting may have been a successful strategy to feed growing populations with the aid of new hunting and seafaring technologies, the plank canoe, and the bow and arrow (Arnold 2007; Blitz 1988; Braje and DeLong 2008; Kennett 2005; Rick 2007). This seems to have been an especially successful adaptation for San Miguel islanders, who may have traded sea mammal meat to the other islanders and mainland peoples (see Colten 1993; Erlandson et al. 2008).

The zooarchaeological record of sea mammal hunting also offers evidence for long-term spatial continuity and ecological resilience across the Northern Channel Islands. The densest ancient populations of sea mammals appear to have been concentrated around San Miguel Island and particularly western San Miguel Island, with decreasing abundances to the east within the archipelago. Modern distributions follow an analogous pattern with western San Miguel Island and Point Bennett serving as the nexus for pinniped breeding and haul-out communities and smaller populations and less frequent occupation on each of the islands to the east. Ultimately, the critical habitat around San Miguel Island has sustained marine mammal populations for millennia, where species appear to have been relatively resilient in spite of heavy predation pressure, especially after ca. 1500 cal BP. Despite a 19th-century blitzkrieg which resulted in the near extinction of many marine mammal species, conservation and recovery efforts have been extremely successful and many modern marine mammal populations seem to be reoccupying similar locations as their ancient ancestors.

Despite the impacts on the distribution of prehistoric pinniped populations, the densest accumulations of pinniped bone are found in sites on San Miguel Island, which supports the largest pinniped populations on the Channel Islands today. The archaeological data suggest that although Native Americans significantly affected the distribution, demographics, and ecology of sea mammals on California's Channel Islands for millennia, the modern pattern of pinniped populations being concentrated toward the western end of the Northern Channel Islands appears to have been maintained for millennia.

A similar pattern may be evident on the Southern Channel Islands. While much less published archaeological data are available, several shell-midden sites on the southern islands suggest that marine mammals were an important prey resource. Today, large breeding populations of northern elephant seals and California sea lions occupy Santa Barbara, San Clemente, and San Nicolas islands, and several Late Holocene shell middens located along the island perimeters have yielded relatively high densities of pinniped bone–suggesting a long-term continuity in sea mammal distribution (see Fagan et al. 2006; Porcasi et al. 2000; Raab et al. 1995; Rick 2001). Similarly, annual migrations of dolphins travel past San Clemente and Santa Catalina islands today and may account for the relatively early and dense concentrations of dolphin bones from sites on these islands–CA-SCLI-43 and CA-SCAI-17 (Porcasi and Fijita 2000).

Throughout the Channel Islands, zooarchaeological data suggest that Guadalupe fur seals were often the focus of the prehistoric marine mammal hunting economy in the Late Holocene. While they have been relatively slow to recover from historic overhunting, only occasionally visiting the Channel Islands today, archaeological data suggest that they were more abundant in the past. While this may be true, they were not necessarily the most abundant pinniped species on the islands. Rather, their breeding and feeding behaviors may have made them more available to ancient hunters on a year-round basis than some other pinniped species.

Archaeological, historical, and modern data suggest some continuity in pinniped populations on the Northern Channel Islands through the Holocene, with most marine mammal populations showing some resilience in the face of intensive predation and dynamic environmental conditions through the Holocene. Despite increasingly intensive human harvesting after ~1500 cal BP and severally overhunted historic populations, most sea mammal species have survived 10,000 years of human hunting on the Northern Channel Islands. The long-term continuity and persistence of Channel Island wind and coastal upwelling may have provided the marine productivity for schooling fish, squid, and other marine fishes that have supported marine mammal communities in the Santa Barbara Channel for millennia. The consistent orientation (NW-SE) of island dunes of Pleistocene and Holocene age suggests that the predominant pattern of strong northwesterly winds that drives upwelling and productivity in the Channel Islands area has operated for many millennia. These winds are especially powerful along western San Miguel and San Nicolas islands and help regulate the body temperatures of large pinnipeds while hauled-out on rocky outcrops and beaches, a pattern that may help explain the resilience and spatial continuity of marine mammal populations along the Northern Channel Islands throughout the Holocene. Kinlan et al. (2005) also argued that Northern Channel Island kelp forests, which contribute significantly to nearshore productivity, have been highly productive throughout the Holocene.

It is increasingly apparent that many nearshore marine habitats have been harvested with intensities sufficient to alter ecosystem structures for millennia (Jackson et al. 2001; Erlandson and Rick 2010; Rick and Erlandson 2008). Disruptions of marine mammal populations, distributions, and densities have been especially severe, beginning at least in the Historic Period and perhaps extending into the deep past. Using archaeological data to help decipher the deeper ecological histories of marine mammal communities can provide indispensable information about the processes involved in the creation of the anthropogenic land- and seascapes so pervasive in today's "natural" world. Building effective management and restoration plans for North Pacific marine mammal communities requires information about historic and ancient harvests and the establishment of reference points (baselines) that account for overfishing or other anthropogenic disturbances (Dayton et al. 1998; Jackson et al. 2001; Pauly 1995; Pauly et al. 1998; Tegner and Dayton 2000). While the relative densities of marine mammal species along the Santa Barbara Channel may have undergone considerable alteration as a result of historic and prehistoric exploitation, archaeological evidence for long-term spatial continuity and ecological resilience provides hope for the restoration of viable, stable, and healthy pinniped and marine mammal communities.

ACKNOWLEDGEMENTS

Our research was supported by the National Science Foundation (multiple grants to Erlandson, Rick, and Braje), Channel Islands National Park (CINP), the Marine Conservation Biology Institute, the National Marine Fisheries Service, Western National Parks Association, and our home institutions. At CINP, we thank Ann Huston, Kelly Minas, Ian Williams, Don Morris, and Georganna Hawley for their support. We compiled data from the research of several Channel Island archaeologists—we are grateful to Jeanne Arnold, Roger Colten, Mike Glassow, Bill Hildebrandt, Terry Jones, Doug Kennett, Judy Porcasi, Mark Raab, Pandora Snethkamp, René Vellanoweth, and Phil Walker. Finally, we thank the production editors and anonymous reviewers for helpful comments and assistance in the revision of this chapter.

REFERENCES CITED

Allen, S. G., D. G. Ainley, G. W. Page, and C. A. Ribic
1971 The Effect of Disturbance on Harbor Seal (*Phoca vitulina*) Haul-Out Patterns at Bolinas Lagoon, California. *U.S.National Marine Fisheries Service Fishery Bulletin* 82:493–500.

Arnold, J. E.
1991 Transformation of a Regional Economy: Sociopolitical Evolution and the Production of Valuable in Southern California. *Antiquity* 65:953–962.
1992 Complex Hunter-Gatherer-Fishers of Prehistoric California: Chiefs, Specialists, and Maritime Adaptations of the Channel Islands. *American Antiquity* 57(1):60–84.
2000 The Origins of Hierarchy and the Nature of Hierarchical Structures in Prehistoric California. In *Hierarchies in Action: Cui Bono?* edited by M. W. Diehl, pp. 221–240. Center for Archaeological Investigations, Carbondale.
2007 Credit Where Credit Is Due: The History of the Chumash Oceangoing Plank Canoe. *American Antiquity* 72(2):196–209.

Arnold, J. E. (editor)
2001 *The Origins of a Pacific Coast Chiefdom: The Chumash of the Channel Islands.* University of Utah Press, Salt Lake City.

Arnold, J. E., and A. P. Graesch
2001 The Evolution of Specialized Shellworking among the Island Chumash. In *The Origins of a Pacific Coast Chiefdom: The Chumash of the Channel Islands,* edited by J. E. Arnold, pp. 71–112. The University of Utah Press, Salt Lake City.

Baldwin, K. P.
1996 *A History of Channel Islands Archaeology.* Master's thesis, Department of Anthropology, California State University, Northridge.

Belcher, R. L., and T. E. Lee, Jr.
2002 *Arctocephalus townsendi. Mammalian Species The American Society of Mammalogists* 700:1–5.

Blitz, J. H.
1988 Adoption of the Bow and Arrow in Prehistoric North America. *North American Archaeologist* 9:123–145.

Braje, T. J.
2010 *Modern Oceans, Ancient Sites: Archaeology and Marine Conservation on San Miguel Island, California.* University of Utah Press, Salt Lake City.

Braje, T. J., and R. L. DeLong
2009 Ancient Sea Mammal Exploitation on the South Coast of San Miguel Island. *Proceedings of the Seventh California Islands Symposium* in press.

Busch, B. C.
1985 *The War against the Seals: A History of the North American Seal Fishery.* McGill-Queen's University Press, Kingston, Ontario.

Carretta, J. V., K. A. Forney, M. M. Muto, J. Barlow, J. Baker, B. Hanson, and M. S. Lowry
2006 U.S. Pacific Marine Mammal Stock Assessments: 2005. NOAA Technical Memorandum NMFS. U.S. Department of Commerce, National Oceanic and Atmospheric Administration, National Marine Fisheries Service, Southwest Fisheries Science Center.

Cassidy, J., L. M. Raab, and N. A. Kononenko
2004 Boats, Bones, and Biface Bias: The Early Holocene Mariners of Eel Point, San Clemente Island, California. *American Antiquity* 69(1): 109–130.

Clementz, M. T.
2002 *The Evolution of Herbivorous Marine Mammals: Ecological and Physiological Tranitions during the Evolution of the Orders Sirenia and Des-*

mostylia. Ph.D. dissertation. University of California, Santa Barbara.
Colten, R. H.
1993 *Prehistoric Subsistence, Specialization, and Economy in a Southern California Chiefdom*. Ph.D. dissertation, University of California, Los Angeles. UMI, Ann Arbor, MI.
1995 Faunal Exploitation during the Middle to Late Period Transition on the Northern Channel Islands. *Journal of California and Great Basin Anthropology* 17(1):93–120.
2002 Prehistoric Marine Mammal Hunting in Context: Two Western North American Examples. *International Journal of Osteoarchaeology* 12:12–22.
Colten, R. H., and J. E. Arnold
1998 Prehistoric Marine Mammal Hunting on California's Northern Channel Islands. *American Antiquity* 63(4):679–701.
Dayton, P. K., M. J. Tegner, P. B. Edwards, and K. L. Riser
1998 Sliding Baselines, Ghosts, and Reduced Expectations in Kelp Forest Communities. *Ecological Applications* 8:309–322.
Delaney, M. J.
1982 *Mammal Ecology*. Blackie and Son, New York.
DeLong, R. L., and S. R. Melin
2002 Thirty Years of Pinniped Research at San Miguel Island. In *The Fifth California Islands Symposium* (CD Publication), edited by D. Brown, K. Mitchell, and H. Chaney, pp. 401–406. U.S. Department of the Interior Minerals Management Service, Pacific OCS Region.
DeMaster, D. P., C. W. Fowler, S. L. Perry, and M. F. Richlen
2001 Predation and Competition: The Impact of Fisheries on Marine-Mammal Population over the Next One Hundred Years. *Journal of Mammalogy* 82(3):641–651.
Ellis, R.
2003 *The Empty Ocean*. Island Press/Shearwater Books, Washington, D.C.
Erlandson, J. M.
1991 The Antiquity of CA-SMI-1: A Multicomponent Site on San Miguel Island. *Journal of California and Great Basin Anthropology* 13:273–279.
Erlandson, J. M., and T. J. Braje
2007 Early Maritime Technology on California's San Miguel Island: Arena Points from CA-SMI-575-NE. *Current Research in the Pleistocene* 24: 85–86.
Erlandson, J. M., T. J. Braje, T. C. Rick, and J. Peterson
2005 Beads, Bifaces, and Boats: An Early Maritime Adaptation on the South Coast of San Miguel Island, California. *American Anthropologist* 107(4):677–683.
Erlandson, J. M. and T. C. Rick
2010 Archaeology Meets Marine Ecology: The Antiquity of Maritime Cultures and Human Impacts on Marine Fisheries and Ecosystems. *Annual Review of Marine Science* 2:165–185.
Erlandson, J. M., T. C. Rick, and T. J. Braje
2009 Fishing up the Food Web? 12,000 Years of Maritime Subsistence and Adaptive Adjustments on California's Channel Islands. *Pacific Science* 63: 4:711–724.
Erlandson, J. M., T. C. Rick, and R. L. Vellanoweth
2004 Human Impacts on Ancient Environments: A Case Study from California's Northern Channel Islands. In *Voyages of Discovery: The Archaeology of Islands*, edited by S. Fitzpatrick, pp 51–83. Praeger, Westport, CT.
2008 *A Canyon through Time: The Archaeology, History, and Ecology of the Tecolote Canyon Area, Santa Barbara County, California*. University of Utah Press, Salt Lake City.
Erlandson, J. M., M. A. Tveskov, and R. S. Byram
1998 The Development of Maritime Adaptations on the Southern Northwest Coast of North America. *Arctic Anthropology* 35:6–22.
Estes, J. A.
1980 *Enhydra lutris. Mammalian Species. The American Society of Mammalogists* 133:1–8.
Etnier, M. A.
2002 Defining and Identifying Sustainable Harvests of Resources: Archaeological Examples of Pinniped Harvests in the Eastern North Pacific. *Journal for Nature Conservation* 15(3):196–207.
Fagan, B., D. R. Grenda, D. Maxwell, A. H. Keller, and R. Ciolek-Torrello
2006 *Life on the Dunes: Fishing, Ritual, and Daily Life at Two Late Period Sites on Vizcaino Point*. Statistical Research, Inc. Technical Series 88, Tucson AZ.
Fritz, L. W., R. C. Ferrero, and R. J. Berg
1995 The Threatened Status of Steller Sea Lions, *Eumetopias jubatus*, under the Endangered Species Act: Effects on Alaska Groundfish Fisheries Management. *Marine Fisheries Review* 57(2): 14–27.
Gales, N., M. Hindell, and R. Kirdwood (editors)
2003 *Marine Mammals: Fisheries, Tourism, and Management Issues*. CSIRO Publishing, Collingwood, Australia.
Gerrodette, T., and W. G. Gilmartin
1990 Demographic Consequences of Changed Pupping and Hauling Sites of the Hawaiian Monk Seal. *Conservation Biology* 4:423–440.

Gifford-Gonzales, D., S. D. Newsome, P. L. Koch, T. P. Guilderson, J. J. Snodgrass, and R. K. Barton
2005 Archaeofaunal Insights on Pinniped-Human Interactions in the Northeastern Pacific. In *The Exploitation and Cultural Importance of Sea Mammals*, edited by G. G. Monks, pp. 19–38. Oxbow Books, Oxford.

Glassow, M. A.
1980 Recent Developments in the Archaeology of the Channel Islands. In *The California Islands: Proceedings of a Multidisciplinary Symposium*, edited by D. M. Power, pp. 79–99. Santa Barbara Museum of Natural History, Santa Barbara, CA.
2005 Prehistoric Dolphin Hunting on Santa Cruz Island, California. In *The Exploitation and Cultural Importance of Sea Mammals*, edited by G. G. Monks, pp. 107–120. Oxbow Books, Oxford.

Glassow, M. A., P. Paige, and J. Perry
2008 The Punta Arena Site and Early and Middle Holocene Cultural Development on Santa Cruz Island, California. Santa Barbara Museum of Natural History Contributions in Anthropology No. 3, Santa Barbara, CA.

Harington, C. R.
1984 Quaternary Marine and Land Mammals and Their Paleoenvironmental Implications—Some Examples from Northern North America. In *Contributions in Quaternary Vertebrate Paleontology: A Volume in Memorial to John E. Guilday*, edited by H. H. Genoways and M. R. Dawson, pp. 5511–525. Carnegie Museum of Natural History Special Publication No. 8, Pittsburgh, PA.

Heizer, R. F.
1978 *Handbook of North American Indians*. Smithsonian Institute Press, Washington, D.C.

Hildebrandt, W. R., and T. L. Jones
1992 Evolution of Marine Mammal Hunting: A View from the California and Oregon Coasts. *Journal of Anthropological Archaeology* 11(4):360–401.

Jackson, J. B. C., M. X. Kirby, W. H. Berger, K. A. Bjorndal, L. W. Botsford, B. J. Bourque, R. H. Bradbury, R. Cooke, J. Erlandson, J. A. Estes, T. P. Hughes, S. Kidwell, C. B. Lange, H. S. Lenihan, J. M. Pandolfi, C. H. Peterson, R. S. Steneck, M. J. Tegner, and R. R. Warner
2001 Historical Overfishing and the Recent Collapse of Coastal Ecosystems *Science* 293(5530): 561–748.

Jones, T. J., and W. R. Hildebrandt
1995 Reasserting a Prehistoric Tragedy of the Commons: Reply to Lyman. *Journal of Anthropological Archaeology* 14(1):78–98.

Jones, T. J., W. R. Hildebrandt, D. J. Kennett, and J. F. Porcasi
2004 Prehistoric Marine Mammal Overkill in the Northeastern Pacific: A Review of New Evidence. *Journal of California and Great Basin Anthropology* 24(1):69–80.

Jones, T. J., and K. A. Klar
2005 Diffusionism Reconsidered: Linguistic and Archaeological Evidence for Prehistoric Polynesian Contact with Southern California. *American Antiquity* 70:457–484.

Kennett, D. J.
2005 *The Island Chumash: Behavioral Ecology of a Maritime Society*. University of California Press, Berkeley.

Kinlan, B. P., M. H. Graham, and J. M. Erlandson
2005 Late Quaternary Changes in the Size and Shape of the California Channel Islands: Implications for Marine Subsidies to Terrestrial Communities. In *Proceedings of the Sixth California Islands Symposium*, edited by D. Garcelon and C. Schwemm, pp. 131–142. Institute for Wildlife Studies, Arcata, California.

Lightfoot, K. G.
2004 *Indians, Missionaries, and Merchants: The Legacy of Colonial Encounters on the California Frontiers*. University of California Press, Berkeley.

Loughlin, T. R., M. A. Perez, and R. L. Merrik
1987 *Eumetopias jubatus. Mammalian Species. The American Society of Mammalogists* 283:1–7.

Lunneryd, S. G., A. Fjälling, and H. Westerberg
2003 A Large-Mesh Salmon Trap: A Way of Mitigating Seal Impact on a Coastal Fishery. *ICES Journal of Marine Science* 60(6):1194–1199.

Lyman, R. L.
1995 On the Evolution of Marine Mammal Hunting on the West Coast of North America. *Journal of Anthropological Archaeology* 14(1):45–77.

McLaren, I. A., and T. G. Smith
1985 Population Ecology of Seals: Retrospective and Prospective Views. *Marine Mammal Science* 1(11):54–83.

Meighan, C.
1959 The Little Harbor Site, Catalina Island: An Example of Ecological Interpretations in Archaeology. *American Antiquity* 24:383–405.

Monks, G. G.
2005 Discussion: Sea Mammals in Zooarchaeology AD 2002. In *The Exploitation and Cultural Importance of Sea Mammals*, edited by G. G. Monks, pp. 167–173. Oxbow Books, Oxford.

Moss, M. L., D. Y. Lang, S. D. Newsome, C. F. Speller, I. McKechnie, A. D. McMillan, R. J. Losey, and P. L. Koch

2006 Historical Ecology and Biogeography of North Pacific Pinnipeds: Isotopes and Ancient DNA from Three Archaeological Assemblages. *Journal of Island and Coastal Archaeology* 1:165–190.

Newsome, S. D., M. A. Etnier, D. Gifford-Gonzalez, D. L. Phillips, M. van Tuinen, E. A. Hadly, D. P. Costa, D. J. Kennett, T. P. Guilderson, and P. L. Koch

2007 The Shifting Baseline of Northern Fur Seal Ecology in the Northeast Pacific Ocean. *Proceedings of the National Academy of Sciences* 104(23):9709–9714.

Ogden, A.

1941 *The California Sea Otter Trade 1784–1848.* University of California Press, Berkeley.

Orr, P. C.

1968 *Prehistory of Santa Rosa Island.* Santa Barbara Museum of Natural History, Santa Barbara, CA.

Pauly, D.

1995 Anecdotes and the Shifting Baselines Syndrome of Fisheries. *Trends in Ecology and Evolution* 10(10):430.

Pauly, D., V. Christensen, J. Dalsgaard, R. Froese, and F. Torres Jr.

1998 Fishing Down Marine Food Webs. *Science* 279(5352):860–863.

Porcasi, J. F., and H. Fujita

2000 The Dolphin Hunters: A Specialized Prehistoric Maritime Adaptation in the Southern California Channel Islands and Baja California. *American Antiquity* 65:543–566.

Porcasi, J. F., T. L. Jones, and L. M. Raab

2000 Trans-Holocene Marine Mammal Exploitation on San Clemente Island, California: A Tragedy of the Commons Revisited. *Journal of Anthropological Archaeology* 19(2):200–220.

2002 Updating Prehistoric Maritime Subsistence at Little Harbor, Santa Catalina Island, California. *Proceedings of the Fifth California Channel Islands Symposium*, edited by D. Browne, K. Mitchell, and H. Chaney, pp. 580–589. Santa Barbara Museum of Natural History, Santa Barbara, CA.

Preston, W. L.

2002 Post-Columbian Wildlife Irruptions in California: Implications for Cultural and Environmental Understanding. In *Wilderness and Political Ecology: Aboriginal Influences and the Original State of Nature*, edited by C. E. Kay and R. T. Simmons, pp. 111–140, University of Utah Press, Salt Lake City.

Raab, L. M., K. Bradford, J. F. Porcasi, and W. J. Howard

1995 Return to Little Harbor, Santa Catalina Island, California: A Critique of the Marine Paleotemperature Model. *American Antiquity* 60(2):287–308.

Riedman, M.

1990 *The Pinnipeds: Seals, Sea Lions, and Walruses.* University of California Press, Berkeley.

Rick, T. C.

2001 Archaeology of Santa Barbara Island: Past Projects and Future Directions. *Pacific Coast Archaeological Society Quarterly* 37(3):57–72.

2004 *Daily Activities, Community Dynamics, and Historical Ecology on California's Northern Channel Islands.* Ph.D. dissertation, Department of Anthropology, University of Oregon, UMI, Ann Arbor, MI.

2007 *The Archaeology and Historical Ecology of Late Holocene San Miguel Island.* Cotsen Institute of Archaeology, University of California, Los Angeles.

2008 An Arena Point and Crescent from Santa Rosa Island, California. *Current Research in the Pleistocene* 25:140–142.

Rick, T. C., R. L. DeLong, J. M. Erlandson, T. L. Jones, T. J. Braje, J. E. Arnold, M. R. Des Lauriers, W. R. Hildebrandt, D. J. Kennett, and T. A. Wake

2009 Where Were the Northern Elephants Seals? Archaeology and Holocene Biogeography of *Mirounga angustirostris. The Holocene*, in review.

Rick, T. C., and J. M. Erlandson (editors)

2008 *Human Impacts on Ancient Marine Ecosystems: A Global Perspective.* University of California Press, Berekeley.

Rick, T. C., J. M. Erlandson, R. L. Vellanoweth, and T. J. Braje

2005 From Pleistocene Mariners to Complex Hunter-Gatherers: The Archaeology of the California Channel Islands. *Journal of World Prehistory* 19:169–228.

Rick, T. C., J. M. Erlandson, T. J. Braje, J. E. Estes, M. H. Graham, and R. L. Vellanoweth

2008 Historical Ecology and Human Impacts on Coastal Ecosystems of the Santa Barbara Channel Region, California. In *Human Impacts on Ancient Marine Ecosystems: A Global Perspective,* edited by Torben C. Rick and Jon M. Erlandson, pp. 77–101. University of California Press, Berkeley.

Rozaire, C. E

1965 Archaeological Investigations on San Miguel Island, California. Los Angeles County Museum of Natural History, Los Angeles.

1978 Archaeological Investigations on San Miguel Island, California. Los Angeles County Museum of Natural History, Los Angeles.

Schoenherr, A., C. R. Feldmath, and M. Emerson

1999 *Natural History of the Islands of California.* University of California Press, Berkeley.

Sharp, John T.
2000 *Shellfish Analysis from the Punta Arena Site, a Middle Holocene Red Abalone Midden on Santa Cruz Island, California.* Master's thesis, Cultural Resources Management, Sonoma State University, Rohnert Park, CA.

Steneck, R. S., M. H. Graham, B. J. Bourque, D. Corbett, J. M. Erlandson, J. A. Estes, and M. J. Tegner
2002 Kelp Forest Ecosystems: Biodiversity, Stability, Resilience, and Future. *Environmental Conservation* 29(4):436–459.

Stewart, B. S., and H. R. Huber
1993 *Mirounga angustirostris. Mammalian Species. The American Society of Mammalogists* 449:1–10.

Tegner, M. J., and P. K. Dayton
2000 Ecosystem Effects of Fishing in Kelp Forest Communities. *ICES Journal of Marine Science* 57(3):5779–589.

Trillmich, F., and K. A. Ono
1991 *Pinnipeds and El Niño.* Springer-Verlag, New York.

Vellanoweth, R. L., T. C. Rick, J. M. Erlandson, and G. Reynolds
2006 A 6000-Year-Old Red Abalone Midden from Otter Point, San Miguel Island, California. *North American Archaeologist* 27(1):69–90.

Wagner, H. R.
1929 *Spanish Voyages to the Northwest Coast of North America in the Sixteenth Century.* California Historical Society, San Francisco.

Walker, P. L., and S. Craig
1978 Archaeological Evidence Concerning the Prehistoric Occurrence of Sea Mammals at Point Bennett, San Miguel Island. *California Fish and Game* 65:50–54.

Walker, P. L, D. J. Kennett, T. L. Jones, and R. DeLong
2002 Archaeological Investigations at the Point Bennett Pinniped Rookery on San Miguel Island. In *Proceedings of the Fifth California Channel Islands Symposium*, edited by D. Browne, K. Mitchell, and H. Chaney, pp. 628–632. Santa Barbara Museum of Natural History, Santa Barbara, CA.

Walker, P. L., and P. Snethkamp
1984 *Archaeological Investigations on San Miguel Island – 1982. Prehistoric Adaptations to the Marine Environment.* Social Process Research Institute, University of California, Santa Barbara. Manuscript on file, Department of Anthropology, University of California, Santa Barbara.

Wilkinson, I. S., and M. N. Bester
1988 Is Onshore Human Activity a Factor in the Decline of the Southern Elephant Seal? *South African Journal of Antarctic Research* 18:14–17.

Yodzis, P.
2001 Must Top Predators Be Culled for the Sake of Fisheries? *Trends in Ecology and Evolution* 16(2):78–84.

13

Perspectives from the Past

ARCHAEOLOGY, HISTORICAL ECOLOGY, AND NORTHEASTERN PACIFIC PINNIPEDS AND SEA OTTERS

Todd J. Braje and Torben C. Rick

OVER THE LAST DECADE a variety of scientists, resource managers, and environmental activists have worked to increase public awareness of the crises facing the world's oceans and the role humans have played in their degradation. Overfishing, pollution, climate change, mismanagement, and other factors have resulted in severely depleted marine ecosystems and anthropogenic seascapes (Dayton et al. 1995; Ellis 2003; Halpern et al. 2008; Jackson 2008; Jackson et al. 2001; Lotze and Worm 2009; Pauly 1995; Pauly and Palomores 2005; Pauly et al. 1998, 2005; Pew Oceans Commission 2003; Pitcher 2001; Woodard 2000; Worm et al. 2006). Steps have been taken to reverse trends, mitigate damage, and restore degraded fisheries and marine ecosystems to a more "natural" balance. To combat these crises, however, it is clear that we need interdisciplinary perspectives and collaboration between ecologists, biologists, resource managers, and social scientists.

Duggins and Davis (1993), Roberts et al. (2001), Halpern and Warner (2002), and other biologists, ecologists, and fisheries managers have documented successful recovery of many marine fish and shellfish with the establishment of reserves that enforce strict no-take zones (see Badalamenti et al. 2002 for contradictory data). These reserves act as the aquatic equivalent of state or national parks, where flora and fauna are protected and preserved and, eventually, act as seed populations for waters outside of reserves. While these measures, and others, have begun to help reduce impacts on global fisheries and reverse the trend of declining biodiversity, the recovery of several North Pacific marine mammals has been much more dramatic and obvious.

Prior to 1973 and the passage of the Marine Mammal Protection and Endangered Species Acts, management of marine mammals was aimed at maintaining a maximum sustainable

FIGURE 13.1. Overview photograph of the large pinniped rookery at Point Bennett on western San Miguel Island, California. Note the abundant seals and sea lions hauled out along the beaches (photo by Braje, January 2006). Less than 500 years ago, a large Chumash village site was located behind the beach, and it is likely this rookery only took shape since marine mammal laws have been enacted to help reestablish North Pacific populations.

yield for commercial harvest (Etnier 2004; Hofman and Bonner 1984). Since about 1980 and the discontinuation of marine mammal hunting, however, direct threats to the populations in U.S. waters have been largely eliminated and management plans have focused on concerns over the reduction of marine mammal prey species and bycatch by commercial fisheries (National Marine Fisheries Service 1995; Sinclair et al. 1994). At the brink of extinction as recently as 50 years ago, North Pacific seal and sea lion populations now number in the hundreds of thousands (Figure 13.1). Although still extirpated from much of its prehistoric range, estimates of the current North Pacific sea otter population exceed 150,000, rebounded from only 13 colonies in the early 20th century (Ellis 2003). Similar population growth has been documented among northern elephant seals *(Mirounga angustirostris)* and Guadalupe fur seals *(Arctocephalus townsendi)*, while others, like Alaskan populations of Steller sea lions *(Eumetopias jubatus)*, are in a state of decline. Given an adequate food base, most pinnipeds are assumed to be repopulating Pacific Coast waters, establishing rookeries, and, perhaps, recovering to "prehistoric levels."

As the studies in this volume have demonstrated, archaeological and historical data document how people hunted, scavenged, and affected pinnipeds and sea otters over long time periods. As is true with finfish, shellfish, and other marine organisms, ancient human hunters, coupled with natural climatic changes, had an influence on marine mammal populations. The research presented here demonstrates that accurate baselines for marine mammal recovery and the development of effective management practices must account for natural and anthropogenic disturbances that began in the deep past. Modern marine mammal distribution, density, and behavior have been influenced by prehistoric and historic human hunting and disturbance, changes in the marine food base and accessibility of breeding and haul-out locations, natural climatic and environmental change, and national and international protection and recovery (Figure 13.2). The recognition of these disturbances and the artificial nature of modern marine mammal populations can create a messy political landscape for environmentalists, resource managers, historical ecologists, and other scientists. The chapters in this volume confront these difficult issues, often challenging assumptions that current marine mammal abundance, behavior, distributions, and rookery locations mirror those of the deep past. In Chapter 1, we raised five primary research questions focused on the historical ecology of northeastern Pacific pinnipeds and sea otters. These questions fell into five broad thematic categories: (1) human impacts on pinnipeds and sea otters; (2) natural versus anthropogenic change; (3) social, cultural, and technological implications of pinniped and sea otter exploitation; (4) analytical techniques and historical ecology; and (5) why is the past important for understanding the present? Here we return to these questions using the ten case studies from this volume as a springboard to help answer them.

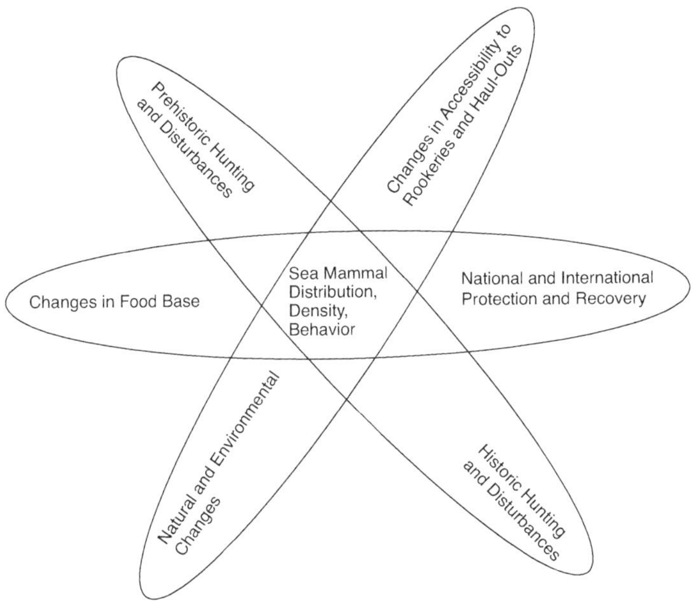

FIGURE 13.2. Model illustrating the ancient and historic processes that have resulted in contemporary marine mammal distribution, density, and behavior.

HUMAN IMPACTS ON PINNIPEDS AND SEA OTTERS • *Change, Continuity, and Sustainability*

How did ancient peoples from southern California to Alaska influence and impact ancient pinniped and sea otter abundance, biogeography, and behavior?

Several of the chapters in this book provide information on long-term trends and developments in pinniped and sea otter *(Enhydra lutris)* biogeography and ecology. In the process, these case studies demonstrate that humans often had an influence on pinniped and sea otters deep in the past. However, the cases studies in this volume also indicate that there are patterns of continuity and sustainability that suggest there was no singular trend in the way ancient peoples interacted with and had an impact on pinnipeds and sea otters in the northeast Pacific. These findings are similar to a recent global overview of the archaeology and historical ecology of marine ecosystems (Rick and Erlandson 2008), suggesting that there was no prehistoric global trend in overconsumption or sustainability.

Some of the chapters in this volume present striking evidence for human overexploitation of pinniped or sea otter populations. Zooarchaeological, ancient DNA, and isotopic analysis, and high-resolution AMS dating by Gifford-Gonzalez (Chapter 10) and a variety of colleagues over the last decade suggest that northern fur seal populations in the Greater Monterey Bay region have undergone dramatic changes during the Holocene. Gifford-Gonzalez found significant evidence for their extirpation from central California by about 1100 cal BP, the combined effects of human hunting and climate change, representing an ancient ecological phase shift vastly different from today. While the archaeological, historical, and modern ecology and biogeography of northern fur seals along the central California coast and the Farallon Islands still pose many questions, Gifford-Gonzalez issues a challenge to researchers to recognize that the patterns we see today are likely very different from the past and can help better inform the future management of these organisms.

Relying on a 9000-year record from central California, Jones, Culleton, Larson, Mellinger, and Porcasi (Chapter 11) investigate the long-term effects of human hunting on sea otter populations. Their study suggests an ancient pattern of human hunting that focused on female otters, perhaps since female otter behavior and feeding patterns made them more accessible to human hunters, which may have resulted in significant population declines by the Late Holocene. While their study demonstrates significant influence of human hunting on sea otter populations, further study is needed to reconcile the nature, scale, and extent of ancient human impacts on California sea otters, as well as the robust population of otters recorded along California at European contact.

Several other papers argue for a middle ground and complex mix of human impacts and natural climate change that have helped shape marine mammal populations across the North Pacific for millennia. Hill's (Chapter 3) historical ecological synthesis of the population densities, migration routes, and human hunting of Pacific walruses indicates that the pattern and intensity of anthropogenic forcing on these animals significantly differs between Alaska, St. Lawrence Island, and the Chukchi coast. Hill found, however, that at least some contemporary and historical migration routes and preferred haul-outs extend into the deep past, as do hydrodynamic processes in the Bering Sea.

Braje, Rick, DeLong, and Erlandson (Chapter 12) also found considerable long-term spatial continuity and ecological resilience of pinniped and sea otter communities along the Northern and, perhaps, Southern Channel Islands. However, they also noted considerable differences in the relative abundance of various species (especially northern elephant seals and Guadalupe fur seals) between prehistoric and modern times (see Rick et al. 2009; Walker et al. 2002). Similar to Gifford-Gonzalez et al., these discrepancies document ancient ecological phase shifts that can help better inform the past, present, and future status of marine mammals.

Whitaker and Hildebrandt (Chapter 9) provide another perspective, arguing for a case of long-term continuity in marine mammal populations, despite human hunting and natural climatic oscillations. In their investigation of the history of pinniped hunting on the northern California coast, Whitaker and Hildebrandt document a relatively continuous and sustainable fur seal hunting economy that persisted for ca. 1500 years until European contact. The case for sustainability further emphasizes the dynamics of ancient human interactions with pinnipeds and sea otters, ranging from degradation to sustainability.

This theme of variability in the degree of human impacts on marine mammal populations can also be seen in breeding behavior. One of the central issues in marine mammal science has been the identification of rookery locations, both historic and ancient, and their reestablishment after historic overhunting and conservation measures. Documenting the location of ancient rookeries has been difficult and controversial, but several studies demonstrate the significant influence people had on ancient breeding and haul out locations (e.g., Hildebrandt and Jones 1992). Gifford-Gonzalez (Chapter 10) uses a variety of zooarchaeolgoical, ecological, and archaeometric evidence to investigate the long-term character of rookeries, from the Farallon Islands to (possible) mainland rookeries along central California. Gifford-Gonzalez suggests that mainland rookeries may have existed and been disrupted by human hunting and other factors, but continued research by a variety of specialists will be necessary to confirm this conclusion. Hill (Chapter 3) found a complex interaction between marine mammal migration routes, human hunting, and climate that both disrupted and maintained breeding locations along the Bering Sea for thousands of years. Finally, Moss and Losey (Chapter 8) identified considerable disconnect between current models of ancient pinniped and sea otter ecology, anthropogenic forcing, and the zooarchaeological record from northern Oregon and southern Washington. They conclude that continued study of existing and new zooarchaeological assemblages and the application of isotopic studies and ancient DNA are essential to test existing models and build new models that explain the interplay between human hunting, climate change, and pinniped and sea otter breeding behavior, population structure, and distribution.

Collectively, the chapters in this volume demonstrate that the degree of human forcing on North Pacific marine mammal populations has ranged from significant to superficial. This variability makes it essential that we build local case studies that investigate the complex interplay between marine mammals, humans, and climate change.

NATURAL CLIMATIC CHANGES OR HUMAN IMPACTS

What were the effects of ancient climate changes on pinnipeds, sea otters, and human society, and how do we differentiate between the effects of human impacts and natural climatic changes?

An important component of investigating anthropogenic alterations to marine ecosystems and pinniped and sea otter populations is distinguishing them from natural environmental and climatic change. This is a complex issue that permeates virtually all of historical and modern ecological research and a variety of ecological studies suggest that an understanding of local and regional climatic patterns is critical to explaining ecological change. As several chapters in the volume demonstrate, it is often difficult to determine whether the appearance or absence of pinniped and sea otter remains in archaeological sites is the product of new human hunting strategies and technologies, human impacts on populations, or environmental fluctuations.

Crockford and Frederick (Chapter 4) tackle the issue of ancient environmental change on prehistoric pinnipeds, implicating Neoglacial sea ice change in the Bering Sea to explain the formation of northern fur seal breeding rookeries along the west coast of North America. Through detailed zooarchaeological analysis, they conclude that Bering Sea pack ice extended much further south than it does today, suggesting a flexible life history for ringed and fur seals in the Arctic. Their findings suggest dramatic environmental changes in the North Pacific during the Neoglacial and will continue to be controversial, but their work demonstrates the potential of archaeological data to help interpret local and regional environmental changes.

Braje et al. (Chapter 12) take a different tack when interpreting the zooarchaeological record of pinnipeds and sea otters on the Channel Islands. They apply modern adult and newborn behavior, breeding habits, and historic/modern distributions to help interpret the archaeological record. Assuming these behaviors have strong evolutionary roots and extend deep into the past, they argue that humans have caused major disruptions to pinniped populations on the Northern Channel Islands. Strategies for resolving different interpretations and teasing apart the signatures of climate change versus human impacts will continue to be an essential part of interpreting the historical ecology of North Pacific marine mammals.

Betts, Maschner, and Lech (Chapter 5) use a 4500-year archaeological sequence in the western Gulf of Alaska to trace fluctuations in otariid abundance related to anthropogenic and climatic influences. They link initial population changes to otariid colonization of the area and the subsequent exploitation of rookeries and haul-outs by humans. Later in the sequence, variability in otariid abundances seems to be linked to regional climatic change. Their study demonstrates the utility of zooarchaeology in elucidating dynamic changes in marine mammal populations that can result from anthropogenic and climatic forcing.

In Chapter 3, Hill found that some modern walrus migration routes and preferred haul-out locations along the nutrient-rich water mass through the Strait of Andyr and the Bering Strait extend into the deep past. Hydrodynamic processes in the Bering Sea that heavily influence marine mammal behavior today have existed for centuries in the far North Pacific and have helped shape human behavior and social structures for centuries. Each of these studies demonstrates that archaeologists should not assume that human hunting or other anthropogenic forces are the primary driver of changes in ancient marine mammal populations or behavior.

What was the role of ancient human technological change, social systems, ideology, and ritual in guiding the types of marine mammals that were taken?

One of the unique contributions of anthropologists and archaeologists in historical ecological studies is their focus on the effects of cultural systems on long-term ecosystem evolution. Marine mammal populations, for example, are influenced by a variety of human cultural developments and function within a diverse array of social, political, and economic systems. In Chapter 9, Whitaker and Hildebrandt document the seemingly sustainable harvest of northern fur seals along the northern California coast for 1500 years. Unlike the case in central California where Gifford-Gonzalez (Chapter 10) links the disappearance of Monterey Bay fur seals with sociocultural elaboration, Whitaker and Hildebrandt point to the development of prestige hunting for pinnipeds and the cultural significance of northern fur seal canine-studded headbands for ceremonial regalia. Hunters, then, focused on the exploitation of adult male animals for the acquisition of large canines, a practice that helped maintain healthy fur seal populations (whether intentional or not) by leaving young and female animals to replenish the population.

The theme of human cultural patterns and marine mammal dynamics is also indicated by Hill's (Chapter 3) compelling evidence for the communal hunting of walruses by Eskimos in the North Pacific that may have afforded the foundation for sociopolitical complexity. Hill argues that communal whaling, long thought to be a marker of increasingly complex cultural organization in the far North Pacific, may have been an outgrowth of the exploitation of the seasonal and predictable migration of walrus populations, which in turn provided valuable subsistence and technological resources to people.

In Chapter 7, McKechnie and Wigen explore the dynamics of cultural and environmental influence on marine mammals in southern British Columbia, noting that pinnipeds were a resource focus along the outer coast of west Vancouver Island and terrestrial mammals were central along the Salish Sea. With regard to marine mammal exploitation, prehistoric hunters focused on northern fur seals, harbor seals, and Steller sea lions along the outer coast and harbor seals along the sea. While this pattern is influenced by environmental factors and shows a correlation with modern distributions of pinnipeds in the area, McKechnie and Wigen also emphasize the importance of human agency and decision making in influencing the pinniped species found in archaeological deposits. These chapters demonstrate that social and natural systems are often intertwined in complex ways and researchers must be diligent in incorporating sociocultural influences in their historical ecological studies of ancient and modern marine mammal populations.

ANALYTICAL TECHNIQUES AND HISTORICAL ECOLOGY

How can various analytical techniques, such as aDNA and stable isotope analyses, be integrated with traditional archaeological data to provide a more comprehensive understanding of ancient pinniped and sea otter populations?

As Lyman points out in Chapter 2, zooarchaeology has entered a new stage of scientific investigation. While traditional zooarchaeological identification and analysis of bone from archaeological deposits remain an important part of investigating the long-term history of marine mammals, archaeologists are employing other techniques such as stable isotopes, ancient DNA, and butchering studies to unravel the dynamic relationships between marine mammals, humans, and climate through time and space (e.g., Burton et al. 2001; Moss et al. 2006; Newsome et al. 2007). Several chapters from this volume provide case studies that integrate traditional zooarchaeolgical analyses with aDNA, stable isotope analyses, or other techniques that provide new perspectives on the long-term histories of pinniped and sea otter populations.

Isotopic studies of archaeological marine mammal remains probably have been the most widely used archaeometric technique. Stable carbon and nitrogen isotopic studies by Gifford-Gonzalez (Chapter 10) and colleagues suggest that northern fur seals may have foraged and bred at latitudes considerably south of modern ranges (see Chapter 2). Whitaker and Hildebrandt (Chapter 9) employ stable carbon and nitrogen isotopic studies to demonstrate that an ancient rookery was not locally exploited along the northern California coast and bachelor herds of northern fur seals from Alaskan breeding populations may have stocked the local population. Jones and colleagues (Chapter 11) integrate zooarchaeology, carbon and nitrogen isotopic studies, and ancient DNA to demonstrate an increasingly intensive pattern of sea otter predation, a decrease in foraging efficiency, and transition from a focus on females to a slight increase in males through time.

Finally, Lech, Betts, and Maschner (Chapter 6) provide a detailed study of pinniped butchering patterns from Sanak Island, Alaska, to elucidate evolving consumption strategies, which seem to coincide with natural sea-surface temperature changes. Their study offers another opportunity for using zooarchaeological remains to link climate, culture, and marine mammals. As archaeological approaches to understanding the historical ecology of pinnipeds, sea otters, and other marine ecosystems continue to evolve and improve, new analytic techniques that push methodological boundaries will help elucidate the complex, long-term histories of North Pacific marine mammals.

WHY IS THE PAST IMPORTANT? • *Integrating Marine Mammal Ecology, Archaeology, and History*

How can reconstructions of ancient human, pinniped, and sea otter interactions that span millennia inform the contemporary management of these organisms and broader marine ecosystems?

The final question raised in Chapter 1 may be the most difficult to answer but, perhaps, is the most important. As the studies in this volume demonstrate, there is increasing momentum and interest in archaeological research on pinnipeds and sea otters. As Lyman noted in Chapter 2, recent research on the topic is pushing itself toward new methodologies, increasingly sophisticated analyses, and the application of results to modern conservation and management protocols. This is an accurate assessment that in many ways serves as a challenge for archaeologists, historians, biologists, and resource managers to continually search for new insight into the ancient and modern dynamics between humans, pinnipeds, sea otters, and marine ecosystems.

While an increasing number of scholars are working to provide deeper historical perspectives to modern management protocols, often by entering into interdisciplinary and collaborative research efforts (e.g., Braje et al. 2009; Burton et al. 2001; Jackson et al. 2001; Moss et al. 2006; Newsome et al. 2007), successful publication of (primarily) archaeological studies in top-tier ecological journals has been infrequent. This is probably due to a variety of factors, but the most difficult obstacle for archeologists to overcome may be the incongruence between ecological and archaeological datasets. Archaeologists will never provide the more detailed perspectives of ancient marine ecosystems in the same fashion ecologists can with modern systems. Taphonomic processes, the selectivity of human hunting and gathering, and limitations in ancient technological capabilities all restrict the types of data that become part of the archaeological record. Partially as a result of these realities and other methodological obstacles (e.g., quantification techniques), many scientists and managers have ignored archaeological and other ancient datasets, suggesting that the absence of modern technologies and reduced human population densities resulted in limited human impacts on past ecosystems.

While there is growing recognition that all humans, regardless of time or space, engage in a dialectic relationship with nature (Crumley 1994), the degree of humanity's influence (especially in the more distant past) is often debated (see Redman 1999). The notion that

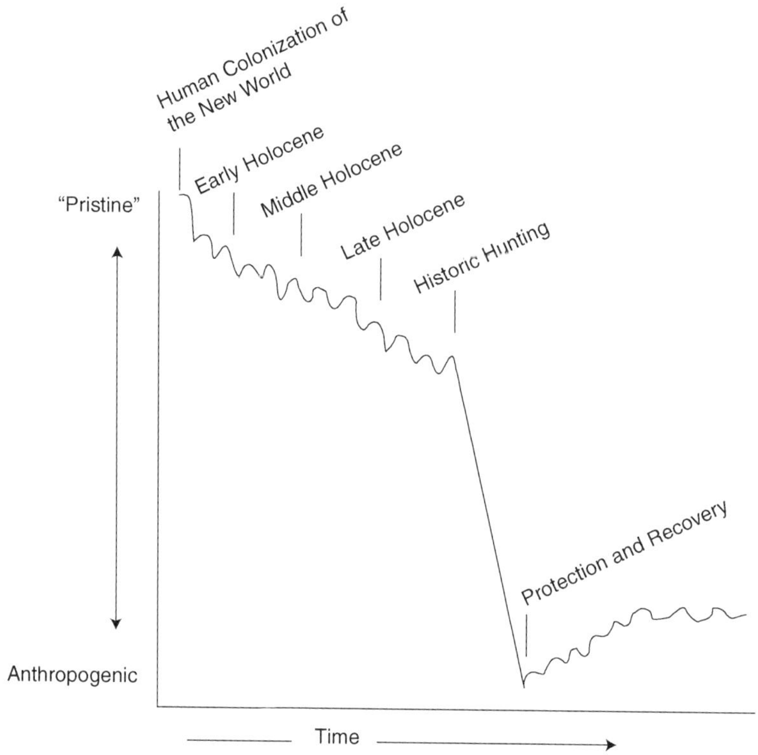

FIGURE 13.3. Model of anthropogenic influences on North Pacific marine mammal distribution, density, and behavior.

dramatic anthropogenic forcing on local and regional ecosystems did not begin until the advent of complex, state-level sociopolitical organization has been shown to severely underestimate the influence of humans (Grayson 2001; Rick and Erlandson 2008, 2009; Lyman 2006; Redman 1999, 2004). However, the target or baseline for effective conservation management and restoration, even when deep historical perspectives are incorporated, are often difficult to set and highly controversial. In North America, for example, there are virtually no post-Pleistocene ecosystems or landscapes (except perhaps far corners of the Arctic and some high elevations) that have not been influenced by human activity, and, thus, could be considered natural or pristine. Should ecological baselines be set at 10,000, 5000, or 1000 years ago, then? Should these temporal baselines remain firm despite local culture histories, where, for example, the chiefly level, maize-based Mississippian societies of the American Midwest are compared with contemporary small-scale hunter-gatherer populations of the Great Basin?

Since archaeological research provides some of the best datasets of ancient ecosystem structure and function, ecological baselines must reflect environments postdating human colonization. Certainly a baseline established prior to the large-scale commercial interests of postindustrial humans is best, and for North America this translates to a pre-Columbian target. Identifying this baseline is difficult and, potentially, highly political. In Figure 13.3, we illustrate the complex history of North Pacific pinnipeds and sea otters since human colonization of the New World. Our model highlights two points. First, a "natural" or "pristine" North Pacific marine mammal ecosystem, unaltered by humans, vanished after the Pleistocene colonization of the New World. Disturbances, either through direct hunting or encroachment, on breeding grounds and haul-outs probably occurred rapidly and had significant effects on ecologically naïve seals and sea lions. Human impacts continued through the Early and Middle Holocene, with significant temporal and geographic variation, until the Late Holocene when new technologies such as more sea-worthy vessels, the introduction of the bow and arrow and composite open-ocean hunting equipment and increased human population densities accelerated the rate of anthropogenic change on North Pacific marine mammals. The most significant human impacts occurred with historic commercial overhunting beginning in the 17th century. After protection and recovery, some pinniped and other marine mammal populations recovered, but under dramatically altered conditions, to establish new distributions, densities, ranges, and behaviors.

Second, no matter what time period we study, conditions will not be natural or pristine, but the deeper the historical perspective, the closer we can come to establishing the "natural" condition or to realizing that some level of human hunting or influence may have existed for millennia. We thus must be explicit about our reference points and the spatiotemporal context of a particular ecosystem. In looking toward the future, we must continue to recognize that change and phase shifts are an important part of all ecosystems in deep and more recent time and that ancient peoples are often a driver of these changes. By understanding the full Holocene history of North Pacific pinnipeds, sea otters, and other marine mammals, we are more likely to identify appropriate management practices by documenting long-term trends in species abundance, composition, and biogeography in association with climatic oscillations and changes in the intensity of human hunting. Ultimately, the only effective way to evaluate the population and behavioral history of North Pacific marine mammals and to decipher if changes are directional, chaotic, or cyclical is to develop a set of long-term observations (see Landres 1992; Lyman and Cannon 2004).

Since nearly all conservation efforts of Pacific pinnipeds and other marine mammals are driven by the Marine Mammal Protection Act of 1972 and the Endangered Species Act of 1973, the application of archaeological data to management strategies becomes especially problematic. By law, management decisions must be carefully balanced between threats to healthy marine mammal populations and threats to commerce and industry. The Marine Mammal Protection Act (16USC1361-1421, Public Law 92-522) requires that this is accomplished within a specific framework: marine mammal populations should be restored to or maintained at an Optimal Sustainable Population, stock-specific determinations should be based on the best scientific data available, and species assessments should include geographic ranges. Unlike national parks and many other federally managed resources that are housed within the Department of the Interior, the National Oceanic and Atmospheric Administration (NOAA) manages U.S. fisheries and sea mammals and is housed within the Department of Commerce. NOAA, then, must balance the commercial interests of domestic fisheries with maintaining healthy populations.

Given that most historic records of marine mammal populations and behavior postdate the massive commercial hunting of the 17th through 20th centuries, archaeologists and historical ecologists are increasingly generating data that stand in stark contrast to widely accepted assumptions about healthy and natural breeding distributions, ranges, relative densities, and behaviors of North Pacific marine mammals. As the chapters in this volume demonstrate, zoological, isotopic, and ancient DNA from archaeological marine mammal assemblages should be included as part of the "best scientific data available" under the Marine Mammal Protection Act. As archaeologists and other scientists engaged in developing long-term histories of marine mammals and continue to produce results that are different from or inconsistent with modern or historic data, controversies over wise management practices surely will arise. In some cases, major contradictions appear between what the archaeological record shows about the human-environmental past and what contemporary observers, the public, and interest groups would like to believe. In our view, however, collaboration between archaeologists, historians, biologists, marine mammal scientists, and other researchers, including the development of transdisciplinary models that integrate the best of modern marine mammal science and historical and archaeological perspectives is the key to better understanding and managing pinniped and sea otter populations in the northeastern Pacific and beyond. It is the role of these researchers to produce objective scientific data, free from political entanglement, with increasingly refined and sophisticated methodologies. Resource managers, conservation biologists, politicians, and the public are responsible for determining the desired future condition of an ecosystem or species and how to best implement scientific data within resource management strategies. Ultimately, archaeologists and other scientists must report data in ways that are useful and understandable to a wide scientific community and publish their results (which include management implications or recommendations) in ecological, biological, or interdisciplinary journals. The chapters in this volume take a decisive step in this direction, and with increased collaboration and improved methodologies, archaeological and ancient datasets will play an increasingly important role in the protection, management, and recovery of marine mammals around the globe.

ACKNOWLEDGMENTS

We thank Mark Raab and Andrew Trites for their helpful comments on an early version of this volume. At UC Press, we are indebted to Blake Edgar, Lynn Meinhardt, Francisco Reinking, and the production staff for helping bring this volume to fruition. A special thanks to all the contributors for their insightful efforts to understand the long and complex history of human, sea mammal interactions along the Northeast Pacific. Finally, numerous conversations with Robert DeLong have been instrumental in shaping our understanding of marine mammal ecology and history.

REFERENCES CITED

Badalamenti, F., G. D'Anna, J. K. Pinnegar, and N. V. C. Polunin
2002 Size-Related Trophodynamic Changes in Three Target Fish Species Recovering from Intensive Trawling. *Marine Biology* 141:561–570.

Burton, R. K., J. J. Snodgrass, D. Gifford-Gonzalez, T. Guilderson, T. Brown, and P. L. Koch
2001 Holocene Changes in the Ecology of Northern Fur Seals: Insights from Stable Isotopes and Archaeofauna. *Oecologia* 128:107–115.

Braje, T. J., J. M. Erlandson, T. C. Rick, P. K. Dayton, and M. B. A. Hatch
2009 Fishing from Past to Present: Long-Term Continuity and Resilience of Red Abalone Fisheries on California's Northern Channel Islands. *Ecological Applications* 19(4):906–919.

Crumley, C.
1994 Historical Ecology: A Multidimensional Ecological Orientation. In *Historical Ecology: Cultural Knowledge and Changing Landscapes*, edited by C. Crumley, pp. 1–16. Santa Fe: School of American Research Press.

Dayton, P. K., S. F. Thrush, and M. T. Agardy
1995 Environmental Effects of Marine Fishing. *Aquatic Conservation: Marine and Freshwater Ecosystems* 5:205–232.

Duggins, D. O., and G. E. Davis
1993 Applications of Marine Refugia to Coastal Fisheries Management. *Canadian Journal of Fisheries and Aquatic Science* 50:2029–2042.

Ellis, R.
2003 *The Empty Ocean*. Washington, D.C.: Island Press/Shearwater Books.

Grayson, D. K.
2001 The Archaeological Record of Human Impact on Animal Populations. *Journal of World Prehistory* 15:1–68.

Halpern, B. S., and R. R. Warner
2002 Marine Reserves Have Rapid and Lasting Effects. *Ecology Letters* 5:361–366.

Halpern, B. S., S. Walbridge, K. A. Selkoe, C. V. Kappel, F. Micheli, C. D'Agrosa, J. F. Bruno, K. S. Casey, C. Ebert, H. E. Fox, R. Fujita, D. Heinemann, H. S. Lenihan, E. M. P. Madin, M. T. Perry, E. R. Selig, M. Spalding, R. Steneck, and R. Watson
2008 A Global Map of Human Impact on Marine Ecosystems. *Science* 319(5865):948–952.

Hildebrandt, W. R., and T. L. Jones
1992 Evolution of Marine Mammal Hunting: A View from the California and Oregon Coasts. *Journal of Anthropological Archaeology* 11(4):360–401.
Hofman, R. J., and W. N. Bonner
1984 Conservation and Protection of Marine Mammals: Past, Present, and Future. *Marine Mammal Science* 1(2):109–127
Jackson, J. B. C.
2008 Ecological Extinction and Evolution in the Brave New Ocean. *Proceedings of the National Academy of Sciences* 105:11458–11465.
Jackson, J. B. C., M. X. Kirby, W. H. Berger, K. A. Bjorndal, L. W. Botsford, B. J. Bourque, R. H. Bradbury, R. Cooke, J. Erlandson, J. A. Estes, T. P. Hughes, S. Kidwell, C. B. Lange, H. S. Lenihan, J. M. Pandolfi, C. H. Peterson, R. S. Steneck, M. J. Tegner, and R. R. Warner
2001 Historical Overfishing and the Recent Collapse of Coastal Ecosystems *Science* 293(5530): 561–748.
Landres, P. B.
1992 Temporal Scale Perspectives in Managing Biological Diversity. *Transactions of the North American Wildlife and Natural Resources Conference* 57:292–307.
Lotze, H. K., and B. Worm
2009 Historical Baselines for Large Marine Animals. *Trends in Ecology and Evolution* 24: 254–262.
Lyman, R. L., and K. P. Cannon
2004 Applied Zooarchaeology, Because It Matters. In *Zooarchaeology and Conservation Biology*, edited by R. L. Lyman and K. P. Cannon, pp. 1–24. Salt Lake City: University of Utah Press.
Moss, M. L., D. Y. Lang, S. D. Newsome, C. F. Speller, I. McKechnie, A. D. McMillan, R. J. Losey, and P. L. Koch
2006 Historical Ecology and Biogeography of North Pacific Pinnipeds: Isotopes and Ancient DNA from Three Archaeological Assemblages. *Journal of Island and Coastal Archaeology* 1:165–190.
National Marine Fisheries Service
1995 *Status Review of Steller Sea Lions (*Eumetopias jubatus*)*. National Marine Mammal Laboratory, Alaska Fisheries Science Center, National Marine Fisheries Service, National Oceanic and Atmospheric Administration, Seattle.
Newsome, S. D., M. A. Etnier, D. Gifford-Gonzalez, D. L. Phillips, M. van Tuinen, E. A. Hadly, D. P. Costa, D. J. Kennett, T. P. Guilderson, and P. L. Koch
2007 The Shifting Baseline of Northern Fur Seal Ecology in the Northeast Pacific Ocean. *Proceedings of the National Academy of Sciences* 104(23): 9709–9714.
Pauly, D.
1995 Anecdotes and the Shifting Baselines Syndrome of Fisheries. *Trends in Ecology and Evolution* 10(10):430.
Pauly, D., V. Christensen, J. Dalsgaard, R. Froese, and F. Torres, Jr.
1998 Fishing Down Marine Food Webs. *Science* 279(5352):860–863.
Pauly, D., and M. L. Palomares
2005 Fishing Down Marine Food Webs: It Is Far More Pervasive Than We Thought. *Bulletin of Marine Science* 76(2):197–211.
Pauly, D., R. Watson, and J. Alder
2005 Global Trends in World Fisheries: Impacts on Marine Ecosystems and Food Security. *Philosophical Transactions of the Royal Society* 306:5–12.
Pew Oceans Commission
2003 America's Living Oceans: Charting a Course for Sea Change, Summary Report. Arlington, VA: Pew Oceans Commission.
Pitcher, T. J.
2001 Fisheries Managed to Rebuild Ecosystems? Reconstructing the Past to Salvage the Future. *Ecological Applications* 1(2):601–617.
Redman, C. L.
1999 *Human Impact on Ancient Environments*. Tucson: University of Arizona Press.
2004 Environmental Degradation and Early Mesopotamian Civilization. In *The Archaeology of Global Change: The Impact of Humans on Their Environment*, edited by C. L. Redman, S. R. James, P. R. Fish, and J. D. Rogers, pp. 158–164. Washington D.C.: Smithsonian Books.
Rick, T. C., and J. M. Erlandson (editors)
2008 Human Impacts on Ancient Marine Ecosystems: A Global Perspective. University of California Press, Berkeley.
Rick, T. C., and J. M. Erlandson
2009 Coastal Exploitation. *Science* 325:952–953.
Rick T. C., R. L. DeLong, J. M. Erlandson, T. J. Braje, T. L. Jones, D. J. Kennett, and P. L. Walker
2009 A 7000-Year Record of Guadalupe Fur Seals (*Arctocephalus townsendi*) on the California Coast. *Marine Mammal Science* 25(2):487–502.
Roberts, C. M., J. A. Bohnsack, F. Gell, J. P. Hawkins, R. Goodridge
2001 Effects of Marine Reserves on Adjacent Fisheries. *Science* 294:1920–1923.
Sinclair, E., T. R. Loughlin, and W. Percy
1994 Prey Selection by Northern Fur Seals (*Callorhinus ursinus*) in the Eastern Bering Sea. *Fishery Bulletin* 92:144–156.

Walker, P. L., D. J. Kennett, T. L. Jones, and R. DeLong
2002 Archaeological Investigations at the Point Bennett Pinniped Rookery on San Miguel Island. In *Proceedings of the Fifth California Channel Islands Symposium*, edited by D. R. Browne, K. L. Mitchell, and H. W. Chaney, pp. 628–632. Santa Barbara Museum of Natural History, Santa Barbara, CA.

Woodard, C.
2000 *Ocean's End*. New York: Basic Books.

Worm, B., E. B. Barbier, N. Beaumont, J. E. Duffy, C. Folke, B. S. Halpern, J. B. C. Jackson, H. K. Lotze, F. Micheli, S. R. Palumbi, E. Sala, K. A. Selkoe, J. J. Stachowicz, and R. Watson
2006 Impacts of Biodiversity Loss on Ocean Ecosystem Services. *Science* 314:787–790.

INDEX

Indexer: Enid Zafran
Composition: Westchester Book Group
Text: 9.5/13 Scala
Display: Scala Sans

www.ingramcontent.com/pod-product-compliance
Ingram Content Group UK Ltd.
Pitfield, Milton Keynes, MK11 3LW, UK
UKHW060822220626
472188UK00008BA/38
9780520267268